Rock Types and Sedimentary Structures

Basic Rock Types

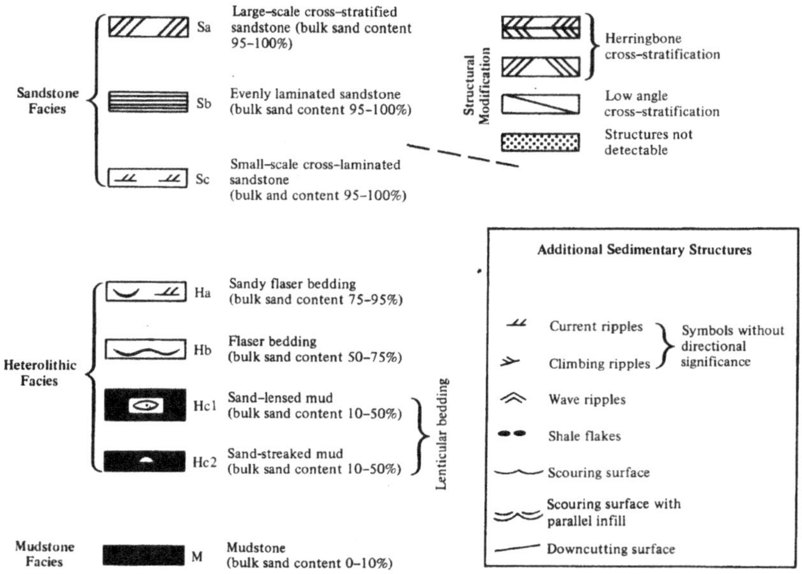

Composite Rock Types

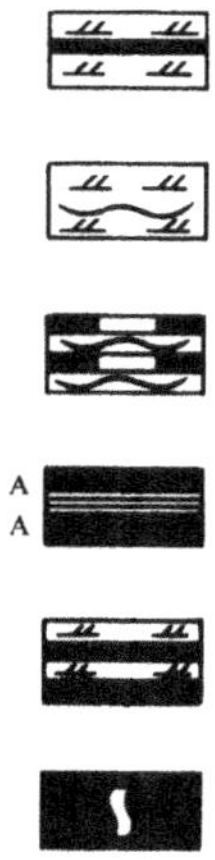

Mudstreaks in cross-laminated sand (Sc) also occurring in Sa, Sb, Ha and Hb. Bulk sand content 75–100%)

Mudflasers in cross-laminated sand (Sc) also occurring in Ha. Bulk sand content 75–100%

Lenticular and flaser bedding in random alternations. Both components up to a few meters thick. Bulk sand content 10–75%

Mudstone (up to a few meters thick) with intercalations of evenly laminated sandstone sheets (up to 2 m thick). Bulk sand content 10–95%

Cross-laminated sand (up to 10 cm thick) and mudbands (up to 4 cm thick) in regular alternation. Bulk sand content 50–75%

Sandy mudstone homogenized by bioturbation. Bulk sand content 10–50%

prepared by Richard Boersma

Guide to Reading the Logs

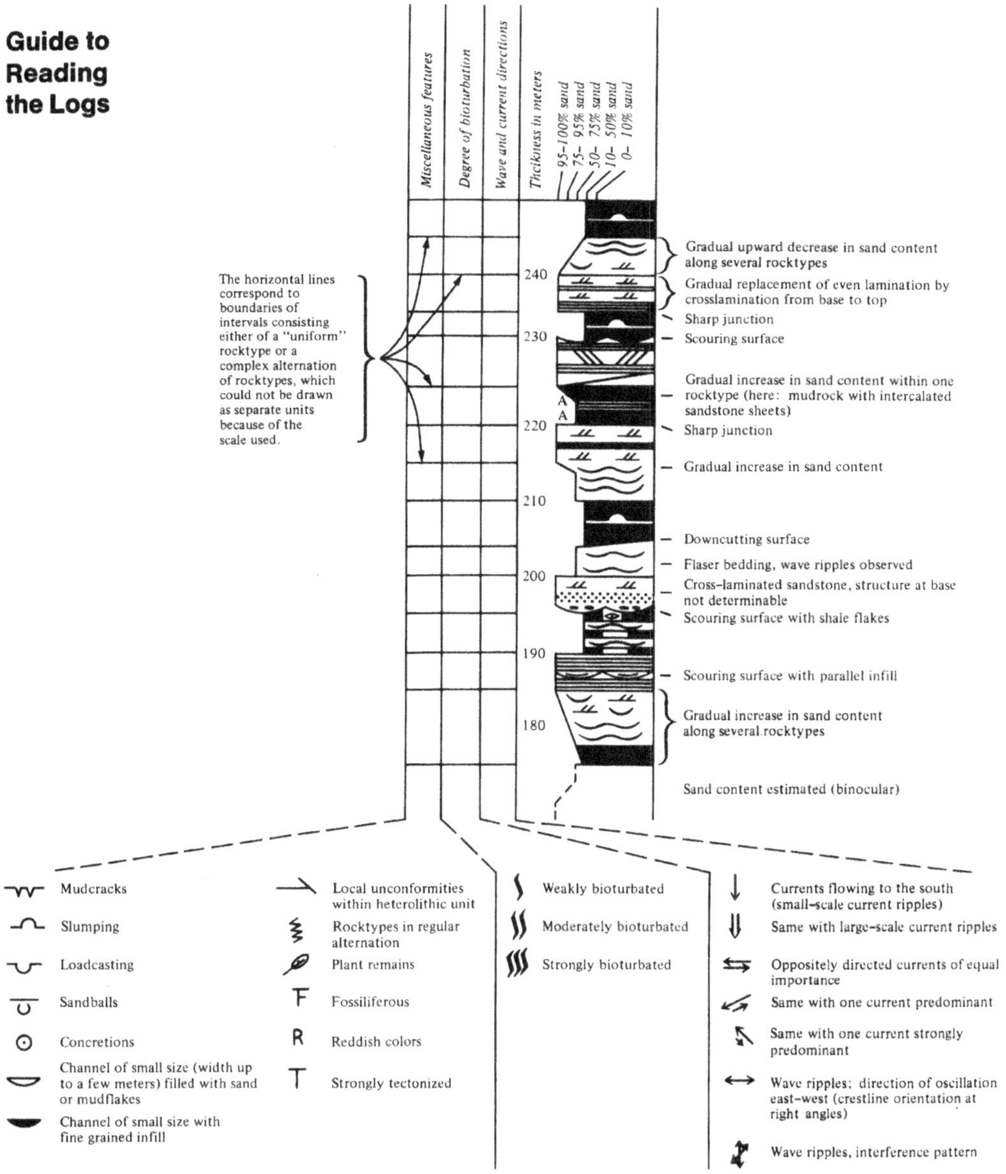

Tidal Deposits

with contributions by
Elizabeth A. Allen University of Delaware
Augustus K. Armstrong United States Geological Survey
John J. Barnes University of Illinois*
F. W. Beales University of Toronto
Moshe Braun Rensselaer Polytechnic Institute
Raymond G. Brown University of Western Australia
Charles H. Carter Ohio Geological Survey
R. Colacicchi Università di Perugia
Robert W. Dalrymple University of Delaware
Richard A. Davis, Jr. Western Michigan University*
Graham Evans Imperial College, London
Alfred G. Fischer Princeton University
Gerald M. Friedman Rensselaer Polytechnic Institute
Robert N. Ginsburg University of Miami
George M. Hagan University of Western Australia
Robert B. Halley State University of New York, Stony Brook*
Lawrence A. Hardie Johns Hopkins University
John C. Harms Marathon Oil Company
Stanley Cooper Harrison Johns Hopkins University
Paul Hoffman Geological Survey of Canada*
Lubomir F. Jansa Bedford Institute of Oceanography
George deVries Klein University of Illinois
R. John Knight McMaster University
John C. Kraft University of Delaware
Naresh Kumar Lamont-Doherty Geological Observatory
Léo F. Laporte University of California, Santa Cruz
Claude Larsonneur Université de Caen
Brian W. Logan University of Western Australia
G. P. Lozej University of Toronto
David B. MacKenzie Marathon Oil Company
James A. Miller Union Oil Company of California
Eric W. Mountjoy McGill University
A. Thomas Ovenshine United States Geological Survey
L. Passeri Università di Perugia
G. Pialli Università di Perugia
B. H. Purser Université de Paris-Sud
J. F. Read University of Western Australia*
Hans-Erich Reineck Senckenberg-Institut
Antonio Rizzini AGIP Direzione Mineraria
John E. Sanders Barnard College
Paul E. Schenk Dalhousie University
Jean F. Schneider Swiss Federal Institute of Technology
Bruce W. Sellwood University of Reading
J. H. J. Terwindt Rijkswaterstaat—Deltadienst, The Netherlands
Allan M. Thompson University of Delaware
Robert W. Thompson California State University, Humboldt
Glenn S. Visher University of Tulsa
Roger G. Walker McMaster University
Harold R. Wanless University of Miami
R. C. L. Wilson Open University, Walton, England
Peter J. Woods University of Western Australia
Isabel Zamarreño Universidad de Oviedo

* See pages 417–419 for present address.

Tidal Deposits

A Casebook of Recent Examples and Fossil Counterparts

edited by Robert N. Ginsburg

Springer-Verlag
New York Heidelberg Berlin

Library of Congress Cataloging in Publication Data

Main entry under title:

Tidal deposits.

Includes index.

1. Marine sediments. 2. Tidal flats. I. Ginsburg, Robert N.

GC380.T52 551.4'5 75-28228

ISBN 978-3-642-88496-2 ISBN 978-3-642-88494-8 (eBook)

DOI 10.1007/ 978-3-642-88494-8

Softcover reprint of the hardcover 1st edition 1975

(Courtesy of Senckenberg Museum, Frankfurt.)

RUDOLF RICHTER

1881–1957

Pioneer in research on tidal deposits

Drawing by Talitha Eck, Ghancea 1946.
From: Abhandlungen Senckenberg. naturforsch. Ges., 1–336, Frankfurt a. M. 1951 (Festschrift zum 70. Geburtstag Rudolf Richter).

PREFACE

One aspect of the post-World War II expansion of sedimentology is a major increase in research on tidal deposits. Before the war only a few researchers studied modern tidal deposits of the North Sea, and still fewer were concerned with analogous fossil deposits. In the past thirty years, however, interest has spread worldwide to sedimentary rocks of all ages; in Reineck's recent bibliography, almost 90 percent of the 500-odd works appeared after 1945.

Descriptions of tidal deposits have become so numerous, and their presentations so varied, that even specialists find them difficult to deal with, and the sheer volume of publications discourages the interest of students and nonspecialists. In discussing this large and impenetrable literature with George de Vries Klein a few years ago, we both recognized the need for an organized collection of representative examples of tidal deposits; we realized that a selection of well-described examples might stimulate synthesis and classification, and would offer students and nonspecialists a valuable introduction to the subject.

The idea of a compilation of examples was received with enthusiasm by our colleagues, and with their help we collected outline descriptions of examples and suggestions for a standard format. From these outlines and suggestions, and with the advice of John Harms and Richard Boersma, we developed an itemized format, adopted standards symbols, and set a maximum length of six pages of typescript and illustration for each contribution.

Our call for contributions was answered with forty-three examples that were preprinted and distributed to some forty-eight participants. Most of these authors joined a week-long workshop at the University of Miami's Fisher Island Station in February 1973. The workshop included field trips to see recent tidal deposits in coastal Georgia and the Bahamas, and featured four days of review of the contributions and group discussions of various problems of tidal deposits. Following this stimulating exchange and review, the authors had a chance to revise their summaries for the present volume.

It is a pleasure to acknowledge the willing help and support of many persons in organizing the compilation and the workshop. George de Vries Klein was the codeveloper of the idea of the compilation, and together we prepared the standard format and solicited the contributions. John Harms and Richard Boersma reviewed the original outlines and made valuable suggestions for the format, and their critical questions during the workshop were a welcome stimulus to all of us.

I thank James Howard and Harold Wanless for organizing valuable field trips during the workshop; Noel James for his continuing advice and encouragement; Lois Keith for arranging the workshop, corresponding with participants, and helping to prepare the contributions for the press. I am grateful to Donald Heuer and James Hitch for producing the preprints on a tight schedule, and to Cynthia Moore and David Gomberg for editorial assistance.

The drawing of Rudolf Richter was provided by Hans Reineck and I thank the Senckenberg Institute for permission to reproduce it.

The expenses of the workshop were supported by Grant No. GA-36225 from the Geology Program of the National Science Foundation, supplemented by contributions from Amoco Production Company, Chevron Oil Company, Elf Oil Exploration and Production Canada Ltd., Esso Production Research Company, Gulf Research and Development Company, Marathon Oil Company, Mobil Oil Research Laboratory, Occidental Petroleum Corporation, Shell Development Company, Sun Oil, Société National des Pétroles d'Aquitaine, Total Leonard Inc., and the Compagnie Française des Pétroles.

I thank all the contributors to this collection for their willing cooperation and care in preparing their summaries.

Robert N. Ginsburg

SYNOPSIS OF 20th-CENTURY RESEARCH ON TIDAL DEPOSITS

The growth of special interest in tidal deposits follows the same round dance of cause and effect as that of several other topics in sedimentology. In each of these—reefs, chalks, deltas—the development of new concepts and approaches is often triggered by studies of Recent examples, studies that, in turn, were stimulated by the problem of interpreting fossil examples. Darwin's visit to the Pacific atolls can be taken as the beginning of the continually expanding research on reefs. In the same way, Rudolf Richter's first exploration of the Wat-Mer, or tidal flats of the German North Sea coast, in the early 1920s marks the beginnings of systematic research on tidal deposits.

For the next twenty years to the beginning of World War II, Richter and his successors at the Senckenberg Institute focused on the processes of deposition, the sedimentary and organic structures, and the fossils of the North Sea tidal flats. Their results were widely used by researchers on ancient sediments, chiefly as guides to interpret the formation of common sedimentary and organic structures, including ripple-marks, mud-cracks, cross-bedding, interlaminated mud, and sand.

The hiatus in research during World War II was followed, just as after World War I, by a major expansion in sedimentology. Part of the new enthusiasm was renewed interest in Recent tidal-flat deposits of the North Sea, at first in the Netherlands by L. M. J. U. van Straaten, soon a renaissance at the Senckenberg Institute under Hans Reineck, and then studies in England by Graham Evans. The appearance for the first time in English of comprehensive works on Recent tidal deposits stimulated further research. One branch led to study of other examples of siliciclastic tidal deposits in the Recent—Bay of Fundy, Atlantic Coast of North America, Baja California—and in the fossil record, ranging from Precambrian to Mesozoic; the other branch led in an entirely new and different direction to the study of carbonate tidal deposits.

The early reports on carbonates, unlike those on siliciclastics, described fossil examples; the interpretations combined the use of sedimentary structures and Walther's law of facies succession to recognize zones of accumulation that were determined by the frequency of exposure or flooding—tidal zones in the broadest sense. This quantum jump in the precision of reconstructing ancient environments stimulated research on a variety of Recent and fossil examples; it led to the discoveries of synsedimentary dolomite, gypsum, and anhydrite, and to the development of a practical model of facies succession.

The meteoric rise of interest in carbonate tidal deposits, the expansion of research on Recent siliciclastics to North America, and to the fossil record worldwide have together produced a substantial literature. An earlier bibliography emphasizing works on the North Sea (Straaten, 1961) listed 240 papers, and the most recent comprehensive one, only twelve years later (Reineck, 1973), cites nearly 500 works.

CONTENTS

SECTION VI

Recognition of Ancient Carbonate Examples: Sedimentary Features and Facies Patterns

Section I

Recent Siliciclastic Examples

As indicated in the preceding synopsis, twentieth-century research on tidal deposits began on the German North Sea coast around 1920, and subsequently spread to the Netherlands, England, and France. Over a period of some fifty years, research in this region has produced a major amount of information on all aspects of tidal deposition.

The North Sea area is particularly appropriate for studying the sedimentology of tidal flats because there is an appreciable tidal range, both mud and sand-sized sediment are available, and there is a variety of animal life. Together these three characteristics have facilitated the development of several major contributions: (1) the overall horizontal and vertical zonation of textures, structures, and organic remains; (2) the origin, description, and zonal distribution of stratification and trace fossils; (3) the net shoreward movement of mud (settling lag and scour lag); and (4) the distinction between the process and record of sedimentation on horizontal surfaces from that in channels. The first two contributions in Section I summarize examples from this classic area; a third by Larsonneur describes a tidal flat in which the sediments are surprisingly rich in lime sand and lime mud.

The North Sea coast is but one of several settings in which holocene tidal deposits occur; several of the other settings are represented in Section I: Harrison and Kraft and Allen describe examples of transgressive tidal-flat deposits in barrier island complexes along the Atlantic Coast; Knight and Dalrymple report the structures and textures of a bay in Nova Scotia where the tidal range is up to 17 meters and drifting ice affects deposition; Thompson emphasizes the importance of an arid climate and the predominance of fine-grained sediment in his description of tidal flats on the margin of the Colorado River Delta; Miller's example from Laguna Madre along the Texas Gulf coast shows that in an arid climate wind can be the major agent of flooding and deposition.

The final two contributions focus on deposition in tidal inlets: The first by Kumar and Sanders reports the sequence of structures formed in a migrating inlet of a barrier island along the Atlantic Coast; the second by Terwindt describes vertical sequences of structures and textures formed in inlets along the Dutch coast and relates the different sequences to variations in the maximum current velocity near the bottom.

In modern environments, where one can measure

Table 1 *Tidal zone terminology*

	Supralittoral[a]	Supratidal[a]	↑		↑	↑
High water						
	Littoral[a]	Intertidal[a]	Intertidal flats[b]	Tidal-flat environment[e]	Tidal flat[d]	Coastal mud flats[c]
Low water						
	Infra- or sub-littoral,[a] neritic (to 100 meters)	Subtidal[a] (lower limit unspecified)				To 12 meters below mean sea level

[a] Gary, M., McAfee, R., and Wolf, C. L., eds. 1972. *Glossary of Geology*. Am. Geol. Inst., Washington, D. C., 857 pp.

[b] Evans, G., 1965, Intertidal flat sediments and their environments of deposition in The Wash. *Quart. J. Geol. Soc. 121*, 209–245.

[c] *Straaten*, L. M. J. U. van, 1954. Composition and structure of recent marine sediments in the Netherlands. *Leidse Geol. Mededel. 19*, 1–110.

[d] Reineck, H.-E. 1967. Layered sediments of todal flats, beaches and shelf bottoms of the North Sea: Estuaries. *Am. Assoc. Advan. Sci. 83*, 191–206.

[e] *Thompson*, R. W. 1968. Tidal flat sedimentation on the Colorado River delta, northwestern Gulf of California. *Memoir 107*, Geol. Soc. Am., Boulder, Colo., 133 pp.

the frequency and range of sea level fluctuations, the terms "tide" and "tidal" are usually reserved for the regular fluctuations produced by the gravitational attractions of the moon and sun—astronomic tides. There is rather general agreement on the boundaries of the tidal zones, but there are differences among some authors on the upper limits of "intertidal" and "tidal flat" (Table 1). The terminology of ancient tidal deposits is reviewed in the introduction to Section II.

1

German North Sea Tidal Flats

Hans-Erich Reineck

Occurrence

Holocene tidal flats on the southern border of North Sea. Length: 450 km. Width: 5 to 7 km, max. 10 to 15 km. Cross section: wedge-shaped. Intersected by rivers and tidal channels.

Hydrography

Tide range: semidiurnal. Spring: 2.6 to 4.1 meters. Neap: 1.8 to 3.1 meters. Tidal current velocities: intertidal flats, 30 cm/second, max. 50 cm/second; smaller channels 100 cm/second; larger channels 150 cm/second.

Sediments

Reworked glacial sediments. Gravels: morainal gravels, shells, mud pebbles. Sand: beach sediments are mainly medium sand; intertidal flat sand is generally fine. Mud: muddy sediments (clayey silt with some fine sand) are found in smaller channels, gullies, mixed flats, mud flats, and supratidal zone. Sediments commonly contain many fecal pellets, Foraminifera, ostracods, spines of the heart urchin (*Echinocardium*), and peat debris.

Distribution of Bedding Types

SUBTIDAL ZONE. The main morphologic features of subtidal zones are channels and sandbars. In the channels bedforms are characterized by megaripple and small-scale cross-bedding. Bipolar directions result in the development of herringbone cross-bedding. Bioturbation is rather low. Shoals are usually medium to fine sand, with well-developed, small-scale ripple cross-bedding. On the channel walls lenticular and flaser bedding and interlayered sand/mud bedding are abundant (Figs. 1–1, 1–2, 1–3). Mud content increases with decreasing channel size, especially in longitudinal cross-bedding, but also in the channel bottom. Longitudinal cross-beds of channels are often muddier than the surrounding intertidal flat sediments. Longitudinal cross-bedding generated on point bars is equally common in the intertidal zone (Figs. 1–4, 1–5). Longitudinal cross-bedding is usually less bioturbated than the surrounding sediments.

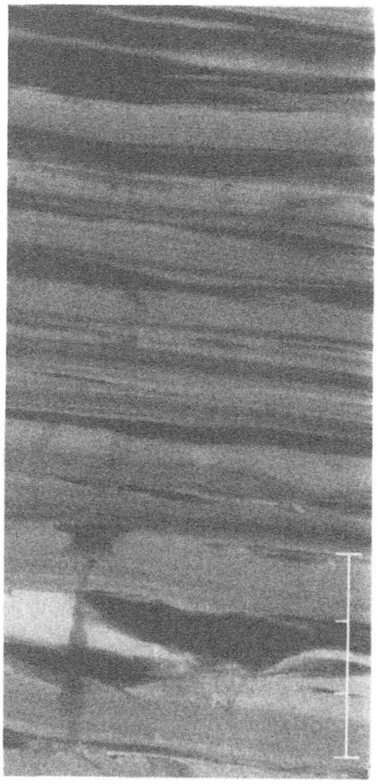

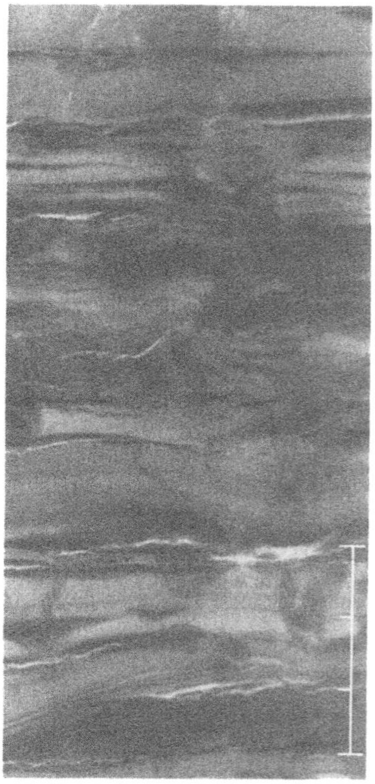

Figure 1-1 (left). Lenticular bedding. Intercalation of flat sand lenses and laminated mud layers from channel wall deposits in the subtidal zone. Water depth, 6 meters. (Radiographic print: dark, sand; gray, mud; scale in centimeters.)

Figure 1-2 (right). Interbedding of mud and sand with horizons of bioturbation structures of bottom-living shells; channel wall deposits of subtidal zone. Water depth, 4 meters. (Radiographic print: dark, sand; gray, mud.)

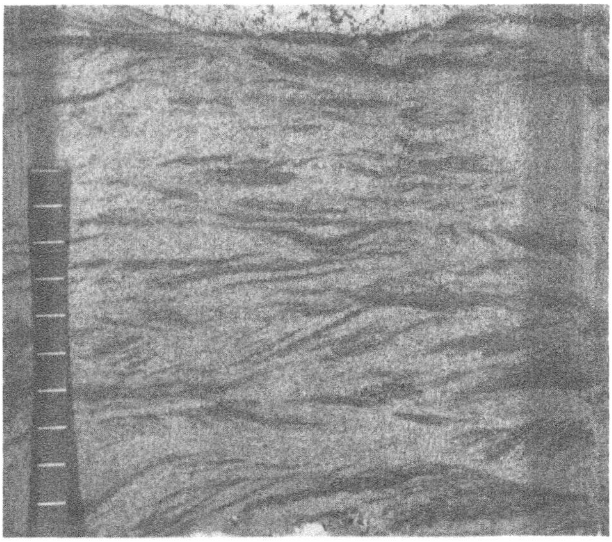

Figure 1-3. Flaser bedding and small-scale ripple bedding from sandy part of mixed flats, showing darker foreset laminae of muddy fecal pellets (scale in centimeters).

Figure 1-4. Tidal-flat surface expression of channel wall deposits showing longitudinal cross-bedding.

Figure 1-5. Longitudinal cross-bedding of filled channel migrating to the right and overlaid by horizontal bedding.

Figure 1-6. Strongly bioturbated mud flat sediments. (Radiographic print: dark, sand; gray, mud.)

Figure 1-7. Small-scale ripple cross-bedding from sand flats. (Radiographic print: dark, sand; gray, mud.)

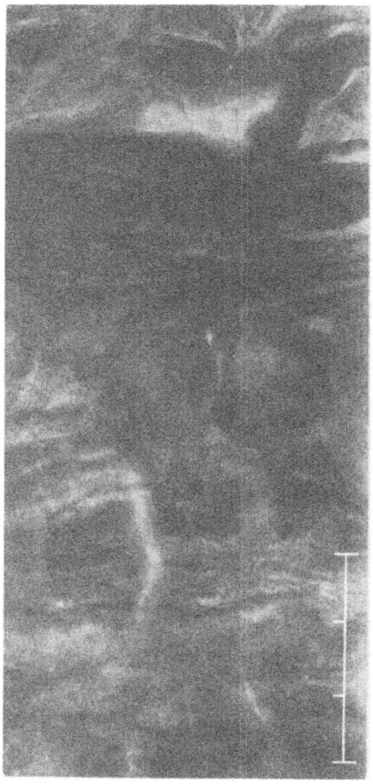

Alternating mud and sand layers are caused by alternating current and slack water conditions. Changes in bedsets are due to changes in weather conditions (wind and wave) (Reineck and Wunderlich, 1969).

INTERTIDAL ZONE. The intertidal zone can be divided into mud flats, located near the high-water line; sand flats, near the low-water line; and mixed flats, a broad transition zone between mud flats and sand flats. The net rate of sedimentation on intertidal flats is rather low, therefore the sediments are more or less strongly bioturbated (Fig. 1–6). Most of the organisms living here are infauna consisting of marine mollusks and polychaetes (Schäfer, 1956).

Sand flats are characterized by well-developed, small-scale ripple cross-bedding (Fig. 1–7) (both current and wave). Less common are laminated sand, flaser bedding, and wavy bedding. Climbing ripple lamination is restricted to the mouths of gullies. The commonest bedding types in mixed-flat environments are flaser and lenticular bedding (Fig. 1–3) in all variations (Reineck, 1972), with thinly and thickly interlayered sand-mud beds. The interlayered sand-mud bedding is especially well de-

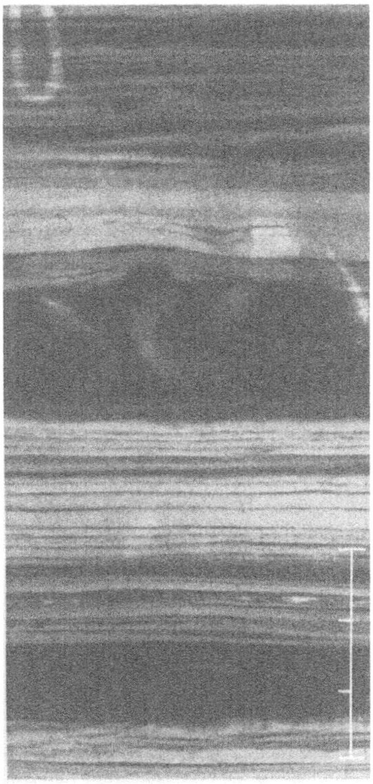

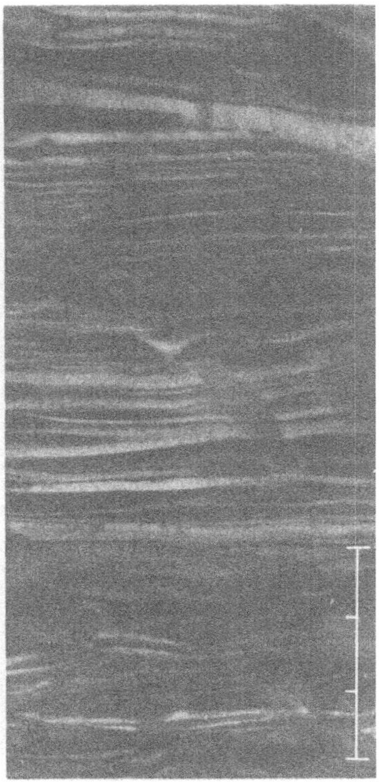

Figure 1-8 (left). Interlamination of mud and sand from mixed tidal flats. U-shaped burrow of *Corophium* in upper part of core. (Radiographic print: dark, sand; gray, mud.)

Figure 1-9 (right). Interlamination of sand and mud from mud flats. The sand laminae are often flat lenses. (Radiographic print: dark, sand; gray, mud.)

veloped in point bar channel deposits (Fig. 1–8) (longitudinal cross-bedding). Horizons of shells in living position are common.

Sand is rare in mud flats, but is often found as thin stripes and intercalations within mud layers (Fig. 1–9). Sometimes the sand is in the form of thin lenticular bedding. Sediments of mud flats are rather strongly bioturbated (Fig. 1–6). The intertidal zone of the German coast has no vegetation, except for some algal mats (Fig. 1–10), sea grass, and *Salicornea herbacea* close to the high-water line. Surface markings and imprints are commonly found on intertidal flats. They are controlled by weather conditions; imprints of rain, hail, and ice needles are produced. During storms the surface is planed off. Strong winds mainly produce wave ripples and occasionally longitudinal ripples in mud. During calm weather, current ripples are abundant, and surface animal tracks are well developed.

SUPRATIDAL ZONE. In the supratidal zone sedimentation takes place during storm tides when the sea level is high enough to push water on the salt marsh above the high-water line. The supratidal zone is vegetated by halophytes. Sandy and muddy layers deposited during storms can be traced over several kilometers. Shell layers, brought by storm waves, are found interbedded with salt marsh sediments. The bedding is uneven, and penetrated by plant stems and roots (Straaten, 1954).

Figure 1-10. Patches of eroded algal mat.

1. North Sea Tidal Flats, Germany

Reineck

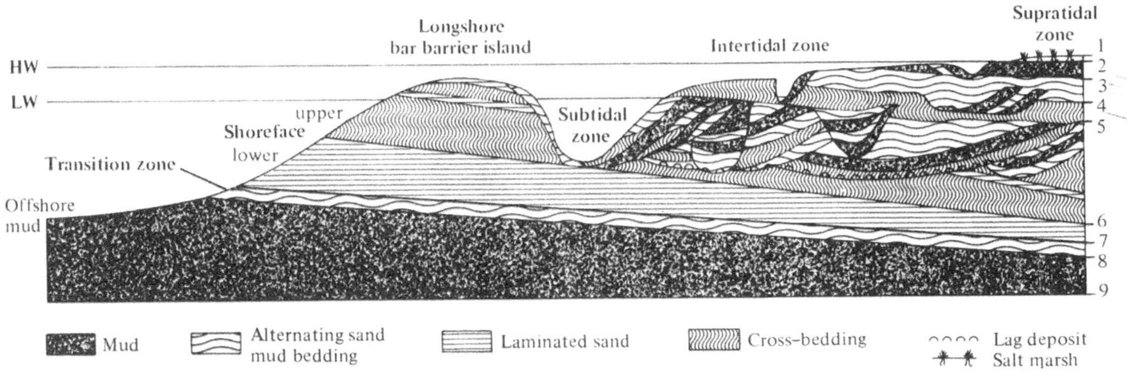

Figure 1-11. Schematic cross sections across the tidal flats of the North Sea of a hypothetical prograding shoreline. Actually, the North Sea tidal flats are resting on Pleistocene and Holocene sediments. (Based on Straaten [1954] and additional works.)

1. Salt marsh (supratidal zone). Very fine sand and mud, interbedded shell layers, plant roots, irregular wavy bedding *2.* Mud flat (intertidal zone). Mud, occasional very fine sand layers, lenticular bedding with flat lenses, strongly bioturbated. *3.* Mixed flat (intertidal zone). Sandy mud, thinly interlayered sand, mud bedding, lenticular bedding, flaser bedding, shell layers, bioturbation strong to weak. *4.* Sand flat (intertidal zone). Very fine sand, small ripple bedding, sometimes herringbone structure, flaser bedding, laminated sand, occasional strong bioturbation. *5.* Subtidal zone. Medium to coarse sand, shells, mud pebbles, megaripple bedding, small ripple bedding, laminated sand, weak bioturbation. *6.* Upper shoreface. Beach bar and ripple cross-bedding and laminated sand, weak bioturbation *7.* Lower shoreface. Bioturbation stronger than in the upper shoreface, laminated sand. *8.* Transition zone. Alternating sand-mud bedding, that is, flaser and lenticular bedding, thinly and thickly interlayered sand-mud bedding, moderate bioturbation. *9* Shelf mud. Mud with storm silt layers, moderate bioturbation.

Summary

Subtidal channels are cut in older sediments. Subtidal deposits are mainly channel and shoal deposits. Channel deposits are characterized by discordances. Even the intertidal zone shows abundant channel deposits with erosional disconformities (Fig. 1–11). The deposits often show strong bioturbation and horizons of molluskan shells in living position. Alternating bedding is abundant, that is, flaser and lenticular bedding, thinly alternating sand-mud layers, and thickly alternating sand-mud layers (Straaten, 1954; Reineck, 1972). Surface markings indicating subaerial exposure are present. Sediments of the supratidal zone are distorted by roots, and bedding is uneven.

References

REINECK, H.-E. 1972. Tidal flats. In Rigby, D. K., and Hamblin, Wm. K., eds. Recognition of ancient sedimentary environments. *SEPM Spec. Publ. 16*, 146–159.

REINECK, H.-E., and WUNDERLICH, F. 1969. Die Entstehung von Schichten und Schichtbänken im Watt. *Senckenbergiana maritima 50*, 85–106.

SCHÄFER, W. 1956. Wirkungen der Benthos-Organismen auf den jungen Schichtverband. *Senckenbergiana lethaea 37*, 183–263.

STRAATEN, L. M. J. U., VAN. 1954. Composition and structure of recent marine sediments in the Netherlands. *Leidse Geol. Mededel. 19*, 1–110.

2

Intertidal Flat Deposits of the Wash, Western Margin of the North Sea

Graham Evans

Occurrence

Holocene intertidal flats fringe the greater part of the coastline of this large embayment on the east coast of England. The more exposed coastlines to the north and east, outside the embayment, have sandy beach-dune shorelines except where coastal bars, barrier spits, or barrier islands have produced locally protected areas behind which intertidal flats have been able to develop.

Physiography

The width of the intertidal flats varies from 0.8 to 6.5 km, and the vertical amplitude is approximately 7.0 meters. The intertidal flats pass seaward into a shallow seafloor which is scoured by tidal currents and usually covered with sand and gravel, but occasionally older clays and peats crop out. Landward, the intertidal flats are bounded by an artificial embankment that is sometimes topped by exceptionally high tides, such as occurred in 1953. Inland of the embankment are extensive, reclaimed intertidal lands that attain a maximum width of 32 km and that form some of the richest and most expensive arable land in England (Inglis and Kestner, 1958). The intertidal zone consists of an inner salt marsh colonized by halophytic plants, a zone of mud flats with an irregular surface (except in its earliest stage of development when it is smooth) and broad sand flats that extend down to low water. In various places, another mud belt occurs near low water—usually bordered by a narrow sand flat immediately adjacent to the low-water mark. The complete series of zones, some of which are often absent, is: (1) higher intertidal zone (salt marsh, higher mud flats); (2) lower intertidal zone (inner sand flat, *Arenicola* sand flat, lower mud flat, lower sand flat) (Evans, 1965). The intertidal zone is dissected by a series of creeks, which are feeder channels to the flats and also drain the area. These creeks have no freshwater sources.

Climate and Hydrography

The area lies in a temperate climatic zone with moderate temperatures (mean daily average 10°C) and rainfall (mean annual average 61.0 cm) with precipitation in excess of evaporation (45.0 cm/year). The predominant wind is offshore from the southwest, but strong northeast inshore winds are common. The seawater temperatures of the adjacent sea range from 3° to 10°C and salinities are approximately 32 to 33‰, although more brackish and saline waters sometimes occur on upper parts of the intertidal zone. The average spring tidal range is 6.5 meters and the average neap tidal range is 3.5 meters. The maximum tidal current velocities vary from 0.9 to 1.2 meters/second in the offshore zone to 0.7 meter/second on the outer intertidal flats, 0.3 to 0.4 meter/second on the inner flats, and 0.1 meter/second on the higher mud flats and salt marshes; minimum tidal current velocities are often close to zero. Waves are usually less than 1.0 meter in amplitude, but may be higher during storms.

Holocene Sequence

The intertidal flat sediments are the youngest deposits of a Holocene sequence, which attains an average thickness of approximately 20 meters (in the buried river valleys it may be even thicker). This sequence consists of peats, brackish water silty clays with subsidiary buried forest beds and thin freshwater marls. Capping this sequence, on its seaward side, is a series of intertidal flat sands, silty sands, clayey silts, and silty clays. Also, these latter deposits infill old channels cut in the earlier deposits; these channel fills, due to compaction and wastage of the peats, now form upstanding meandering ridges called roddons. The deposition of this intertidal flat series commenced about 2000 BP, as a result of the so-called Romano-British transgression. Since that time, most of the coast has been regressive. The intertidal flats have prograded seaward, and this natural accretion has been substantially aided by man's engineering works on the drainage network and land reclamation. In many places the natural accretion has been very slow and has only attained measurable rates when man has interfered with natural processes (Willis, 1961).

Sediments

Sediments range from sands to silty sands to clayey silts to silty clays. The sand and coarse silt usually occur as well-sorted layers interlaminated with silty clay and clayey silt layers. Bioturbation has frequently mixed these various types. The sands are mainly quartz with small amounts of feldspar (orthoclase 0 to 15 percent; plagioclase, 2 to 3 percent; and a few grains of microcline), mica (0 to 4 percent), and rock fragments, including coal. Admixed with these are molluskan shell debris, plant fragments, foraminiferal tests, echinoderm plates and spines, bryozoan tests, diatom tests, calcispheres, and various aggregates, including fecal pellets and other minor components. Pyrite is commonly found developing within the sediment (Love, 1967). On the average, the accessory minerals consist of augite (18 percent), amphiboles (16 percent), and garnet (11 percent), with small amounts of apatite, epidote, alterite, tourmaline, rutile, staurolite, zircon, and kyanite (Chang, 1971). This assemblage differs slightly from that found in the sediments of the beaches and dunes of the neighboring North Sea coast and of the adjacent seafloor. However, these differences are largely the result of differences in grain size rather than differences in source. The sand appears to have been derived from the adjacent seafloor and neighboring coasts to the north and east of the Wash by reworking and erosion of the tills and outwash of the last Weichselian ice sheet in the area.

The clays consist of illite (40 percent), illite-montmorillonite (30 percent), kaolinite (20 percent), and chlorite (10 percent). Because these clays are very similar in composition to those supplied by the rivers draining into the area and to those being eroded from the neighboring coasts and adjacent seafloor, it is impossible to assign a definite source to them (Shaw, 1961).

SALT MARSH. Laminated silty clays and clayey silts with subsidiary laminae of coarse silt and sand, with sand and silt decreasing inland and away from the creeks (Fig. 2–1); iron oxide and hydroxide-coated mud cracks are common; plant debris and plant root structures coated with iron oxide and hydroxide are abundant (Love, 1967). The deposits of shallow pools (salt pans) are intensively bioturbated, mainly by worms and crabs and certain *in situ* surface-feeding gastropods (*Littorina* sp., *Hydrobia* sp.) and burrowing bivalves (*Scrobicularia* sp., *Macoma* sp.) (Fig. 2–2).

HIGHER MUD FLATS. Laminated and interlaminated silty clays and clayey silts with sand and coarse silt more abundant than in the marsh deposits (Figs. 2–2, 2–3); both symmetrical and asymmetrical ripple marks are present paralleling the shoreline with the latter commoner; iron oxide and hydroxide-coated mud cracks are abundant; shallow scours oriented perpendicularly to the shoreline and infilled with clasts of silty clay and shell debris are common; poorly developed stromatolitic lamination is present. Scattered bivalves (*Scrobicularia* sp., *Macoma* sp., *Mya* sp.) and gastropods are present, particularly on the seaward parts; considerable bioturbation by worms and occasionally crabs occurs mainly on the seaward parts.

Figure 2-1. Exposure produced by the migration of a creek, showing the upper part of the progradational intertidal flat sequence.

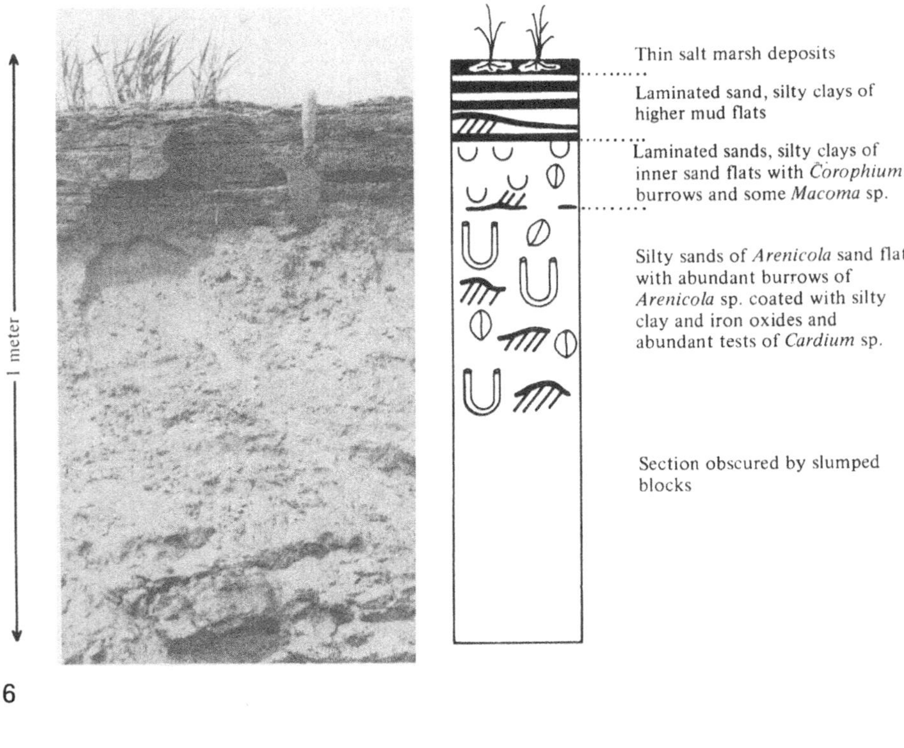

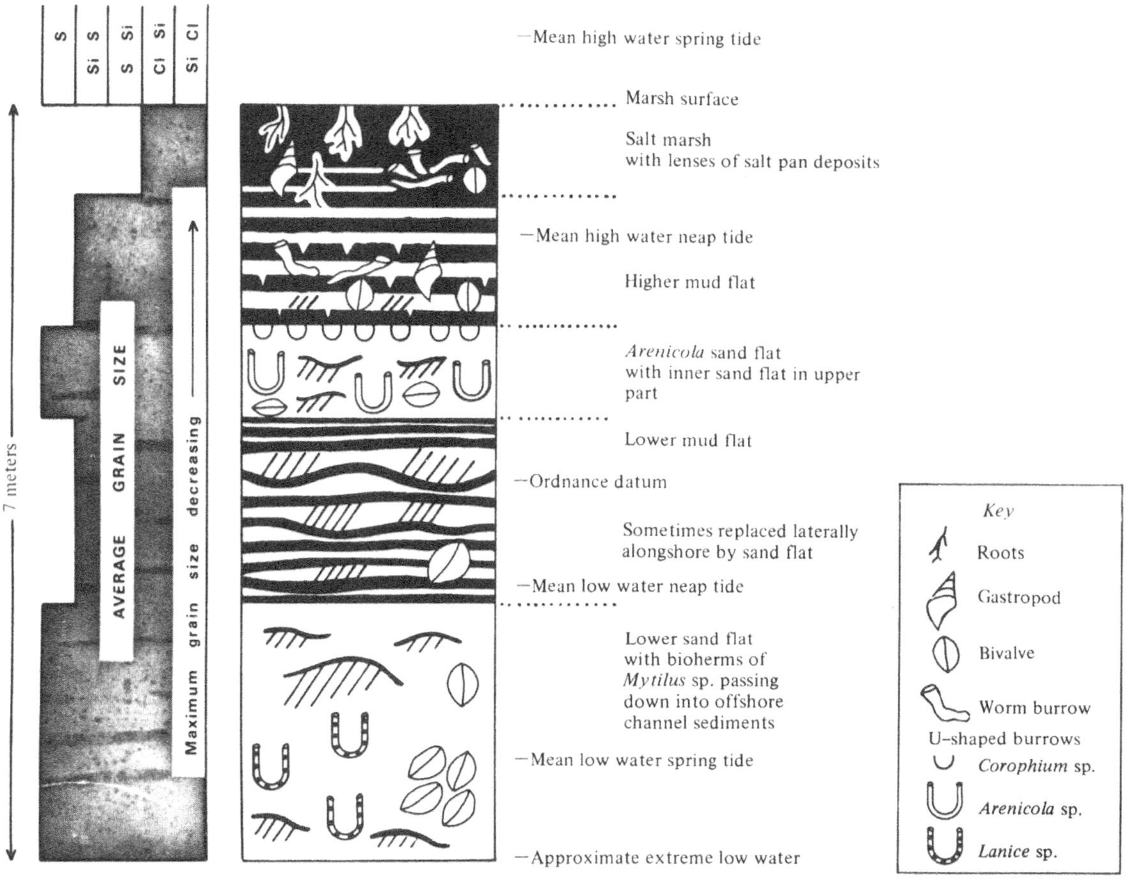

Figure 2-2. Sediment log of hypothetical intertidal flat sequence, assuming that each zone progrades seaward over the adjacent zone (note that the lower mud flat is not always present). This sequence is cut by the deposits of the creeks, which may in places entirely replace it.

INNER SAND AND *Arenicola* SAND FLAT (Figs. 2–1, 2–2, 2–3). Laminated or ripple-laminated sands and silty sand with silty clay and clayey silt drapes and some flat lamination; abundant ripple marks, both symmetrical and asymmetrical with the latter commoner, generally paralleling the shoreline; abundant bioturbation by crustaceans (*Corophium* sp.) (Fig. 2–3), on the inner sand flat and by worms (*Arenicola* sp.) (Fig. 2–3) on the seaward *Arenicola* sand flat; some burrowing bivalves, particularly on landward parts (*Scrobicularia* sp., *Macoma* sp., *Mya* sp.) and abundant *Cardium* sp. on the seaward part; thin lenses of molluskan debris are sometimes present.

LOWER MUD FLATS (Figs. 2–2, 2–3). Interlaminated ripple-laminated sand with silty clay and clayey-silt; elsewhere ripple-laminated sands with silty clay and clayey silt drapes; abundant ripple marks symmetrical and asymmetrical, parallel, oblique, and at right angles to the shoreline; abundant small scours ori-

Figure 2-3. Cores illustrating the sediments of the various zones: salt marsh (*1*), higher mud flats (*2*), inner sand flats (*3*), *Arenicola* sand flats (*4*), lower mud flats (*5*), lower sand flats (*6*). The key to the annotation on the core photographs is *a*(*2,6*) silty clay drapes on ripple marks; *b*(*2*) algal banding?; *b*(*3*) silty-clay-lined *Corophium* sp. burrows; *d*(*1*) silty-sand-infilled plant root cavities; *f*(*2*) burrow; *f*(*4*) silty-clay-lined *Arenicola* sp. burrows; *g*(*2*) silty-clay-lined worm burrows.

entated perpendicularly to the shoreline infilled with silty clay and clayey silt clasts and shell detritus, some *in situ* mollusks and also some bioturbation, but, although locally common, this is rarer than in the adjacent *Arenicola* and inner sand flat deposits.

Lower sand flats (Figs. 2–2, 2–3). Laminated ripple-bedded sands and silty sands with silty clay and clayey silt drapes; elsewhere, festoon- and planar cross-bedding sands dipping alongshore with silty clay and clayey silt drapes; abundant ripple marks parallel and perpendicular to the shoreline; scattered burrowing bivalves and surface-dwelling gastropods and low biohermal mounds of bivalves (*Mytilus* sp.); very noticeable abundance of agglutinated worm tubes (*Lanice* sp.) near low-water mark.

Creeks and adjacent areas. The floors of the creeks consist mainly of ripple cross-laminated sands with silty clay drapes; interbedded layers of clasts of silty clay and clayey silt and molluskan shell debris; asymmetrical ripple marks parallel to the shoreline; abundant evidence of contemporaneous slumping and soft sediment intrusion; slip-off banks of meandering creeks consist mainly of laminated and ripple cross-laminated sands and silty sands with silty clay and clayey silt laminae; sets of climbing ripples and herringbone cross stratification are common only in these creek deposits; large "ball and pillow" structures are found in some locations; iron oxide and hydroxide-coated mud cracks are abundant; creek meandering produces a sequence of festoon-bedded sediments oriented obliquely to the shoreline; some burrowing mollusks (*Macoma* sp., *Scrobicularia* sp., *Mya* sp.) and some bioturbation by worms and crabs.

The Vertical Sequence

Both historic evidence and repeated surveying indicate coastal progradation (Inglis and Kestner, 1958). This progradation has produced a generally fining-upward sequence of sediments, sometimes interrupted by the deposits of the lower mud flats, which reflect the transition from the relatively high energy zones found near lower water marks to the lower energy conditions found on the salt marshes (Fig. 2–2). The sediments of the zones thus form a series of flat sheets, each of which is, of course, diachronous. Only the upper part of the sequence has been proven in pits, and it is possible that there has been in some places a vertical *in situ* build-up of sediments rather than a horizontal progradation. This fining-upward sequence—approximately 7.0 meters thick, of laminated and interlaminated sand, silty sands to silty clays—is interrupted by deposits of meandering creeks that often show large-scale trough cross-stratified units due to sideways infilling. Available evidence suggests that the creeks are restricted to meander belts and do not rework the whole intertidal flats; however, meander belts of adjacent creeks may unite in some locations to completely replace the flat sheets of deposits of the various zones with creek deposits (Willis, 1961; Evans, 1965).

Without man's interference, the broad coastal plain of reclaimed intertidal flats would have been covered by swamp in which peat would have accumulated so that the natural top unit of the sequence would be peat or, in the geologic column, coal. The whole sequence has an abundant but poorly diversified infauna and epifauna with trace fossils produced by burrowing worms and *in situ* mollusks forming the dominant element; mixed with these is a derived, displaced, and *in situ* restricted microfauna. If the offshore channel became infilled and the intertidal flat built seaward over it, the intertidal flat sequence of sediments would be underlain by up to 10 meters of offshore channel sediment, which consists of festoon cross-bedded sands with silty clay drapes interbedded with gravels composed of mainly shells and silty-clay clasts.

References

CHANG, S. C. 1971. A study of the heavy minerals of the coastal sediments of the Wash and adjacent areas. M. Phil. thesis, Univ. of London, p. 223.

EVANS, G. 1965. Intertidal flat sediments and their environments of deposition in the Wash. *Quart. J. Geol. Soc. London 121*, 209–245.

INGLIS, C. C., and KESTNER, F. J. T. 1958. Changes in the Wash as affected by training walls and reclamation works. *Proc. Inst. Civ. Eng. 11*, 435–466.

LOVE, L. G. 1967. Early diagenetic iron sulphide in Recent sediments of the Wash, England. *Sedimentology 9*, 327–357.

SHAW, H. F. 1973. Clay mineralogy of Quaternary sediments in the Wash embayment, eastern England. *Mar. Geol. 14*, 29–46.

WILLIS, E. H. 1961. Marine transgression sequences in the English Fenlands. *Ann. N.Y. Acad. Sci. 95*, 368–376.

3

Tidal Deposits, Mont Saint-Michel Bay, France

Claude Larsonneur

Occurrence

The Bay of Mont Saint-Michel is deep inside the gulf of Saint-Malo in the western part of the English Channel, between Brittany and the Cotentin Peninsula (Fig. 3–1B). It constitutes a wide-open basin slanting regularly toward the northwest and opening on to the sea along a distance of 20 km, between the Pointe du Grouin and the cliffs of Carolles. The elegant outline of Mont Saint-Michel towers above the bay, which is characterized by an exceedingly wide beach (up to 15 km) furrowed by a shifting dendritic pattern of channels at low tide. Toward the south, it merges into some low coastal lands reclaimed from the sea and now cultivated: the Dol marshes, the polders of Mont Saint-Michel (Fig. 3–1D).

Climate and Hydrography

The climate is a temperate, oceanic one: moderate temperatures (annual average 11.5°C at the Pointe du Grouin), frequent and fairly well distributed rainfall (750 to 800 mm/year). Sea-

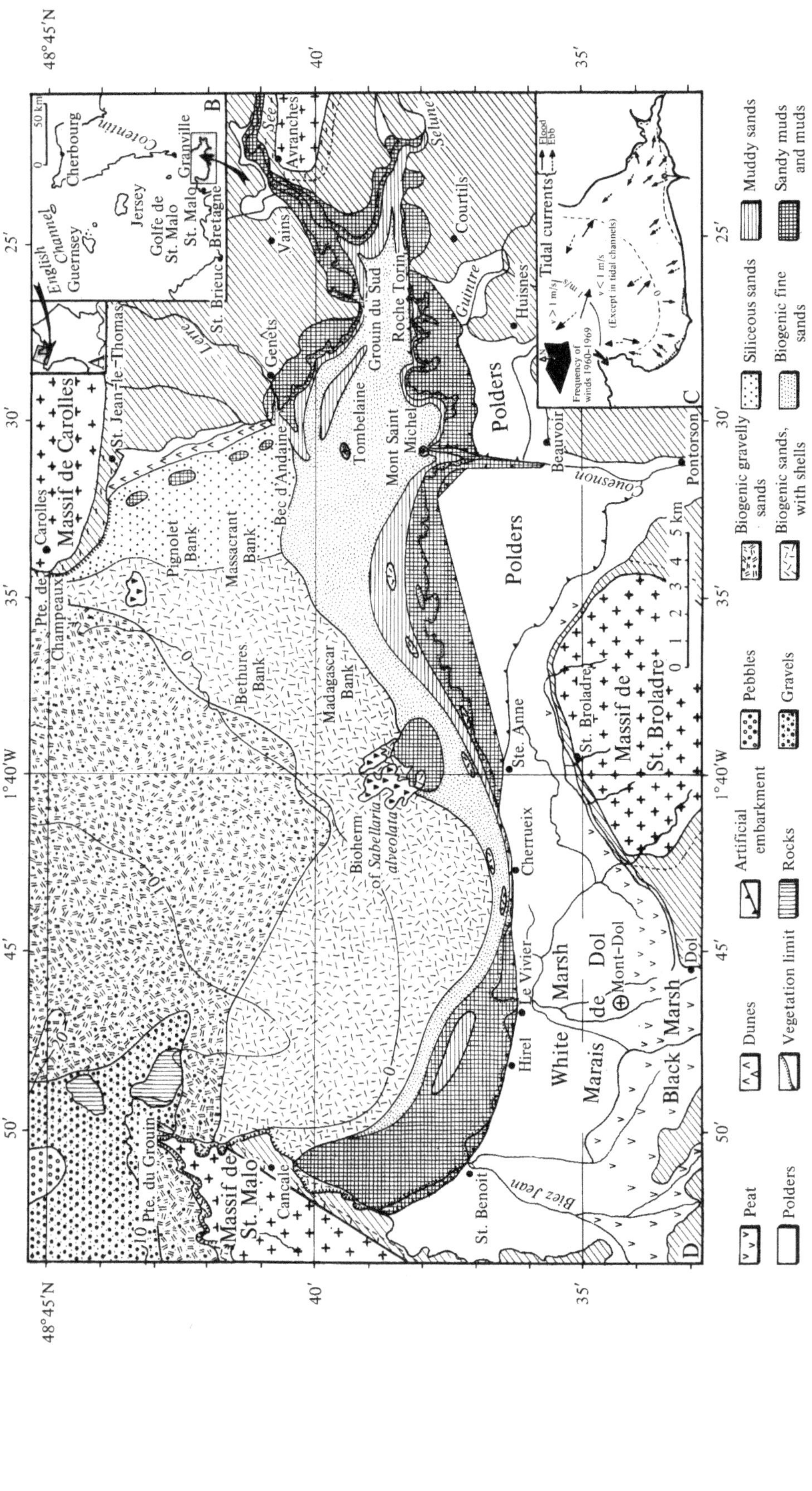

48°45′N
40′
35′
50′
45′
1°40′W
30′
25′
Massif de Carolles
Carolles
Pte. de Champeaux
St. Jean-le-Thomas
Pignolet Bank
Massacrant Bank
Bethures Bank
Madagascar Bank
Bec d'Andaine
Genêts
Vains
Avranches
Sée
Sélune
Courtils
Guintre
Huisnes
Tombelaine
Grouin du Sud
Roche Torin
Mont Saint Michel
Polders
Beauvoir
Couesnon
Pontorson
Ste. Anne
St. Broladre
Massif de St. Broladre
Cherrueix
Bioherm of Sabellaria alveolata
Le Vivier
Hirel
White Marsh
Marais de Dol
Mont-Dol
Black Marsh
Dol
St. Benoit
Biez Jean
Massif de St. Malo
Cancale
Pte. du Grouin
English Channel
Guernsey
Jersey
Golfe de St. Malo
St. Malo
St. Brieuc
Bretagne
Granville
Cotentin
Cherbourg
50 km
5 km
Tidal currents
Flood
Ebb
v > 1 m/s
v < 1 m/s (Except in tidal channels)
Frequency of winds 1960–1969
A
B
C
D
Peat
Polders
Dunes
Vegetation limit
Artificial embarkment
Rocks
Pebbles
Gravels
Biogenic gravelly sands
Biogenic sands, with shells
Siliceous sands
Biogenic fine sands
Muddy sands
Sandy muds and muds

water temperatures vary from 7°C in February to 18°C in August. Salinity fluctuates between 33 and 35‰ (slightly higher in summer) and turbidity is generally high owing to the fine particles that are reworked by each tide. The rivers flowing into the bay—principally, the See, the Selune, and the Couesnon—have a small flow (20 cubic meters/second for the Selune in middle water), but they form ebb and flood channels that direct the dynamics of sedimentation.

The semidiurnal tide is characterized by its exceptional range, reaching 15.3 meters (about 5 meters in neap tides). Since the beach slopes gently seaward, especially in the east part (1‰), the intertidal flats are very large (200 km^2). The velocity of tidal currents (Fig. 3–1C), more than 1 meter/second north of the Pointe du Grouin, gradually decreases at the upper beach, except in tidal channels where it remains constantly high (1 to 2 meters/second and higher during spring tide). Generally these currents are reversing and the flood to the southeast is faster than the ebb. However, giratory currents exist in the western part of the bay. Wave action is weak or moderate, rarely violent; it is mainly the result of local winds. In this region, dominant winds, which are also the strongest, blow from the west (Fig. 3–1C). The western part of the bay is under the prevailing influence of rather frequent, strong northeasterly winds (Phlipponneau, 1956; Mathieu, 1966).

Sedimentary Environments and Sediments

Typical sedimentary sequence

The different facies of this sequence (Figs. 3–1 and 3–2) follow each other from the depth of about −20 meters to the estuarian area situated between Cherrueix and the Bec d'Andaine. It includes deposits decreasing in grain size from the open sea to the coastline, corresponding to the gradual decrease in velocity of the tidal currents (Bourcart and Boillot, 1960; Dolet, Giresse, and Larsonneur, 1965; Giresse, 1969; Mathieu, 1966).

1. Supratidal zone (Figs. 3–3, 3–4, 3–5). This zone, situated above +12 meters, is flooded only at high-water spring tides (30 percent of the tides) and is invaded by halophyte plants whose density increases as one gets nearer the salt marsh, which is locally separated from the higher mud flats by a small cliff. The deposits (sandy muds), called *tangue*, contain from 15 to 75 percent of fine particles (<0.05 mm); the median varies from 0.03 to 0.09 mm; the carbonate content (skeletal debris) fluctuates from 40 to 50 percent. These deposits show finely laminated bedding and lenticular bedding, sometimes weakly

Figure 3-2. Typical sedimentary sequence.

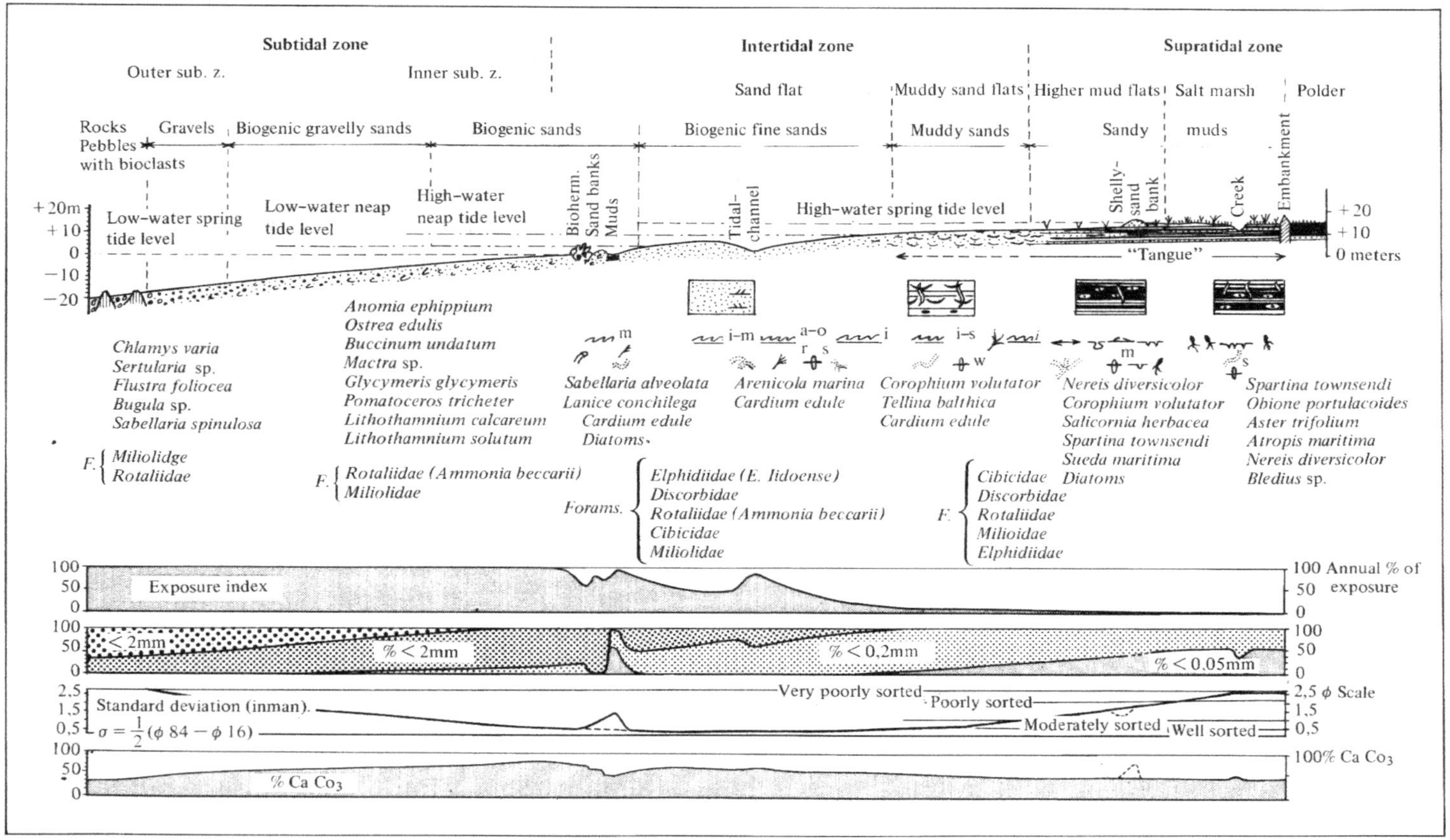

Figure 3-3. Salt marsh connected to the higher mud flats by a small cliff eroded by wave action at Genêts.

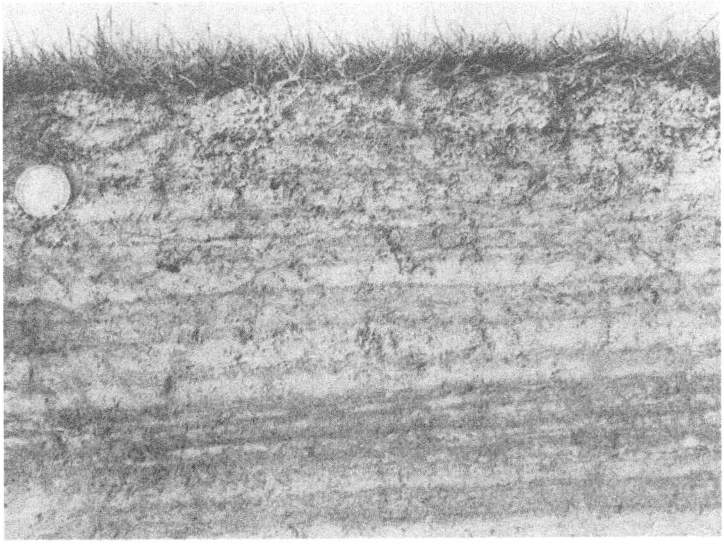

Figure 3-4. Vertical section in a salt marsh. Laminated sand-mud bedding and lenticular bedding, moderately bioturbated (estuarian zone).

Figure 3-5. Dune lying on an old salt marsh near Bec d'Andaine at the limit of the estuarian zone.

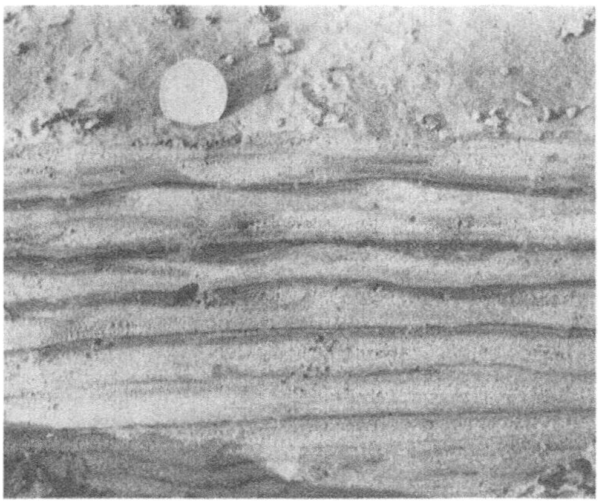

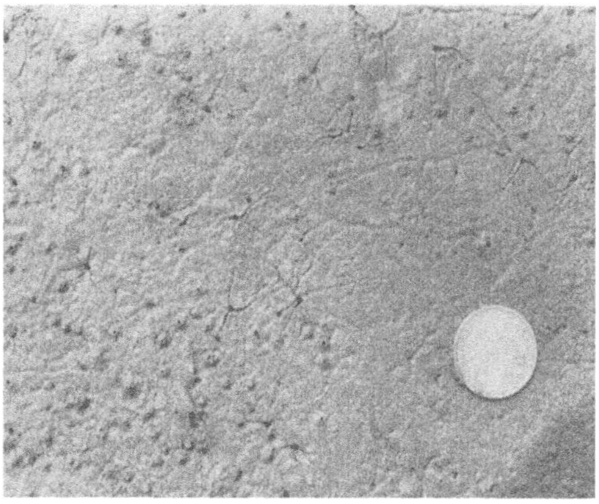

Figure 3-6 (left). Higher mud flats. Lenticular bedding connected with thick lenses and interlaminated sand-mud bedding (estuarian zone).

Figure 3-7 (right). Muddy sand flats inhabited by *Corophium volutator* and *Tellina balthica* (estuarian zone).

bioturbated. Mud cracks and surface trails are abundant. Locally, the tangue is covered by shelly sand banks accumulated by storms along the shoreline.

2. *Intertidal zone* (Figs. 3–6 through 3–12 inclusive). In the upper part of this zone, the muddy sand flats make up a strip of sediments intermediate between the tangue and the sands. This strip extends from +10.5 to +12 meters and it is flooded by 30 to 60 percent of the tides. The deposits contain from 5 to 15 percent of fine particles (<0.05 mm), their median can reach 0.12 mm. Ripple marks are common, and they show irregular stratification and flaser bedding. Burrowing organisms are very abundant, mainly *Corophium volutator, Tellina tenuis*, and *Cerastoderma-Cardium edule* (Bajard, 1966).

Figure 3-8. Muddy sand flats. Flaser bedding and finely laminated bedding (estuarian zone).

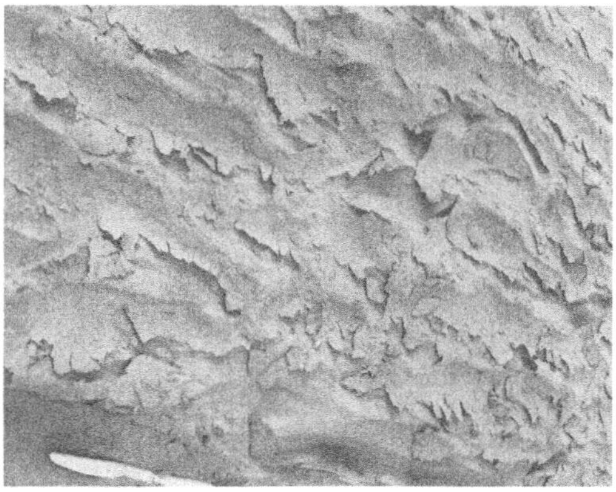

Figure 3-9 (above left). Muddy sand flats that have remained emerged for a few days. Ripple marks deformed by the rain action and partly covered again with eolian sands (estuarian zone).

Figure 3-10 (above right). Muddy sand flats during a period of neap tides. Thin mud curls and fine sands showing forms of eolian erosion (estuarian zone).

The wide sandy areas that occupy the middle and lower parts of the intertidal flats consist of fine sands and biogenic sands (carbonate fraction varying between 50 and 70 percent). These deposits show symmetrical, asymmetrical, linguloid, and interference ripple marks. Oblique as well as rhomboidal ripple marks appear on the channel bank, while megaripples can be seen on coarse sands. Generally the sands are homogeneous and well sorted. They contain many *Arenicola marina* and *Cerastoderma-Cardium edule.* The Foraminifera are principally represented by *Elphidium lidoense.*

A bioherm of *Sabellaria alveolata* exists in the lower intertidal flats off Sainte-Anne. The organic structures are associated with biogenic sand banks accumulated in high hydraulic dunes, with biogenic fine sands, including *Cerastoderma-Cardium edule* and *Lanice conchilega,* and with muddy sediments deposited in the comparatively sheltered depressions (Mathieu, 1966).

Figure 3-11 (below left). Oblique ripple marks on the left of a tidal-channel-linguloid ripple marks appear on the bottom (estuarian zone).

Figure 3-12 (below right). Sand flats. Rhomboidal megaripples and asymmetrical ripple marks (eastern part of the bay).

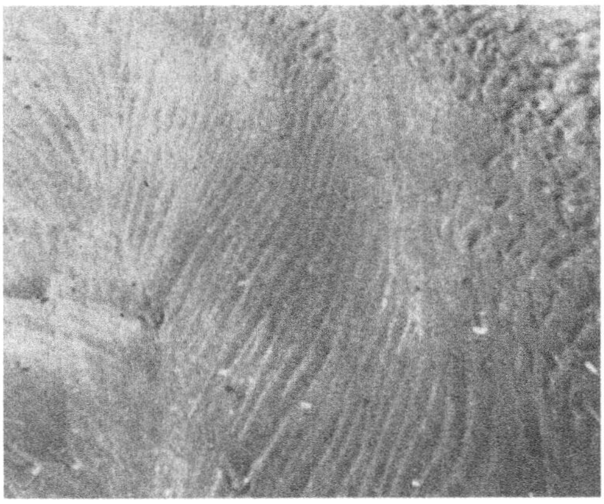

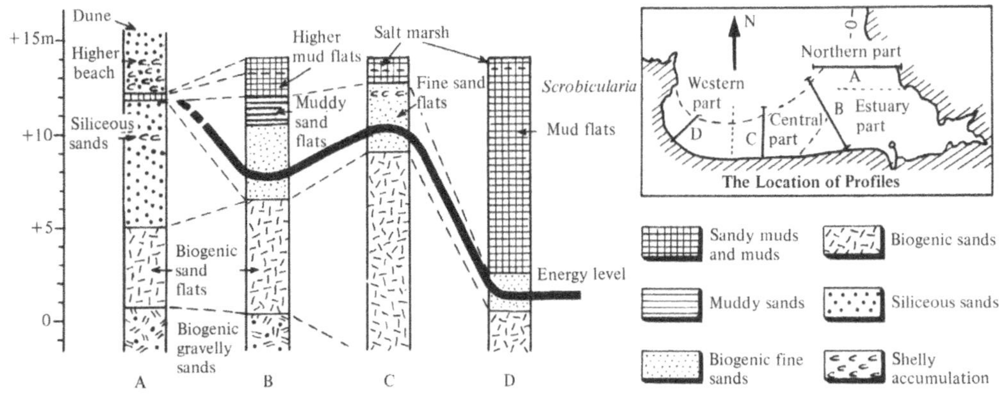

Figure 3-13. Profiles showing the succession of sedimentary environments in the principal parts of the bay.

3. Subtidal zone. Biogenic sands, becoming gravelly toward the open sea, then biogenic gravelly sands (over 15 percent gravel) follow the intertidal deposits. These deposits frequently contain shells of pelecypods (*Ostrea edulis, Anomia ephippium, Glycymeris glycymeris* especially) and locally some debris calcareous algae (*Lithothamnium calcareum* and *L. solutum*). *Ammonia beccarii* is the dominant Foraminifera. The influence of tidal currents increases with depth, inducing the formation of a series of sediments increasing in grain size and in lithoclastic content. Simultaneously, the sedimentary layer thins and, in some places, exposes the bedrock below. Attached organisms are dominant.

The elements of this typical sequence occupy surfaces varying in size according to the distribution of the hydrodynamic energy. For instance, the stronger influence of currents and waves off Cherrueix reduces the extent of the muddy deposits on the upper beach (Figs. 3–1 and 3–13). The sediments are arranged according to energy levels as shown in Figure 3–13.

Other sedimentary environments

North of the Bec d'Andaine, the estuarian zone gives way to a zone bordered with dunes of siliceous sands (only 25 to 45 percent of organic debris, Fig. 3–5). However, near the shore some depressions retain muddy deposits (Fig. 3–1). Beyond, great banks separated by channels appear on the sand flats. This part of the bay is the most exposed to wave action. In contrast, in the western part of the bay, weak turbulence leads to muddy sediments over the greater part of the intertidal zone (*Scrobicularia* mud flats; Figs. 3–1, 3–2, 3–14, 3–15). The deposits are furrowed by a series of channels perpendicular to the coastline and inhabited by an abundant fauna, mainly *Scrobicularia plana,*

Figure 3-14 (above left). *Scrobicularia* mud flats carved off by tidal channels perpendicular to the coastline (western part of the bay).

Figure 3-15 (above right). Detail of Fig. 3-14. 5. The irregular surface of the mud shows abundant trails of *Nereis diversicolor* and some mud cracks.

Hydrobia ulvae, and *Nereis diversicolor*. The Foraminifera association frequently found in estuaries is dominated by *Ammonia beccarii* and *Elphidium lidoense*.

Holocene Succession

This succession can be summarized as follows (Fig. 3–16):

Heterogeneous fluviatile deposits (sandy clays, clayey sands with gravels and pebbles) containing some lower postglacial peat lenses (Boreal).

Marine sediments (shelly sands, fine sands, muddy sands,

Figure 3-16 (below). Vertical section showing the Holocene deposits.

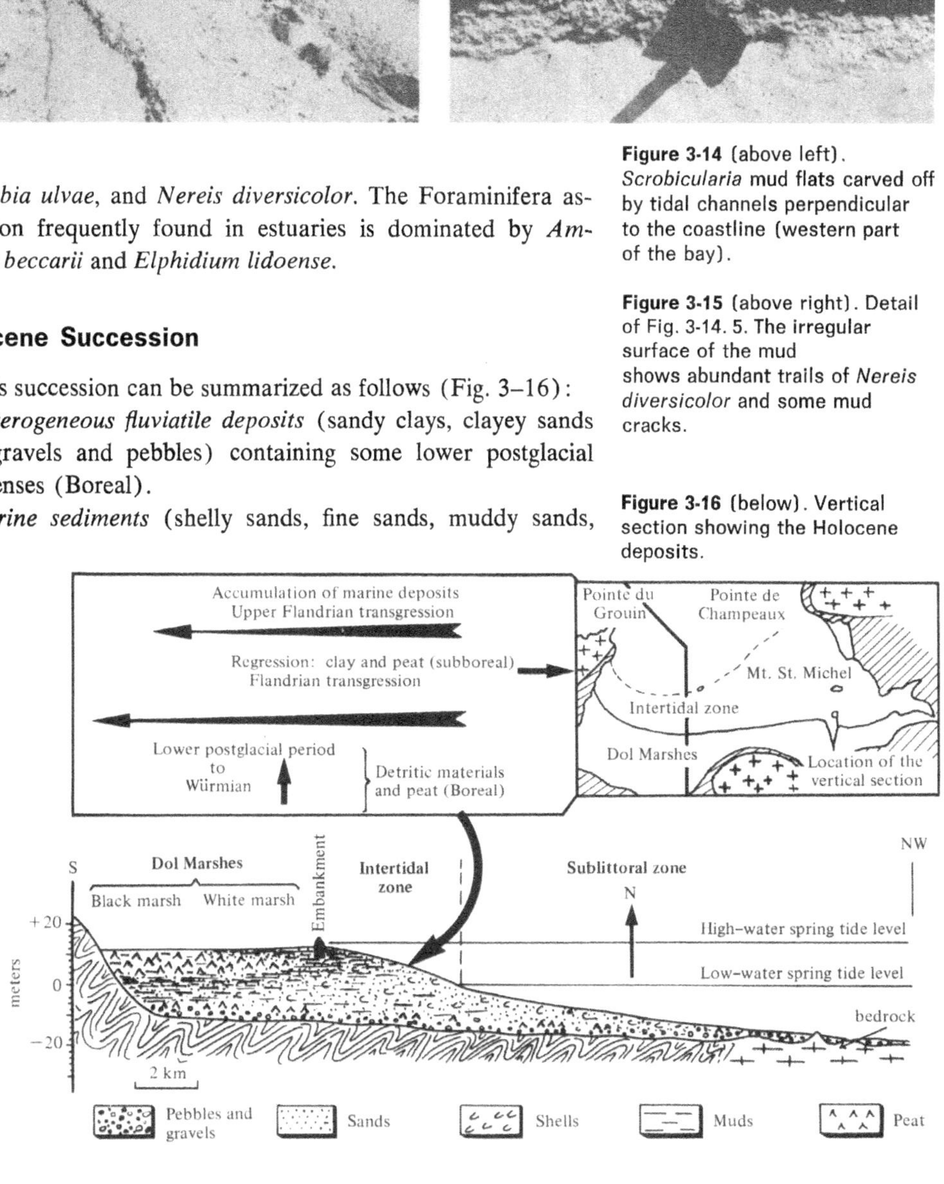

tangue) brought by the Flandrian transgression, which overflowed the present coastline and progressed up to the hills of the hinterland.

Peaty deposits (subboreal) in the Dol marshes, which appeared at the end of a regressive sequence partly resulting from the progressive filling up of the bay.

Muddy deposits (tangue) accumulated in the white marsh part of the Dol marshes during the Upper Flandrian transgression, early Christian era. These muddy sediments are overlain by fining-upward, regressive marine deposits that filled up the estuaries and produced shorelines. Since the Lower Flandrian the rate of sediment accumulation in the Bay is estimated at one million cubic meters per year. Most of this sediment was brought in from the sea by waves and currents; only a small amount came from local erosion of sandy cliffs.

References

BAJARD, J. 1966. Figures et structures sédimentaires dans la partie orientale de la baie du Mont Saint-Michel. *Rev. Geog. Phys. Geol. Dyn. VIII*, 39–112.

BOURCART, J., and BOILLOT G. 1960. La répartition des sédiments dans la baie du Mont Saint-Michel. *Rev. Geog. Phys. Geol. Dyn. III*, 189–199.

DOLET, M., GIRESSE, P., and LARSONNEUR, C. 1965. Sédiments et sédimentation dans la baie du Mont Saint-Michel. *Bull. Soc. Linn. Norm. 6*, 51–65.

GIRESSE, P. 1969. Essai de sédimentologie comparée des milieux fluvio-marins du Gabon, de la Catalogne et du Sud-Cotentin. Thèse, Caen, 730 pp.

MATHIEU, R. 1966. Contribution à l'étude du domaine benthique de la baie du Mont Saint-Michel: Sédiments actuels, microfaune, écologie. Thèse, Paris, 287 pp.

PHLIPPONNEAU, M. 1956. La baie du Mont Saint-Michel. *Mem. Soc. Geol. Min. Bretagne XI*, 7–65.

VERGER, F. 1968. *Marais et wadden du littoral français*. Bordeaux, Biscaye imp., 539 pp.

4

Tidal-Flat Complex, Delmarva Peninsula, Virginia

Stanley Cooper Harrison

Occurrence

Atlantic Ocean coast of Virginia, facing a gently sloping continental shelf (Fig. 4–1). Length: 100 km. Width: 3 to 12 km. Thickness: 30 meters maximum on Atlantic Ocean and thinning to 0 meter against the Pleistocene mainland.

Climate

Pleasant, temperate climate characterizes the area. Mean annual precipitation, 112 cm, exceeds evaporation (U.S. Weather Bureau, 1959). Temperature means: January 0° to 2°C, July-August 30° to 31°C. Hurricane winds may approach 240 km/

This report is part of a 1971 Ph.D. dissertation submitted to The Johns Hopkins University, Baltimore, Maryland. Financial support for the study was furnished by the Coastal Studies Institute, Louisiana State University, under Contract N00014-69-A-0211-0003, Project No. NR 388 002, Geography Programs, Office of Naval Research; and by The Johns Hopkins University. Field facilities were furnished by the Eastern Shore Laboratory of the Virginia Institute of Marine Science.

hour and, if coincidental with high astronomic tides, onshore winds may double the tide range (to 3 to 4.0 meters), producing extensive flooding and coastal changes.

Hydrography

Inlets: spacing; 9 km average, range 2.5 to 14 km. Depths: major, 15 to 25 meters and 1 to 2 km wide with maximum (non-storm) current velocities 100 to 200 cm/second; minor, 1 to 10 meters and 100 meters to 1 km wide with maximum veloci-

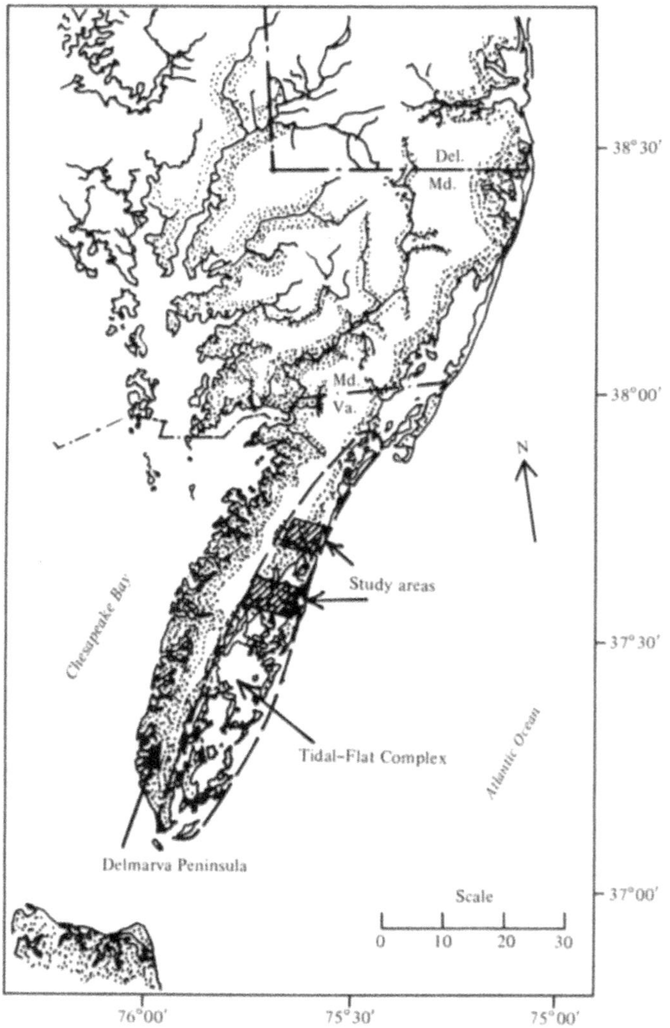

Figure 4-1. Map of the Delmarva Peninsula showing the location of the study areas.

ties of 60 to 125 cm/second. Internal circulation pattern involves two or more inlets connected by branching channel systems. Tides: moderate-low (perhaps more typical of ancient epeiric seas). Semidiurnal with a small inequality; mean range 1.06 to 1.25 meters; spring range 1.5 to 2.1 meters at coastal stations with a range decrease of 10 to 20 cm at mainland stations.

Sediment Source and Sedimentary Dynamics

Sediment is derived from erosion of the seaward side of the complex and from littoral drift from farther north along the Delmarva Peninsula. Pebbles and cobbles indicative of the Potomac River drainage system suggest sediments of an ancestral river system are being reworked by the modern transgression (Harrison, 1972, pp 50–51, 96–97).

Mechanical sedimentation of the incoming material is controlled by a continuous spectrum of winds, waves, tides, and currents ranging from normal to catastrophic (storm) conditions. Normal processes of surf and tidal current exchange produce foreshore laminations and spit accretions on the islands and transport sand and fine-grain sediment into the protected areas of the complex. The abundant silt and clay-size particles, in contact with organic debris, are ingested by organisms to form pellets or other organic-inorganic agglomerates that are subsequently current-deposited in tidal flats, bays, and marshes to form muds (Harrison, 1972). A genetic sequence is built vertically and laterally from a bay floor through subtidal channel levees to intertidal levees (tidal flats) to marshes. Storms disrupt the normal processes by pumping large quantities of sand through existing inlets, over the islands, or through new inlets to form deltaic sand wedges extending into the complex or sand ridges on the complex margin. Reestablishment of normal processes results in the storm sand wedge being covered with muds.

Geomorphic-Sedimentary Subenvironments

Three gross geomorphic environments that parallel the coast can be recognized: an offshore marine, the tidal-flat complex, and the Pleistocene mainland. Within the tidal-flat complex, vertical tide fluctuations are used to define geomorphic subenvironments. Spatial relations are illustrated in Figures 4–2 and 4–3. In contrast with many siliciclastic tidal flats these geomorphic-sedimentary subenvironments do not parallel the coast but flank symmetrically the inlet-channel elements.

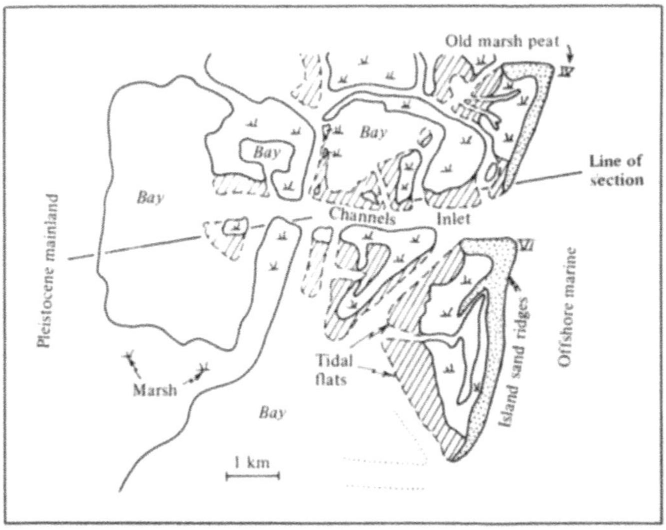

Figure 4-2. Plan view of the major geomorphic subenvironments.

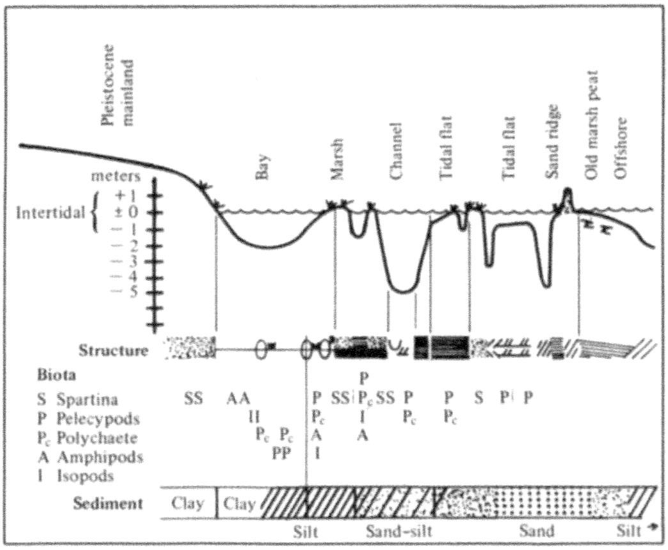

Figure 4-3. Cross section showing the distribution of grain size, sedimentary structures, and biota in the subenvironments.

Offshore marine. A gentle seaward slope of 1 to 2 meters/km is characteristic. Coarse to medium sands form sand waves near shore. Shell debris and mud balls eroded from the tidal-flat complex at the swash line are locally abundant; surf clams, burrowing shrimp, and worms are minor inhabitants; mud content increases seaward.

Inlets. Large-scale reversing sand waves or megaripples (herringbone) characterize major inlets. Shell debris is minor and the base of the channel may be a smooth surface incised into older sediments. Minor inlets have abundant shell and boulder to silt-size mud clasts derived from erosion of the nearby complex, incorporated in cross-bedded and laminated sands. Flood-oriented cross-bedding predominates in areas of rapid transgression.

CHANNELS. The channels are 600 meters wide and 20 meters deep near inlets and narrow to 2 meters wide and 1 meter deep in the marshes. Maximum current velocities range from 11 to 60 cm/second. The channel boundaries are sharp (erosional) in sandy (high current) areas, but comformable to adjacent subenvironments in distal areas. Channel sediments are coarser than sediments in adjacent environments but channels do not migrate. Cross-bedded sands dominate near the inlets and laminated muds (microcross-bedded) characterize the more distal parts of the channels. Shells of the pelecypods *Ensis* sp., *Solen* sp., and *Tagelus* sp. are locally common along the sandy channel boundaries, whereas *Crassostrea* sp. and *Mercenaria* sp. increase in muddy areas. The burrows of *Ensis* sp., *Solen* sp., *Tagelus* sp., *Callianassa* sp., and *Arenicola* sp. characterize sandy channels and adjacent tidal flats; muddy sediments are churned by isopods, amphipods, and various polychaete worms. A few anomalously wide, deep, or bluntly terminated channels are found behind some beach ridges and evidently were connected to earlier inlets or seaward extensions of existing inlets; these channels are filling with mud, in sharp contrast with their older sandy walls.

BAYS. Bays are of two types: Large, open, relatively deep bays occur in the widest part of the complex and may have developed from a protected lagoon, whereas small shallow (less than 50 cm water cover at low tide) bays encircled by marsh characterize the narrow portions of the complex. The sediments, physical structures, and biota vary in the different settings. Cross-bedded sands characterize the seaward margin of the open bays with extensively burrowed muds in more distal portions. Rippled and laminated fine sands and silts with scattered burrows typify intermediate areas. Marsh-encircled bays are floored with intensively burrowed fine muds.

TIDAL FLATS. Extensive areas of true intertidal flats occur between low tide level and the midtide level at which marsh grasses become important. The most extensive areas (2 km by 6 km) are marginal to large channels (channel levees) extending to large bays. Other important areas of intertidal flats occur as deltas behind temporary (storm) inlets and landward of washover areas on the coastal sand ridges. Flood-oriented sand waves or megaripples of coarse sand characterize the deltas and large shoals in front of the inlets. Rippled medium and fine sand is found along the main channel margins, but, more distally, mud dominates and burrowing increases significantly and may destroy all of the laminations formed during the tidal exchange. Laminated sands and algal mats are found in the intertidal washover areas landward of the coastal sand ridges. Desiccated algal mats locally form "jelly rolls," "turnovers," and "cookie chips" (Harrison, 1972, pp. 56–58)—indicative of their intertidal origin. The biota of the tidal flats and bays are similar. *Callianassa* sp., *Arenicola* sp., *Ensis* sp., *Solen* sp., and *Tagelus* sp. occupy individual, discrete burrows in the high current areas of sandy sediments. As the sediment becomes siltier, *Crassostrea* sp., *Mercenaria* sp., and many burrowing worms become important. Isopods, amphipods, worms, and *Mercenaria* sp. are very abundant in clayey sediments.

SALT MARSHES. *Spartina alterniflora* forms over 95 percent of the halophyte plants in the marsh. This plant, which starts to grow at the midtide level, covers extensive areas on channel margins, fringes of bays and tidal flats, and narrow strips along the mainland side of the complex and on the landward side of the coastal sand ridges. Although the marsh may initiate growth in sands, fine silts and clays are the typical sediment in more mature marshes. Faint laminations can be recognized in marsh sediments along channel margins, but extensive deposits behind those levees are more typically structureless. The burrowing decapod *Uca* sp. is very common on the edge of the marsh with the pelecypods *Modiolus* sp. and *Crassostrea* sp. locally common. *Littorina* sp., the surface-feeding gastropod, is very abundant.

ISLAND-SAND RIDGES. The seaward margin of this complex is formed by 11 islands of sand that overlie other elements of the tidal-flat complex. The maximum thickness of the sand is about 9 meters but averages closer to 2 meters. Foreshore (beach) laminations are common on the front of the complex; horizontal and mainland dipping laminations and small-scale cross-bedding are found on the crest and on the mainland side of these sand ridges. Spillover lobes and eolian dunes form local, high-angle sand waves. Abundant shell debris and mud boulders from erosion in the swash zone litter the islands.

The Stratigraphic Record

This complex is being rapidly transgressed, at rates up to 10 meters/year. If such rates continue the only record of this complex will be the deeper scours at the inlets or at the intersections of major channels, which will be filled with coarse, cross-bedded sand. If this tidal flat is totally preserved it will be a landward thinning wedge with an unconformity, locally erosional, at its base. In gross aspect the seaward side will be composed dominantly of sandy sediments, with clay or fine silt typical of the landward side. Current structures will dominate the seaward section and burrows or structureless peat-mudstones the landward wedge. The sediments will fine upward and the section will be capped by the structureless peat muds (Fig. 4–4).

In general the basal sedimentary unit just above the unconformity will be a structureless peat or burrowed mud, probably only a few tens of centimeters thick. Succeeding that unit will be the open bay (lagoon) sequence, consisting of cross-bedded sandstones on the seaward side and burrowed muds on the landward side. Higher in the section the lithology will be more complex, reflecting differentiation into bays, channels, subtidal-intertidal levees (tidal flats), and the succeeding marsh. Intercalated within those units will be thick-bedded (30 to 50 cm), cross-bedded sand wedges deposited as storm sheets. These storm wedges will thin landward. Mud-filled channels may be found above these sand wedges because of disruption of channel patterns by the storm deposits.

Figure 4-4. Diagrammatic cross section showing the expected stratigraphic record.

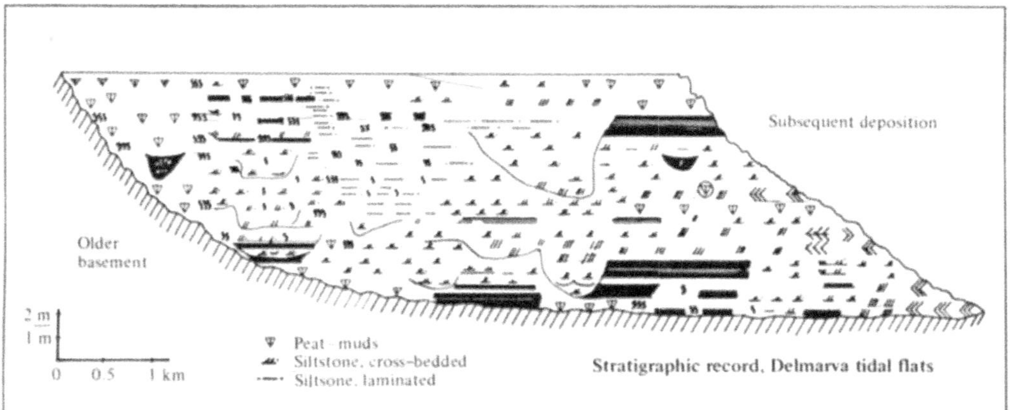

In transverse view, the channels, bays, and tidal flats will show a vertical succession to the structureless peat muds. Channels show little evidence for migration and will be distinguished from adjacent deposits by coarser texture, shell lag deposits, and greater abundance of current structures. Channels may become smaller near the top of the wedge and burrowed sediments will increase away from the channel. Low-angle (up to 15°) laminations may connect channels and adjacent tidal flats. Structureless muds or peats will cap the section, and, as a lithologic unit, will be extremely variable in thickness. Total thickness of a complete sequence will range between 5 and (rarely) 30 meters. Postdepositional erosion may truncate the upper members of the clastic wedge.

References

HARRISON, S. C. 1972. The sediments and sedimentary processes of the Holocene tidal flat complex, Delmarva Peninsula, Virginia. Coastal Studies Inst., Louisiana State Univ., Baton Rouge.

U. S. WEATHER BUREAU. 1959. *Climates of the States (Virginia)*. Dept. Commerce, Washington, D.C.

5

A Transgressive Sequence of Late Holocene Epoch Tidal Environmental Lithosomes Along the Delaware Coast

John C. Kraft and Elizabeth A. Allen

Studies of Holocene coastal sedimentary environments provide a direct key to the interpretation of sediments deposited in coastal and nearshore marine environments of the Atlantic continental shelf during the past 100,000,000 years (Fig. 5–1). Sequences of sediments and patterns and distributions of sedimentary structures can be used to form models for the interpretation of transgressive sequences along the entire Atlantic Coast.

The patterns of distribution of Holocene coastal sediments are complex. This complexity results from lateral and vertical movement of successive tidal environments of deposition over a deeply incised pre-Holocene landscape. The sediments infilling the estuaries and forming around eroding headlands of Pleistocene materials include characteristic shoreline deposits, such as spits, dunes, and baymouth barriers; an intermeshing network of tidal deltas and washover bars; nearshore marine erosional-depositional sands and gravels; and lagoons or estuaries with fringing *Spartina, Distichlis,* and *Phragmites* marshes that form the leading edge of the transgressive units.

The transport agents within the sedimentary environments developed in a coastal lagoon-barrier system such as illustrated are highly complex. Sediment is eroded directly from the Pleistocene surface being submerged by coastal erosion and relative sea level rise. Some sediment moves down the tidal creeks from the inland surface of erosion. Tidal currents move sediments in water wedges of variable salinities back and forth across the lagoon. The eroding marsh shorelines of the lagoons provide sediment to a thin, muddy sand area in the nearshore area of the coastal lagoons. During the frequent storms along the coast, very large amounts of sand and gravel are washed across the barrier. The sands, gravel, and shells of washover barriers often show cross laminae that are foreset landward and truncated sharply, indicating the nature of the strong currents that move sediments landward across the marshes.

During these storms some of the beach-berm material is washed back into the shallow marine area. As a result, it can be seen that an extremely complex pattern of current movement, sediment erosion, and sediment transport and deposition is involved in the relatively small coastal area depicted. As would

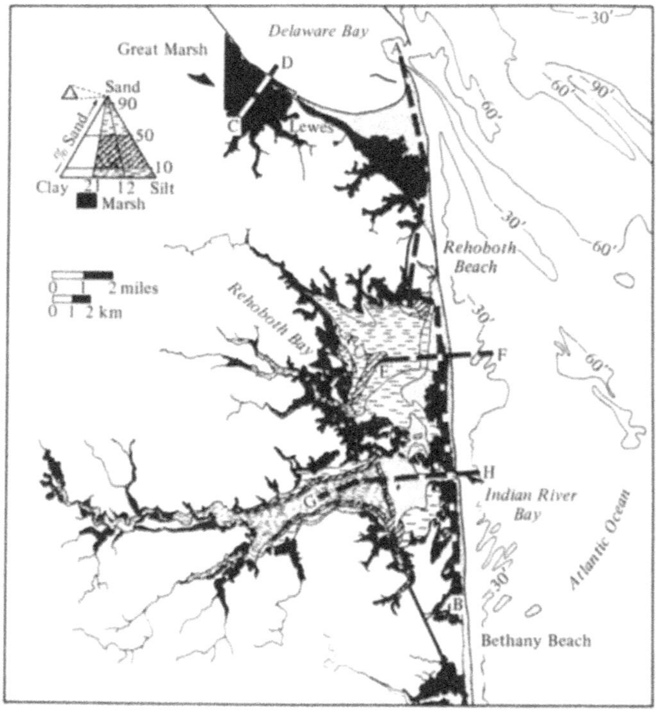

Figure 5-1. Map index to Holocene sedimentary environments of the Delaware coast and locations of cross sections in text (dashed lines). Sediment facies patterns are represented by a triangle ratio diagram. This map emphasizes the relationship between the coastal Holocene sedimentary environments and the geomorphic surface being inundated (nonpatterned land area) as sea level rises and the late Holocene transgression proceeds.

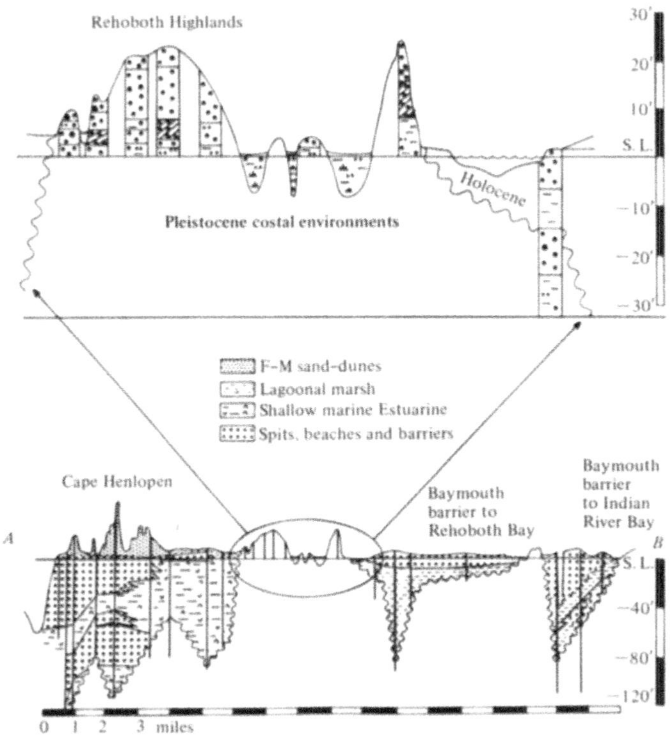

Figure 5-2. A cross section of Pleistocene coastal sediments in the Lewes-Rehoboth Canal at Rehoboth Beach, related to an interpretive cross section showing Holocene coastal sedimentary environments overlying the deeply incised pre-Holocene coastal typography in a section parallel to the Atlantic Coast. The valleys shown here are part of the trellis-drainage system believed to be tributary to the ancestral Delaware River Valley, which lies to the east of the line of section. (After Kraft, 1971a.)

be expected the sediments become quite difficult to correlate laterally. Lagoonal washover beaches are often an anomaly completely surrounded by *Spartina* marshes on the landward side and extremely muddy sands grading into dark gray muds on the lagoon side.

The top of Figure 5–2 shows details of the Pleistocene sediments exposed in the Lewes-Rehoboth Canal and on Thompson's Island. Beach and dune sands are most abundant in the Pleistocene section. Steeply foreset sands and gravels are interpreted to be products of tidal delta environments within the beach-barrier system. At four localities, clearly identifiable and correlative exposures of lagoon-marsh muds have been noted. The evidence for interpreting these sediments as probable marsh muds is the distribution of what appear to be oxidized roots and stems throughout the section. No fauna are present in in the Pleistocene section; however, it is most likely that calcareous skeletons have been leached out of the sediments. Rare open *Mercenaria* sp. molds indicate the magnitude of the carbonate leaching. The Pleistocene sedimentary elements are laterally adjacent to similar Holocene-Recent coastal environment

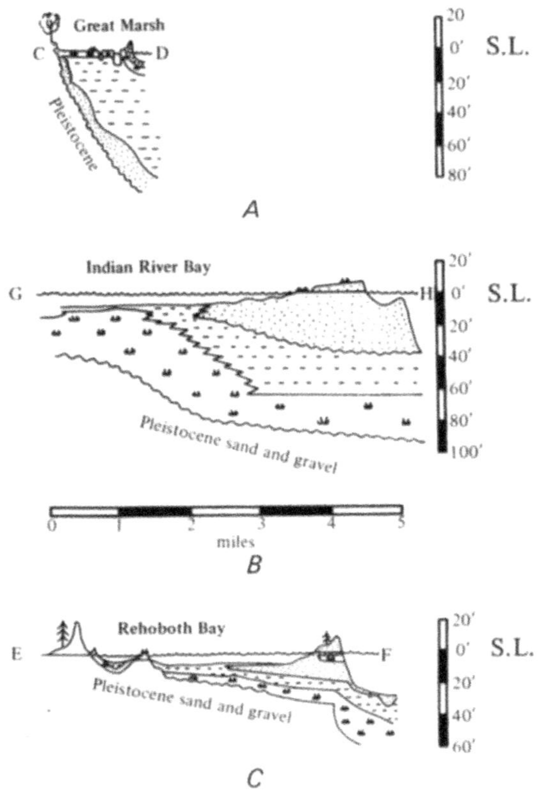

Figure 5-3. *A.* The Great Marsh was formerly a coastal lagoon-tidal creek system that silted, perhaps in the very recent past (just before the Europeans settled in the area), and became covered with a *Spartina* salt-marsh flora. (Tidal range 4.5 ft.)

B. The leading edge of the marine transgression in the Rehoboth Bay transect is bounded by a fringing *Spartina* salt-marsh against a highland composed of Pleistocene coastal sediments. This is followed in turn by a geographic sequence of lagoon-muds, barrier-lagoon sands, back-barrier marsh and forest, dunes, and the beach-berm system. Drillhole evidence from the barrier shows that a similar vertical sequence of environments may be anticipated. As the transgression proceeds, coastal erosion on the submerged beach face destroys part of the depositional record of the Holocene coastal sediments. With continuing rise of sea level, the greater portion of the record might be preserved under a nearshore lag deposit of sand. (Ocean tidal range 4.5 ft, lagoon tidal range less than 1 ft.)

C. The section through Indian River Bay is in the vicinity of an actively forming tidal delta. This lagoon has formed along the axis of a deeply incised pre-Holocene valley. Accordingly, for the kind of transgressive coast under study, the marsh, lagoon, and barrier sections are exceptionally thick. (Ocean tidal range 4.5 ft, lagoon tidal range 1.5 to 0.5 ft.) (After Kraft, 1971b.)

sediments. The stratigraphic record will be very complex should the sedimentary environmental lithosomes be preserved in this fashion.

Figure 5–3 shows three schematic cross sections of the leading edge of the Holocene transgression. The diagrams are based on interpretation of cores in each area. They indicate some of the many sediment variations that exist at the edge of the transgression, depending upon the local topography and general conditions of sediment source and supply.

The cross section in Figure 5–4 shows a detailed interpreta-

Figure 5-4. An interpretive cross section of a baymouth barrier south of Dewey Beach showing the vertical sequence of Holocene sediments and associated radiocarbon dates. The date of 2360 years BP at 18 ft below sea level establishes the rapid transgression of Atlantic coastal washover barriers across the low-lying adjacent shoreline. The somewhat lenticular cross section of the baymouth barrier should be considered to be undergoing erosion in the emerged and submerged beach face and then washed over landward and upward in space and time. Part of this stratigraphic record may be preserved in the nearshore area under shallow-marine sands and gravels. The marsh surface overridden by the barrier sands may crop out at low tide, undergo erosion, and be redeposited as reworked material. (After Kraft, 1971a.)

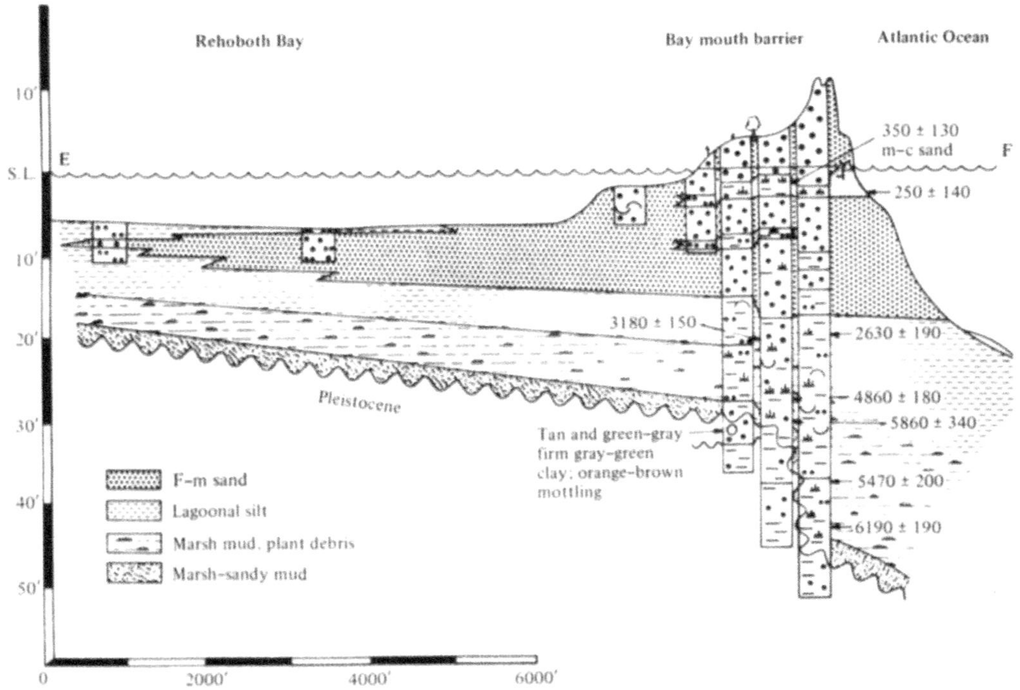

tion of the baymouth barrier between the Rehoboth coastal lagoon and the Atlantic Ocean, based on drillhole borings through the barrier sequence. Study of the three deep borings on the baymouth barrier shows that the drill penetrated, first, the dune-washover sands of the surface of the barrier, then barrier-back marsh muds and sands, then the central elements of the baymouth barrier, including possible lagoon sands of the barrier-back and other barrier-back marshes; eventually it broke through into lagoon muds characterized by a *Crassostrea virginica* fauna. Shortly thereafter, the drillhole entered *Spartina* marsh muds and *Spartina* peats. Just before reaching the pre-Holocene transgression surface, the drill encountered a marsh-fringe deposit of sandy mud with rootlets. The pre-Holocene surface of Pleistocene coastal sediments was characterized by oxidized light tan, light green, orange, and yellow color, mottling and a relatively stiff or hardened sediment sequence.

Radiocarbon dates from the marsh muds throughout the section and from tree trunks in the barrier-back sequence are shown along the boreholes. These radiocarbon dates are critical in constructing a local curve of relative rise in sea level based on a position of the living salt-marsh surface. The dates, their position, and the materials from which they are derived are important in forming concepts of coastal change for projecting into the past and into the future.

Figure 5–5 is a core log through the linear baymouth barrier south of Dewey Beach (located on profile *EF* of Fig. 5–4). The first 10 ft are coarse to medium-grain washover sands deposited over marsh sediments during large storms such as hurricanes and northeasters. These washover sediments range from fairly clean-washed sands to dirty sands mixed with plant debris. The marsh mud and peat below the washover sands represent a fringing marsh on a barrier-back. The uneven bedded sands mixed with plant fragments indicate washovers in which the marsh plants have pentrated the sands and reestablished the marsh surface.

Immediately under the barrier-back marsh are tidal delta sequences consisting of poorly sorted sands and gravels with high-angle foreset beds or fine to coarse barrier-back lagoon sands with occasional silt streaks. At the base of the subtidal transgressive-barrier sands the sediment changes sharply to fine-grain blue-gray lagoonal muds. This section is heavily mottled and burrowed and contains *Crassostrea virginica* and other mollusks. The next underlying unit, the salt-brackish water marsh mud and peat, represents an early Holocene marsh at the leading edge of the marine transgression. At the base of

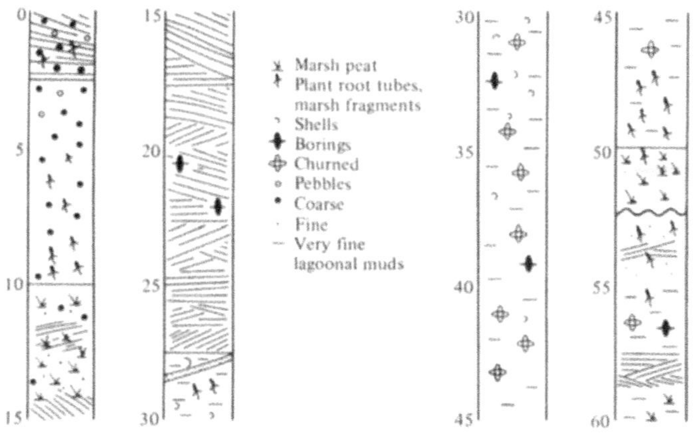

Figure 5-5. Core log from drillhole core located on Dewey Beach barrier transect.
0–2.5: Light tan sands from the washover fans, medium to coarse grain, low-angle lamination with concentrations of plant fragments. *2.5–5:* Clean washover sand with no bedding apparent. *5–7.5:* Washover sands, plant fragments. *7.5–10:* Washover sands, heavy concentrations of plant fragments. *10–12.5:* Barrier-back marsh peat, high concentration of sand changing to unevenly bedded sands with reworked marsh fragments. *12.5–13.75:* Barrier-back marsh peat, high concentration of sand. *13.75–27.5:* Medium- to fine-grain lagoonal and tidal delta sands, high-angle–low-angle cross bedding, thin lamination of muds intermixed, occasional burrowing infilled with mud. *27.5–37.5:* Gray lagoon muds with sand lenses, some plant fragments, heavy mottling, and burrowing of sediments. *37.5–40:* Gray lagoon muds, heavily mottled. *40–45:* Gray lagoon muds with some shell fragments, mottled. *45–50:* Gray lagoon muds with abundant interfingering of marsh fragments and plant rootlets. *50–52:* Lagoon-marsh fringe with occasional sand lenses. *52–55:* Light gray, hard, muddy sand, Pleistocene. *55–57.5:* Hard mud, burrowed, churned, mixed, and oxidized, Pleistocene. *57.5–59:* Coarse sands, cross-bedded, bright orange color, possible Pleistocene soil zone, heavy organic layers. *59–60:* Fine-grain, blue lagoon muds, hard, mottled, Pleistocene.

the Holocene section is a marsh fringe of dark-brown muddy sand and roots. Beneath the land surface being inundated and undergoing transgression is hard marsh mud and peat, oxidized sands, and deep-blue lagoonal muds.

The Holocene section therefore represents a landward transgression of the Atlantic Ocean over a Pleistocene surface. The Holocene lagoonal environment was transgressed by barrier-back lagoonal sands and tidal delta sands. Upon these protected subtidal sands, a barrier-back marsh developed. The marsh surface was subsequently washed over by barrier sands during severe storms, which raised the surface above sea level and drove it landward in the transgressive sequence. As illustrated in the core log the landward sequence of horizontal environmental elements is repeated in the vertical stratigraphic sequence (see also Figs. 5–1 and 5–4).

Depending upon local topography and the rate of rise of sea level, the stratigraphic record of nearshore tidal deposits can be extremely complex. Both of these factors control the amount of erosion and thus the completeness of the sedimentary column. Nevertheless, a definite pattern of succession of coastal environments exists and can be used to reconstruct the geologic history of transgressive shorelines.

References

KRAFT, JOHN C. 1971a. Sedimentary facies patterns and geologic history of a Holocene marine transgression. *Geol. Soc. Am. 82*, 2131–2158.

——— 1971b. *A Guide to the Geology of Delaware's Coastal Environments*. Coll. of Marine Studies, Univ. Delaware, Newark.

6

Intertidal Sediments from the South Shore of Cobequid Bay, Bay of Fundy, Nova Scotia, Canada

R. John Knight and Robert W. Dalrymple

Occurrence

The deposits described here are located along the south shore of Cobequid Bay (Fig. 6–1), the eastermost extension of the Minas Basin between Burncoat Head and Salter Head, Nova Scotia (lat. 45°20′ N; long. 63°30′ W to 63°45′ W) (Pelletier and McMullen, 1972). Comparable environments occur to the west along the south shore at Walton (Swift and McMullen, 1968) and further west along the north side of Minas Basin at Economy Point and Five Islands (Klein, 1970).

Climate and Hydrography

The Minas Basin has a temperate maritime climate. The severest storms occur during June to November, but winter

The writers gratefully acknowledge the critical reading and suggestions of Dr. G. V. Middleton. Appreciation is expressed to the Atlantic Geoscience Center, Bedford Institute and McMaster University for logistical and financial support during the field season.

Figure 6-1. *A.* Location of field area in Cobequid Bay, the easternmost extension of Minas Basin. *B.* Airphoto mosaic of Cobequid Bay. Selmah Bar is located immediately west of the airport. Noel Bay is located in the left bottom corner. North is to the top. Scale is approximately 1 inch = 4 mi. (After Atlantic Tidal Power Board, 1969.)

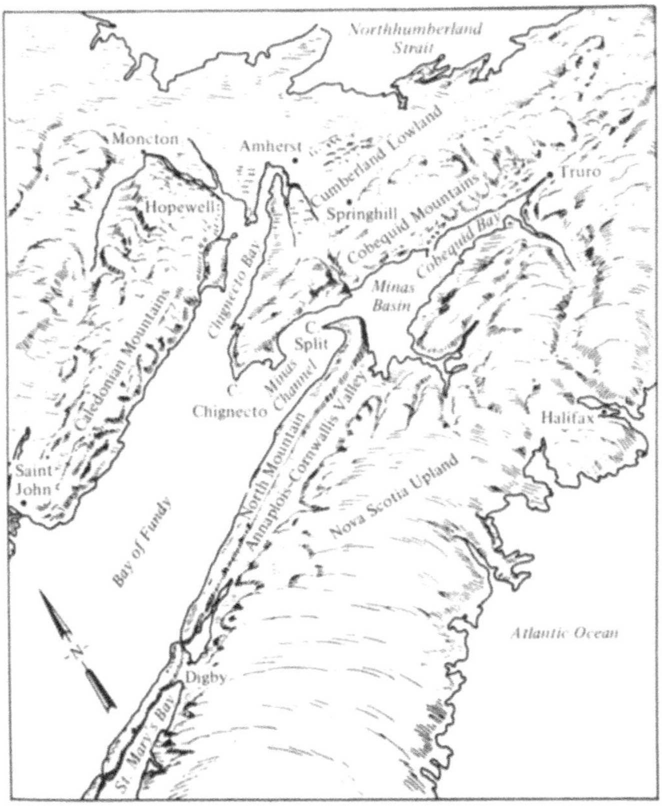

A

B

storms often produce violent gales. January is the coldest month, with a mean daily average temperature of −6°C; July is the warmest month, with a mean daily average temperature of 20°C. Mean annual precipitation for the area is about 110 cm. Westerly winds prevail throughout most of the seasons; however, complete reversals are not uncommon during a tidal cycle (Ann. Meteorological Summary, 1967).

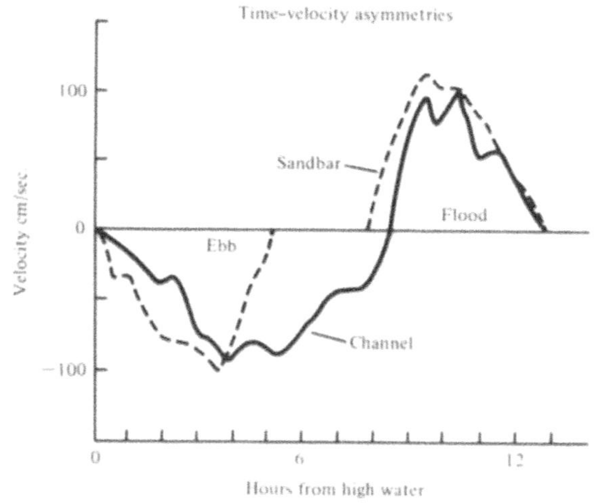

Figure 6-2. Comparison of channel and sandbar time-velocity asymmetries from a sandbar in Cobequid Bay.

Tides within the study area are semidiurnal and have a mean range of 11.5 meters with a maximum spring tide range of approximately 15 meters (Tide Tables, 1971, 1972).

Bottom tidal currents attain speeds between 30 and 115 cm/second. Measurements of tidal current velocities, directions, and rates of water level rise and fall over the sandbars and channels display strong time asymmetries (Fig. 6–2). Topo-

Figure 6-3. Falling water marks on megaripple slip-face and superimposed late-stage emergent-runoff linguoid and straight current ripples. Megaripple crest is slightly planed-off. Scale in centimeters and decimeters.

Figure 6-4. Oblique airphoto of Noel Bay with Noel Bay Bar in background. Meandering dendritic tidal channels dissect the mud flats. Tidal marsh is located around the periphery of the bay. View is to the northwest.

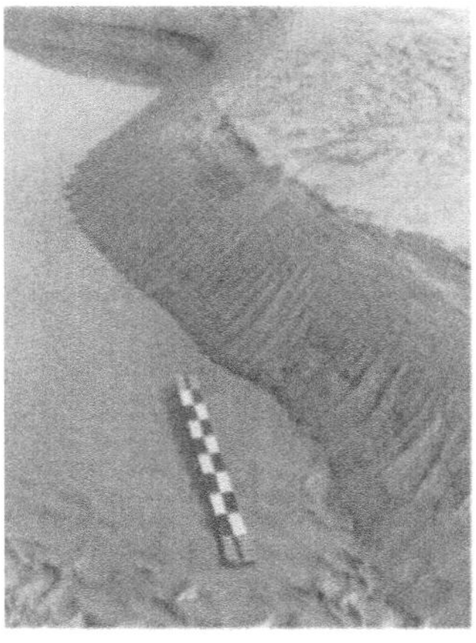

graphic shielding and vertical position give marked areal and temporal variations in tidal asymmetries and magnitudes.

In summer, winds blow mainly from the south and southwest; during the rest of the year they blow from the west and northwest. Wave activity is particularly well developed along the shores owing to the alignment of the basin geometry with the direction of the prevailing westerly winds. Wave-generated currents develop in response to the oblique angle of attack by wave fronts to the shore.

In late stages of ebb-current runoff, and after emergence, open channel flow develops in both sandy and muddy areas. In sandy areas this flow is confined to megaripple and giant ripple troughs (Fig. 6–3). The muddy areas are drained by meandering dendritic creeks of variable depth (Fig. 6–4).

Drifting ice plays an active role in sediment transport. During the winter months ice forms in the upper portions and along the edges of the bay. This ice drifts back and forth along a path formed by tidal current and wind stresses. The drifting ice collects its load by direct incorporation of muddy water and by grounding on tidal flats and sandbars at low tide.

Vegetation is restricted to the upper intertidal zone and supratidal areas where marsh grasses predominate.

Sediments and Sedimentary Structures

SHORE MARGINS. Most of the sediment in the Minas Basin intertidal zone is derived from sea cliffs up to 20 meters tall composed of Carboniferous and Triassic bedrock and Pleistocene glacial till and outwash sand. Maximum erosional rates for Triassic sedimentary rocks from reconnaissance studies range from 1 to 2 meters/year from the combined action of waves, currents, and ice.

GRAVEL BEACH. The gravel beach (Fig. 6–5) is caused by erosion of materials from the shore margin to form a wave-cut bench and beach. Marginal gravels of boulder and cobble size and of varying lithologies occur at the high-water mark. Triassic basalts and sedimentary rocks and granite and metamorphic rocks derived from Pleistocene fluvioglacial deposits also occur. Sediment size decreases seaward.

At high slack water, mud settles out of suspension to leave a mud matrix in the gravel framework or a thin veneer of mud overlying the gravel surface. Thickness of the mud veneer varies from a few to a maximum of 30 cm. Both bedrock and Pleistocene deposits frequently outcrop along the gravel beach. The

Figure 6-5. View north across gravel beach, Selmah Bar area. Horizontal facies succession includes coarse marginal gravels (foreground), mud veneer over lag gravels, followed by harrow marked sands and gravels. Megaripples are visible on sandbar in middle background.

width and lithology of the gravel beach appear to be closely controlled by the adjacent shore or underlying lithologies. Low tide drainage across the gravel beach is by open channel flow in the gravel base or by open channels cut through the veneered surface muds exposing a gravel or bedrock channel floor.

Primary structures are evident only toward the lower end of the gravel beach; they include current lineations, imbricate pebbles and cobbles, scours, swash bars, and ripple marks. Some minor amounts of bioturbation are evident on the mud-veneered areas.

SAND BEACH. The sand beach occurs immediately adjacent to and directly on Triassic sedimentary deposits. Grain size decreases seaward from medium- to coarse-grain sand to silts. Ledges of bedrock outcrop are common and these help to trap sand in the beach position.

Interference and current ripples, current lineations, scours, and imbricate pebbles are common.

MUDFLATS. The mud flats (Fig. 6–4) occur in the upper part of the intertidal zone and are composed of muds and silts. Surface muds are brown. With depth, the mud becomes blue-gray owing to organic reduction of iron oxides to sulfides. Deposition of these muds and silts occurs at high slack water from suspended sediments settling out and erosion of adjacent salt marsh deposits. There is little input of suspended sediment from rivers in the area (Atlantic Tidal Power Programming Board, 1969). Mud flat sediments progress from muds to silts with ripples in the seaward direction. The mud flat sediments are laminated (Fig. 6–6) in laminae of 1 to 2 mm. Some bioturbation is evident by shell and worm organisms. Silty deposits overlying some mud flat areas occur as interference or ladder-back ripples with small mud pellets in the troughs. Some silt and fine-grain sand occurs as small lenses within the mud laminae.

The mud flats slope slightly seaward and are drained by deeply entrenched dendritic drainage systems up to 2.5 meters deep, exposing internal structures of the mud flat deposits (Fig. 6–6).

SANDBARS. The sandbars (Figs. 6–1 and 6–7) are separated from the shore beach environments by a channel system floored by coarse sands, gravel lag and/or bedrock. The sand on the sandbars ranges from 0.90 to 2.5 ø and is quite well sorted.

Figure 6-6. Parallel and wavy laminated muds and silts of mud flats. View is to the east. Scale in centimeters and decimeters.

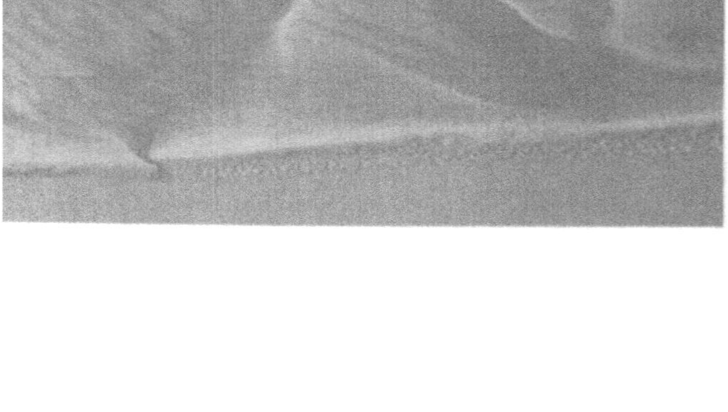

Figure 6-7. Oblique airphoto of Selmah Bar. Megaripples (wavelengths 0.5 to 12 meters) and giant ripples (wavelengths greater than 12 meters) extensively developed. Most of large ripples here are flood oriented and show ebb modifications at low tide. A poorly developed gully system is present in the left foreground. View is to the south.

Figure 6-8. Sinuous megaripples on the north side of Selmah Bar. Ripples show planed-off crests and superimposed current ripples. Ripple wavelengths 8 to 10 meters and amplitudes 60 to 70 cm. Scale is 1 meter.

Figure 6-9. Superimposed current ripples on small-scale, ebb-oriented megaripples, Selmah Bar. Megaripple wavelengths approximately 1.5 meters. Amplitude in scour pits approximately 25 to 30 cm.

The sandbars are generally linear in an east-west direction and asymmetric in cross section, north-south. The steeper sides of these sandbars have slopes up to 18°.

Source materials are derived from local bedrock and Pleistocene deposits both along the shore and from the floor of the basin. The Pleistocene deposits provide most of the fine- to medium-grain materials and the Triassic bedrock supplies the coarser sands. The dominant mode of sand transport is by tidal currents; wave action appears to provide minor modifications.

The sandbars occur in the middle to lower part of the intertidal zone, being covered by 5 to 15 meters of water at high tide, but exposed at low tide.

A wide range of sedimentary structures and bedforms occurs.

Megaripples and giant ripples (Figs. 6–3, 6–7 to 6–9) have wavelengths from 1.5 to 38 meters and amplitudes from 10 to 150 cm. Scour megaripples and planed-off megaripples occur. Other bedform features include plane bed, falling watermarks, interference ripple marks, and current ripple marks that occur independently as well as superimposed on larger bedforms in varying directions. Bidirectional planar and trough cross stratification is evident within the larger bedforms. Low-angle reactivation surfaces are common. Some mud drapes and bioturbation appear in the megaripple troughs, but these are destroyed by each successive flood tide.

TIDAL MARSH. The tidal marsh (Fig. 6–4) occurs in the supratidal zone and undergoes submergence only at spring tide intervals. Some of these marsh areas have been diked in the past. The Cobequid Bay marsh deposits appear to be a mixture of clay and silt overlain by a organic horizon as thick as 30 cm. The surface is covered in salt marsh grasses. Sediments below the organic zone are brown and become blue-gray with increased depth. No stratification is evident, but remnant organic concentrations do occur.

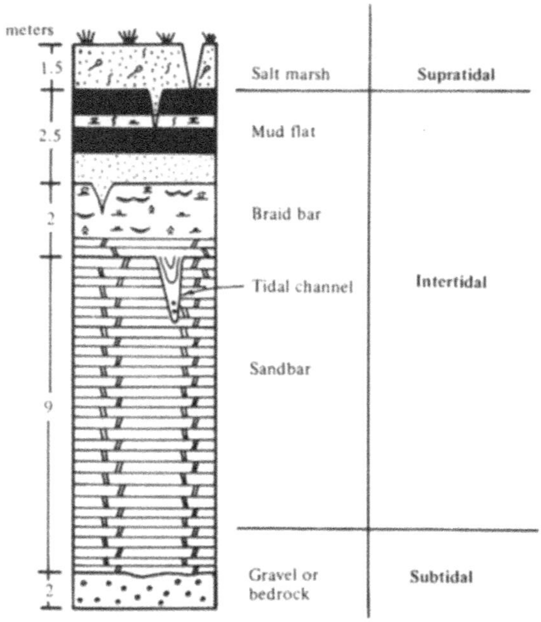

Figure 6-10. Hypothetical regressive vertical sequence of sedimentary facies in the study area.

Vertical Sequence

The vertical sequence in Figure 6–10 represents a simple hypothetical regressive composite of the various environments in the study area. The proposed sequence assumes that sufficient sediment is available to cause the eventual infilling of the bay and does not necessarily represent any present vertical succession in the area. The sequence assumes almost complete preservation of sand bars as indicated by the 9 meters of cross-stratified sands.

The sequence generally fines upward, reflecting a transition from the relatively high energy lower intertidal and subtidal environments to the lower energy conditions of the upper intertidal and supratidal zones. The sediments are cut by a series of incised channels that may or may not show any trough cross stratification due to sideways infilling, but may exhibit scoured-erosional bases and channel outlines overlain with a coarser sediment lag that fines upward and that may itself be cross stratified.

References

ANNUAL METEOROLOGICAL SUMMARY. 1967. Dept. Transport, Meteorological Branch, Halifax, Nova Scotia.

ATLANTIC TIDAL POWER PROGRAMMING BOARD. 1969. Feasibility of tidal power development in the Bay of Fundy. Ottawa, Ontario (restricted circulation).

KLEIN, GEORGE. 1970. Depositional and dispersal dynamics of intertidal sand bars. *J. Sediment. Petrol. 40*, 1095–1127.

PELLETIER, B., and MCMULLEN, M. 1972. Sedimentation patterns in the Bay of Fundy and Minas Basin. *In* Gray, T. J., and Gashus, O. K., eds., *Tidal Power*. New York, Plenum Press, pp. 153–187.

SWIFT, D. J. P., and MCMULLEN, R. M. 1968. Preliminary studies of intertidal sand bodies in the Minas Basin, Bay of Fundy, Nova Scotia. *Can. J. Earth Sci. 5*, 175–183.

TIDE TABLES. 1971, 1972. *Canadian Hydrographic Survey*. Ottawa, Ontario.

7

Tidal-Flat Sediments of the Colorado River Delta, Northwestern Gulf of California

Robert W. Thompson

Occurrence

Barren mud and salt flats occur on the seaward margin of the Colorado River delta, in the northwestern Gulf of California, flanking the lower 50 km of the Colorado River course, and extend southward from the river mouth for 60 km along the coastal plain of Baja, California (Fig. 7–1). The sedimentology and geologic history of this southern extension were studied in some detail by Thompson (1968), and the results show that intertidal and shallow subtidal sediments here differ markedly from those of other extensive tidal flats described in the literature (for example, Straaten, 1954; Evans, 1965; Reineck, 1967).

Climate and Hydrography

The northwestern Gulf of California is characterized by a hot, arid climate and extreme tides. Annual rainfall averages about 5 to 7 cm, most of which falls in occasional downpours

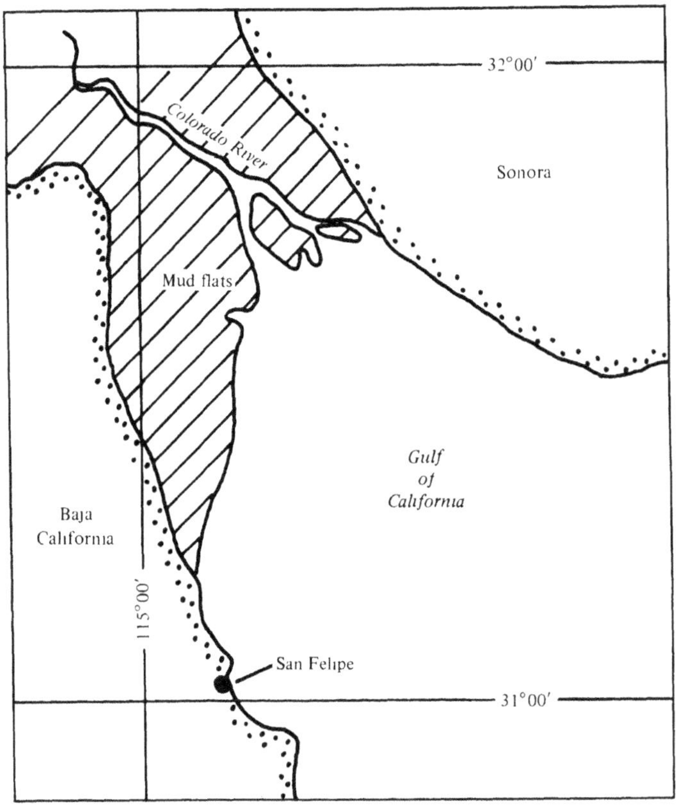

Figure 7-1. Regional setting.

in the late summer. Evaporation rates are the order of 200 to 250 cm/year, hence conditions are conducive to extreme desiccation in all environments not frequently inundated by tides. Tides in the northwestern gulf are mixed, with a mean range of 4 to 5 meters and a spring range of 6 to 8 meters, the values increasing from south to north. A maximum spring range of about 10 meters has been recorded near the mouth of the Colorado River. Spring tidal currents reach velocities of 150 cm/second near the river mouth and in restricted tidal channels of the southern mud flats; across the open intertidal flats, maximum tidal currents are the order of 25 to 50 cm/second. Net transport of sand along beaches in the area is northerly, toward the head of the gulf, in response to larger, although less frequent, waves of southeasterly approach.

Morphology

Three main morphologic units are distinguished within the coastal mud flats (Fig. 7–2):

1. The high flats, which constitute a nearly horizontal plain situated near extreme tide level. Parts of the high flats, in particular the salt pans, are still flooded by extreme spring tides, whereas other parts are supratidal and essentially inactive. The high flats abut seaward-dipping alluvial fans of the Piedmont Plain on the west and are fringed on the seaward side by a zone of semicontinuous beach ridges.
2. The intertidal flats, which dip seaward at gradients of about 0.1° to 0.2° from spring higher high water (SHHW) to spring lower low water (SLLW) level. Parts of the intertidal flats occur between the younger (seawardmost) beach ridges.
3. The subtidal flats, which extend from SLLW to a depth of about 11 meters at an average gradient of 0.05°.

From a morphologic view, the mud flats are most remarkable for their uniform smoothness. Significant small-scale relief features include only the beach ridges in the upper intertidal zone, and tidal channels that dissect a very restricted part of the high flats near the south end. Seaward from the subtidal flats, the

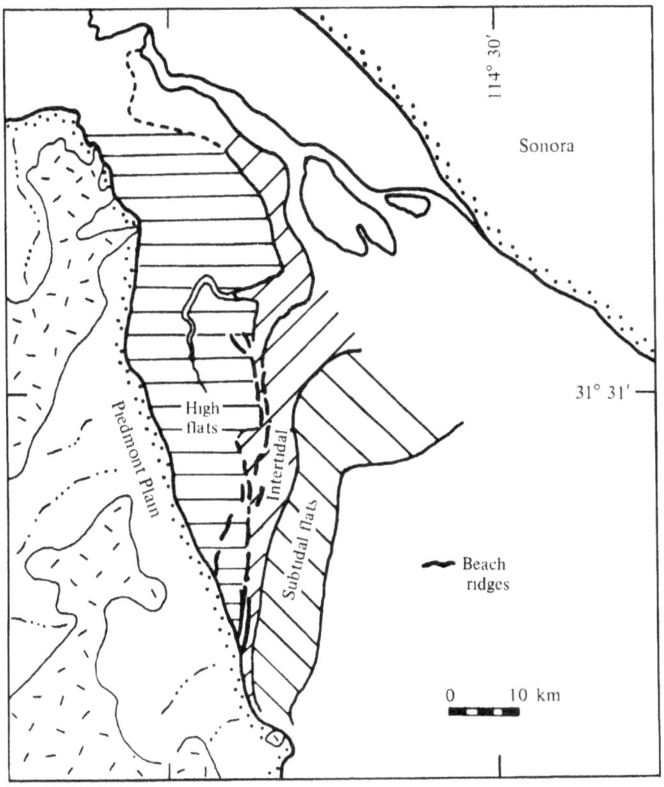

Figure 7-2. Morphologic units of the tidal flats.

Gulf floor is characterized by elongate tidal current ridges and intervening troughs that extend parallel to the Gulf axis.

Sedimentary Facies

Surface and near-surface sediments show a distinct zonation into sedimentary facies that corresponds approximately to the morphologic subdivisions (Fig. 7–3). Characteristics of the sedimentary facies are summarized in Figure 7–4. Among the more important processes responsible for the sediment zonation are tidal currents, wave action, benthic organism burrowing, and desiccation and attendant evaporite crystallization.

Sediments on the subtidal flats and ranging up into the lower intertidal zone are subtly laminated, brownish-gray silty clays (facies E). Thin (<1 mm) laminae of coarse silt-very fine sand occur interspersed at irregular intervals of 1 to 10 cm within the silty clays. These muds have settled from suspension in tidal currents, and variation in the intensity of these currents

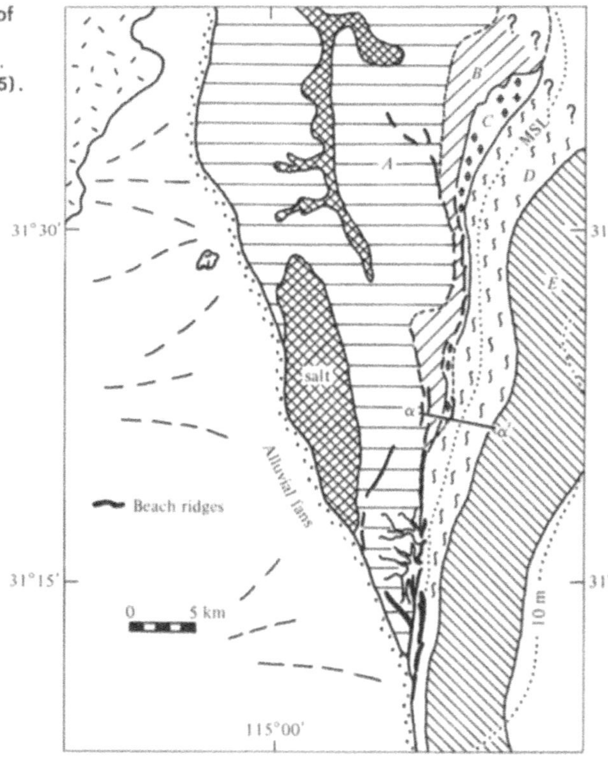

Figure 7-3. Aerial distribution of sedimentary facies. *A–E:* Sediment facies (see Fig. 7-4). *a-a'*: cross section (see Fig. 7-5).

with concomitant slight reworking is most likely responsible for the lamination.

Between the levels of mean low and mean high water (MLW-MHW), pale brown to brownish-gray silty clay and clayey silt prevail (facies D and C). Structure of these sediments is determined largely by the effects of burrowing organisms concentrated over this interval. Discrete, small-scale burrows and disturbed (wavy) lamination, probably produced by various species of mollusks, echinoids, and polychaete worms, typify the sediments from near mean low water to neap high water. Details regarding distribution of these organisms and their effects are masked by recent reworking in this zone which relates to cessation of mud supply from the Colorado River. Above neap high water, intense burrowing by fiddler crabs (*Uca monilifera* and *Uca coloradensis*) has resulted in a lithologically mottled sediment of mixed shell, sand, silt, and clay.

Very well-laminated clayey silt (facies B) prevails on the intertidal flats between mean high and spring high-water levels. These sediments are characterized by alternation of clean silt, clayey silt, and silty clay in laminations varying from 1 mm to a maximum of about 8 mm, but averaging 2 to 3 mm in thickness. The clean silt laminations, which are best developed in the lower part of this zone, originate under the combined influences of small waves and tidal currents that sweep back and forth across the tidal flats with increasing intensity during rising spring tides. Clays of the finer laminae settle near the high-water mark of each tide, which recedes across the mud flats during waning spring tides, and from thin sheets of water left covering the flats during ebb tide. Lack of burrowing organisms above mean high water allows excellent preservation of the lamination.

Several progressive changes are evident in the sediments in going from near mean high water to spring high-water level:

1. Texture changes from clayey silt to silty clay
2. Color changes from brownish-gray, typical of slightly reducing conditions in zones further seaward, to moderate brown or reddish-brown, indicative of oxidizing conditions.
3. Lamination changes from somewhat lenticular or wavy to more uniform and of finer scale
4. Gypsum and halite are significant constituents of the mud
5. Deformed (warped) and ruptured laminae due to shrinkage cracks and evaporite crystallization become increasingly common.

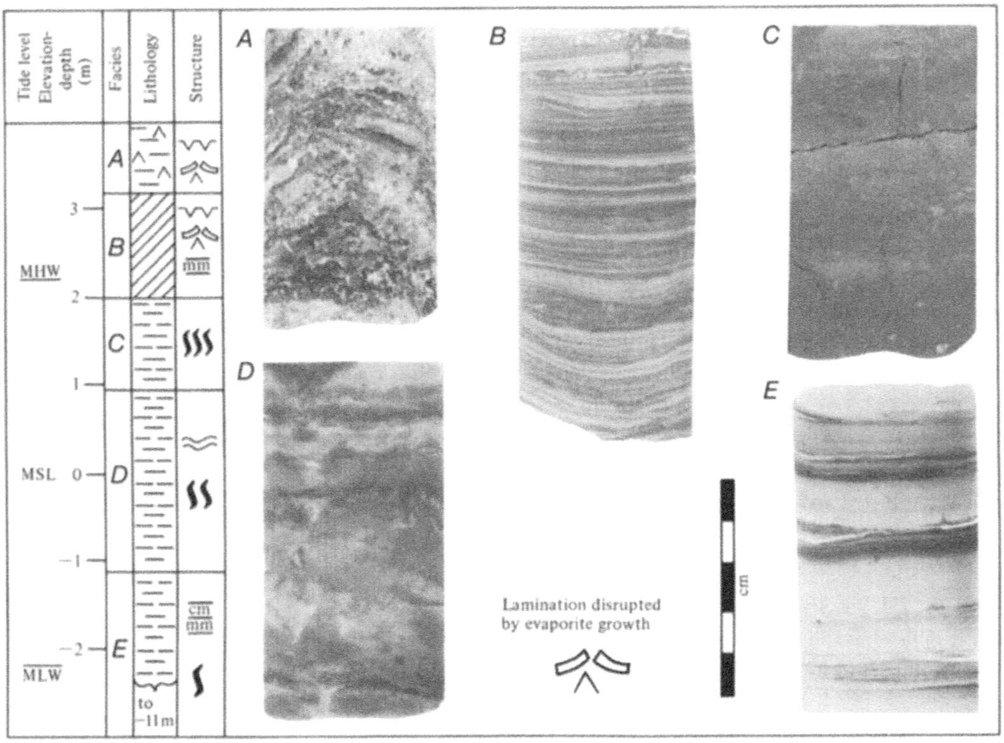

Figure 7-4. Characteristic facies of a regressive tidal-flat sequence. *A*, chaotic clay-evaporite mixture; *B*, brown laminated silt; *C*, gray-brown mottled silty clay; *D*, gray, burrowed silty clay; *E*, gray, laminated silt and clay. (Photos all x-radiographs except *C*. See Figs. 7-3 and 7-5 for distribution of facies.)

These changes relate to waning velocities of the tidal currents and to more prolonged intervals of subaerial exposure at higher tide levels.

Once the flats have aggraded to near the extreme tide level, only minor amounts of the finest silt and clay are furnished by tidal floodings, which occur only a few times each year. Intervals between flooding are marked by excessive evaporation whereby gypsum and halite become concentrated in the near-surface muds. Crystal growth of these minerals, along with shrinkage and mud-crack formation, combine to destroy totally any lamination and yield a very porous, chaotic, mud-evaporite mixture (facies A) that caps the tidal-flat accumulation. In several areas, salt crystallization has so elevated the mud-flat surface that it is now essentially deactivated.

Geologic History

Borings confirm that tidal-flat facies as described above underlie a major portion of the coastal mud flats and indicate they have grown by depositional regression across the toes of preexisting alluvial fans during the late stages of Holocene sea level rise. Silt and clay comprising the flats were supplied by the Colorado River and were dispersed to this site largely by tidal currents. Depositional regression by mud-flat accretion was possible at this open coastal site because of high rates of mud supply and low wave energy.

Seaward growth of the mud flats was temporarily interrupted by intervals of erosional transgression on at least two and probably three occasions during the last 3000 years BP. During these intervals, mud supply from the Colorado River was relatively slight so that waves had sufficient time to rework previous mud-flat deposits and to concentrate the coarser (>0.062 mm) components. Reworked coarse sand- to gravel-size shell remains initially accumulated as low bars near neap high-water level and, by berm accretion, later grew above spring high-water level to form beach ridges. Fine terrigenous sand and some shell were concentrated into a semicontinuous veneer covering former silt and clay facies of the lower intertidal zone. Alternating conditions of high and low mud supply since about 3000 years BP have resulted in development of the zone of beach ridges and intervening mud flats that fringes the high flats. Cross sections through this zone show lenses of intertidal mud-flat deposits of different ages that are separated horizontally and vertically by coarse deposits formed during erosional transgression (Fig. 7–5).

Figure 7-5. Cross section of interbedded transgressive sand and regressive mud facies. For location, see Fig. 7-3.

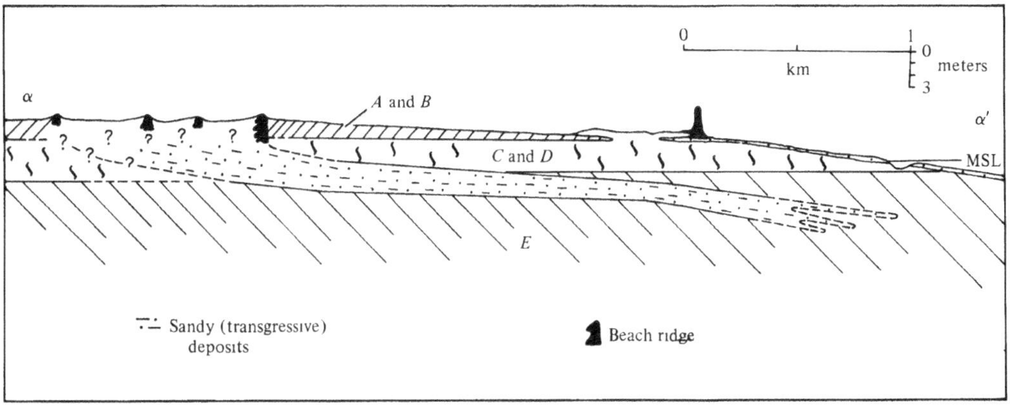

Comparison With Other Tidal Flats

A model of sedimentary facies expected from tidal-flat deposition was proposed recently by Klein (1971) for purposes of recognizing ancient tidal-flat deposits and estimating paleotidal range. Information concerning components of this model has come largely from tidal flats that have accumulated in temperate regions in either the lee of large barrier islands or protected coastal embayments. In these circumstances, restriction of tidal access to and from the tidal flats yields substantial currents and the development of elaborate tidal channel systems that largely control the sedimentation patterns. Furthermore, substantial sand is transported to the tidal flats under the combined influence of littoral drift and tidal currents.

The resulting tidal-flat deposits are predominantly sand and show a distinct upward-fining in texture through three main facies: (1) well-cross-bedded and cross-laminated sand at the base (subtidal); (2) interlaminated sand and mud in the middle (intertidal); (3) organic-rich clay or peat at the top (supratidal). The sediments are generally dark gray-black and the remains and traces of organisms are conspicuous throughout. Evaporites are negligible in the sequence. In the geologic record, such deposits might be anticipated in a stratigraphic complex of interbedded lenticular sandstones, dark gray shales or mudstones, and coals that accumulated in associated barrier island, shallow neritic, and deltaic environments.

With only this image of tidal-flat deposits, recognition of ancient analogs of the Gulf tidal flats in the geologic record would be improbable. Little sand has reached these tidal flats because of the tidal circulation system, and this, combined with low wave energy and copious supply of silt and clay, has resulted in insignificant barrier development. Tidal currents therefore are little restricted and move over the flats as a broad uniform flow with little tendency to develop channel systems. This is responsible for several of the more obvious differences between the Gulf tidal flats and those summarized by Klein:

1. Suspension-deposited silty clays in the subtidal and lower intertidal zones rather than cross-bedded sands
2. Typically uniform horizontal lamination and bedding on the intertidal flats rather than lenticular (wavy) or flaser bedding
3. Predominance of silt and clay and lack of a significant upward-fining of texture.

Other conspicuous differences between the gulf tidal flats and those summarized by Klein include abundance of evaporites in the upper intertidal and supratidal zones and absence of marsh; virtual lack of biogenous traces and predominance of brown, oxidized sediments in the upper intertidal zone. These differences relate primarily to the extreme aridity in the gulf setting.

The gulf tidal-flat sediments, with time and burial, would likely evolve into a sequence of red siltstones and claystones with significant lenticular beds and seams of gypsum and halite (Walker, 1967). Such sediments, interfingering with and overlain by alluvial fanglomerates and eolian sands, would more likely be interpreted in the geologic record as the product of an arid, inland basin rather than a paralic tidal-flat sequence, particularly considering the prevailing models of tidal-flat accumulations. Ancient red bed sequences of the western interior of the United States might be examined profitably in this light.

References

EVANS, G. 1965. Intertidal flat sediments and their environments of deposition in the Wash. *Quart. J. Geol. Soc. London 121*, 209–245.

KLEIN, G. DeV. 1971. A sedimentary model for determining paleotidal range. *Geol. Soc. Am. Bull. 82*, 2585–2592.

REINECK, H.-E. 1967. Layered sediments of tidal flats, beaches, and shelf bottoms. *In* Lauff, G. H., ed., *Estuaries.* Am. Assoc. Advan. Sci. Publ. No. 83, pp. 191–206.

STRAATEN, L. M. J. U. VAN. 1954. The composition and structure of recent marine sediments in the Netherlands. *Leidse Geol. Mededel. 19.*

THOMPSON, R. W. 1968. Tidal flat sedimentation on the Colorado River delta, northwestern Gulf of California. *Geol. Soc. Am. Memoir 107.*

WALKER, T. R. 1967. Formation of red beds in modern and ancient deserts. *Geol. Soc. Am. Bull. 78*, 353–368.

8

Facies Characteristics of Laguna Madre Wind-Tidal Flats

James A. Miller

Occurrence

Laguna Madre Flats is a predominantly terrigenous, wind-tidal flat forming a land bridge between Padre Island and the mainland on the south Texas coastal plain (Fig. 8–1). The flats are approximately 20 km long and 10 km wide.

Climate and Hydrography

The climate is semiarid, with evaporation estimated to exceed precipitation by more than 50 cm/year. Lagoon waters adjacent to the wind-tidal flat are usually hypersaline, commonly exceeding 50‰ in the summer months. During normal conditions, wind tides have an amplitude of 30 to 50 cm and may cover areas of more than 50 km^2. Sedimentation is dominated

Technical and logistical support for this study was provided by the U.S. Geological Survey, Office of Marine Geology, Corpus Christi, Texas; from the Marine Geology Fund at The University of Texas at Austin; and from the Union Oil Company of California.

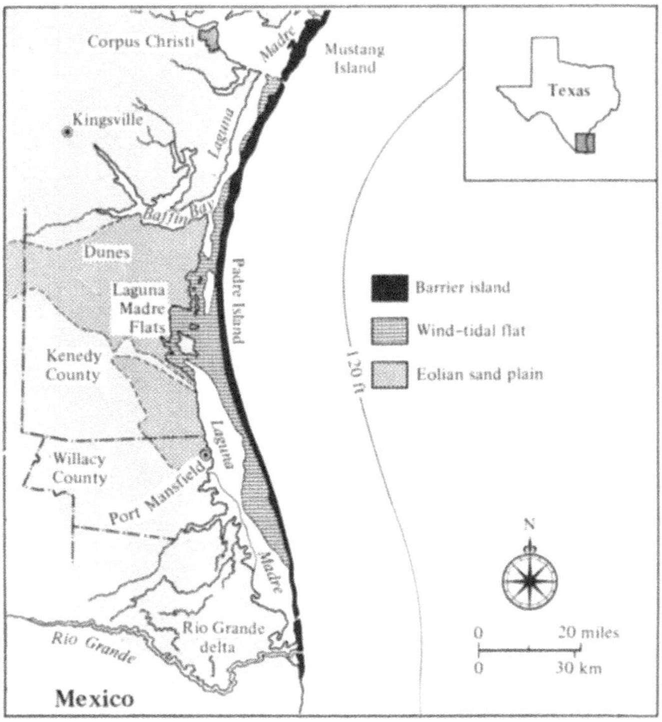

Figure 8-1. Location map of the south Texas coast showing the distribution of major depositional environments. (Modified after Hayes, 1967.)

by eolian and wind-tidal processes. Slight differences in elevation on the wind-tidal flat markedly affect the frequency with which areas are flooded, thus creating a complex of facies within the wind-tidal environment.

Rate of Accumulation

^{14}Carbon dates of algal-mat material show that the rate of sedimentation in the land bridge area has been gradually decreasing (50 mm/100 years to 25 mm/100 years or less) during the past 2,500 years. The algal mat dates have also shown that the sedimentation rate was not uniform throughout the land bridge. Rather, there were several small basins that have been dominated by different proportions of eolian and wind-tidal sedimentation.

Depositional Facies

Fisk (1959) produced a topographic map of the Laguna Madre Flats that reveals the slight but significant differences in elevation that determine the frequency with which an area is flooded. These differences, plus variations in the supply of sand and mud from Padre Island and Laguna Madre, produce a variety of sedimentary facies for what is actually one environment (Figs. 8–2 and 8–3).

Light olive-gray clay and blue-green algal-mat sequences are well developed on the frequently flooded wind-tidal flat adjacent to the south lagoon. The clay bands and algal mats are contorted and disrupted by desiccation and burrowing. Tiny granules of

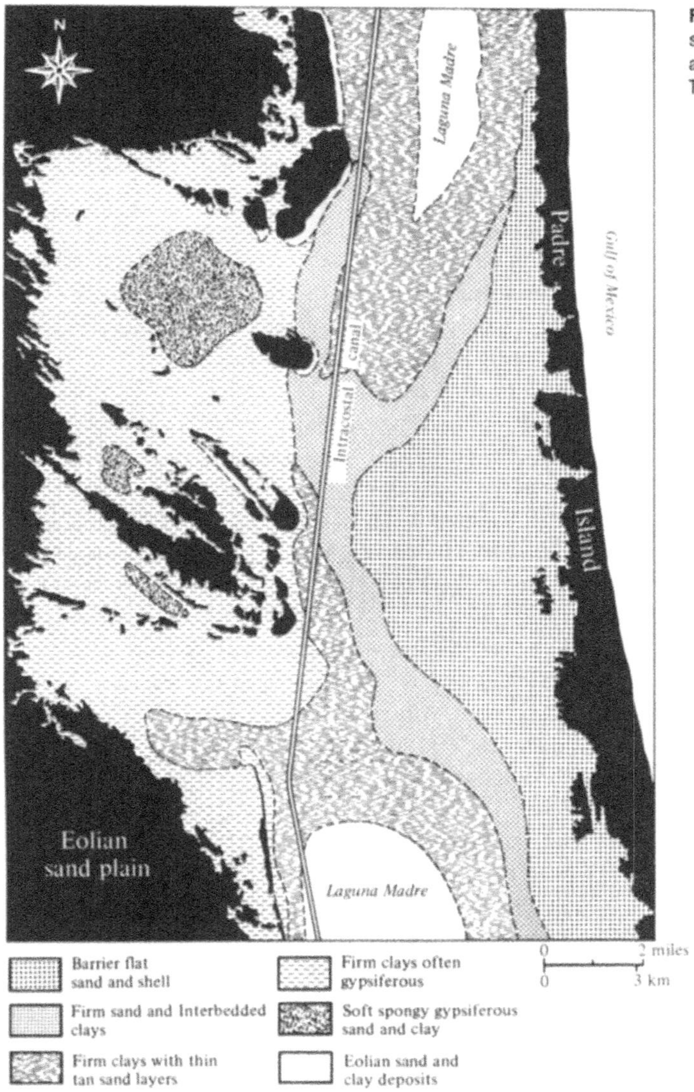

Figure 8-2. Distribution of sedimentary facies, Padre Island and Laguna Madre Flats, Texas. (Modified after Fisk, 1959.)

Figure 8-3. Aerial view of the northern portion of the Laguna Madre wind-tidal flat. Color variations denote slight differences in elevation. Scale in inches.

carbonate material form thin bands in association with the algal mats; gypsum is generally absent (Fig. 8–4). Beneath the clay and algal sequence there is light to medium gray lagoonal sand with poorly defined clay laminae. Areas closer to the mainland, which are subject to less flooding and higher rates of eolian sediment accumulation, produce facies in which light gray or buff-colored sand becomes a major component, with poorly developed algal mats and massive and indistinct clay bands. There may be some disruption of bedding by worm and insect burrows and root mottling. Thin stringers of sucrose gypsum may be associated with the clay and algal laminae.

Shallow depressions on the surface of the wind-tidal flat pond water when the area is inundated by wind tides or heavy rains. Most of these ponds are less than 25 cm deep and appear to be the final stage in the filling of natural depressions on the Laguna Madre Flats.

When the basins are flooded, algal mats thrive and clay settles from suspension and is incorporated in the algal mat or forms discrete clay bands. After the flooding, water in the basins begins to evaporate, and unless the water is diluted by rain or subsequent flooding, salt pans may develop. Most of the salt pans are less than 100 meters in diameter with crusts less than 2 cm thick, but pans of more than 800 meters in diameter with crusts 5 to 10 cm thick have been observed.

Unlike the Arabian Gulf and Baja California, halite is normally the only mineral found in these brine pans, and the pans are not regularly recharged by successive floodings. The salt crusts are strictly empheral, and cores from the salt pans reveal that none of the halite is being preserved. The only evidence of a salt pan's existence is a 2 to 8 cm layer of highly organic black clay with decaying grass and filamentous green algae.

Figure 8-4. Interlaminated clay and algal sediments with minor amounts of sand and sucrosic gypsum. Note burrowing, desiccation cracks, and distortion of algal mats.

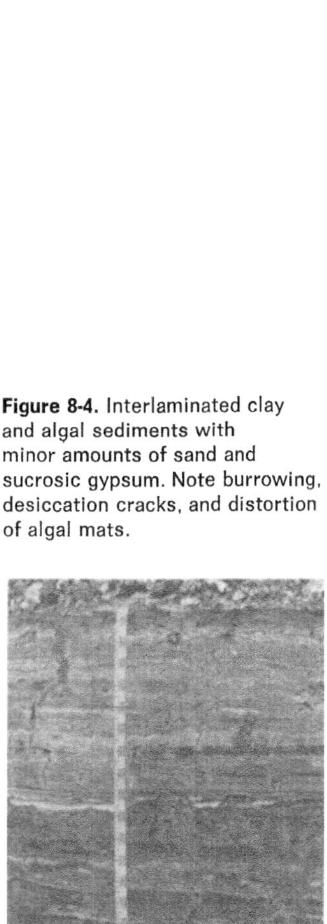

Diagenetic Features

Clay and algal-mat sequences are well developed on those frequently flooded portions of the wind-tidal flat. The clay and algal laminae are commonly contorted and disrupted by desiccation, gypsum crystallization, and burrowing (Fig. 8–5). Gray subtidal sand containing gypsum sand crystals and poorly defined clay laminae underlies the clay and algal zone.

In the more restricted and arid portions of the flats, gypsum becomes a prominent sediment constituent, but no anhydrite has been detected in any of the samples. Investigations by Masson (1955), Fisk (1959), and myself indicate that nearly all of the gypsum develops in or beneath the surface sediments by precipitation from interstitial brines. As might be expected, the gypsum occurs in a variety of forms, dependent upon the conditions of precipitation and the composition of the surrounding sediments. Table 8–1 summarizes the various types of gypsum and their mode of occurrence.

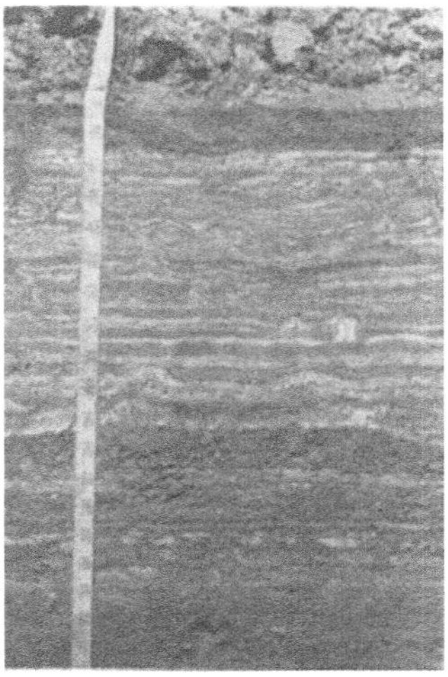

Figure 8-5. Slightly contorted interlaminae of clay, algal mats, and sucrosic gypsum. Most deformation results from desiccation and the growth of gypsum crystals. Scale in inches.

Table 8–1 *Gypsum types and modes of occurrence*

Gypsum type	Mode of occurrence
Sucrose gypsum (1–2 mm selenite)	
1. Thin bands, stringers and pods of sucrose gypsum	Forms in association with clay laminae and algal-mat sequences.
2. Nodules of sucrose gypsum	Forms as burrow fillings in muddy sand near the margin of the flats, or fills voids formed by the desiccation of algal mats or trapped gas bubbles.
Biconvex lens-like senenite crystals	
1. Small crystals (≦1 cm diameter) a. Individual crystals b. Crystal aggregates (multiple penetration twins)	Generally occur in rather dry sandy sediments in areas adjacent to the margin of the flats; crystals are often unoriented, but many are parallel to bedding; crystal aggregates may develop as burrow filling; only minor amounts of sediment are included in the crystals.
2. Large crystals (>1 cm diameter) a. Amber crystals (individual) b. Amber crystals (aggregates)	Generally occur in clay and algal-rich sediments of middle and lower wind-tidal flat; crystals contain little or no included sediments; they are most abundant at depths of approximately 1 m in sediments with high interstitial water content; crystals may conform to the primary bedding or be discordant.
c. Sand crystals	Gray colored, containing large quanities of included sand and shell; occur below the water table in lagoonal sand facies; usually occur at depths of greater than 2 m; crystal aggregates average 6–8 cm in diameter, although aggregates with a diameter of 25 cm have been collected.

Authigenic aragonite, dolomite, and magnesium calcite are present in varying quantities throughout the land bridge area. Overall, these minerals constitute less than 5 percent of the total sediment, but local concentrations may be nearly 100 percent pure carbonate.

The distribution of the carbonates and the relative proportions in which they occur are quite complex. Most of them are found in association with clay and algal materials, with the algal mats seeming to be the more important. Aragonite formation appears to be favored in sediments that are predominantly sand, with moderate to minor quantities of clay, algae, and gypsum. Conversely, dolomite seems favored in sediments dominated by clay, algae, and gypsum, and containing only minor amounts of sand.

The carbonates occur in a variety of forms: (1) thin lime bands (2 mm or less) formed within or immediately beneath active algal mats; (2) carbonate bands 2 to 12 mm thick, which may be from direct precipitation or early diagenetic formation; (3) fragments of lime bands that have been broken up and reworked by waves and currents; (4) cement or matrix in sandy and shelly sediments; (5) calcareous and dolomitic marls closely associated with clay, algae, and the amber selenite crystals; (6) a precipitate or alteration product in burrows, mud clasts, and clay bands.

The gray or salt-and-pepper sand in the Laguna Madre sediments is primarily the result of black iron sulfide (hydrotroilite?) and microcrystalline pyrite coating many of the quartz sand grains.

Although some bacteriologic work has been done in the area (Fisk, 1959; Sorensen and Conover, 1962), the extent of bacterial activity in Laguna Madre sediments is largely speculative. My observations, geochemical data, and mineralogic analyses indicate that bacteria probably play an active role in the production of interstitial gases and that they may strongly influence local Eh and pH conditions. Furthermore, it is quite likely that the formation of carbonates, gypsum, and iron sulfides is promoted by the action of ammonifying and sulfate-reducing bacteria.

References

Fisk, H. N. 1959. Padre Island and the Laguna Madre Flats, coastal south Texas. 2nd Geog. Conf., Coastal Studies Inst., Louisiana State Univ., 103–152.

HAYES, M. O. 1967. Hurricanes as geological agents: Case studies of Hurricanes Carla, 1961, and Cindy, 1963. *Rept. Investigations No. 61*, Bureau Econ. Geol. Univ. Texas, Austin, pp. 1–56.

MASSON, P. H. 1955. An occurrence of gypsum in southwest Texas. *J. Sed. Petrol. 25*, 72–79.

SORENSEN, L. O. and CONOVER, J. T. 1962. Algal mat communities of *Lyngbya confervoides* (C. Agardh) Gomont. *Publ. Inst. Mar. Sci. Univ. Texas 8*, 61–74.

9

Inlet Sequence Formed by the Migration of Fire Island Inlet, Long Island, New York

Naresh Kumar and John E. Sanders

Occurrence

Fire Island inlet, Long Island, New York, and up to 8 km east of the inlet under Fire Island. Present location of Fire Island inlet is long. 73°18′ W and lat. 40°38′ N (Fig. 9–1). The inlet is 4.8 km long and 1.1 km wide at present; its long axis trends in an east-west direction. Maximum depth: 10 meters.

Hydrography and Winds

Tide range: 1.23 meters, semidiurnal. Spring tides are 20 percent greater than the mean tides. Tidal currents in the inlet: 1.2 meters/second for the ebb and 1.05 meters/second for the flood current. Wind direction: east and southeast in winter and spring, west and northwest in summer and autumn. Waves: 72 percent of all deep-water waves approach from the sector east-northeast through south-southeast. Wave heights decrease as the water shoals across the continental shelf. Greatest reported wave heights in deeper water are 7.5 to 9 meters; mean wave

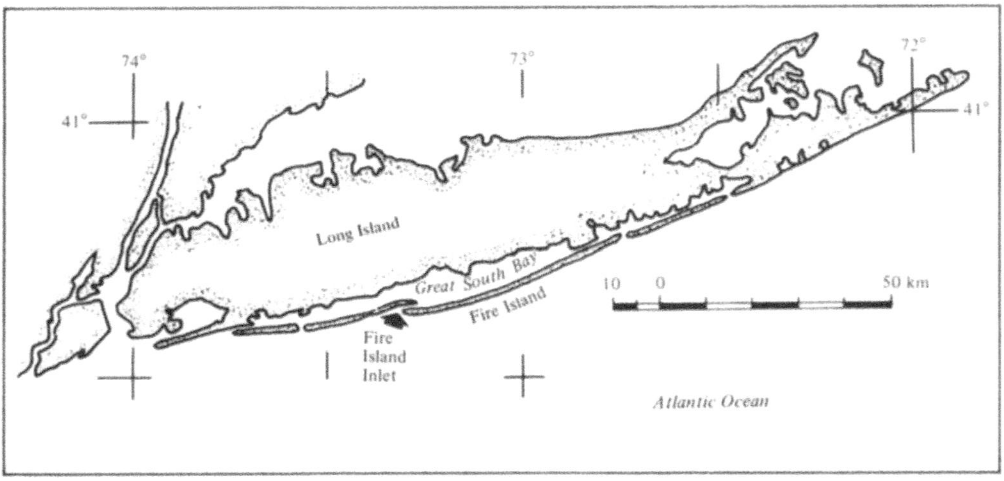

Figure 9-1. Location of Fire Island inlet, south shore of Long Island, New York.

heights of 36 cm during the period 1950 to 1954. Hurricanes: tropical hurricanes and extratropical storms, the "northeasters," frequent the area.

Mechanism of Inlet Sequence Formation

Inlets typically occur on submerging coasts. Depending upon the direction from which the waves approach the coast, the tidal inlets may migrate systematically in one direction. Longshore currents, powered by waves, deposit sediment on one bank of the inlet, which causes the tidal currents to erode the other side.

Fire Island inlet has migrated 8 km westward during the 115-year period between 1825 and 1940, in the manner described above.

Hydraulic conditions vary in the different parts of the inlet from the bottom of the channel up to the spit. Hence the sedimentary structures and textures also differ in sediments deposited at different depths in the inlet. During the lateral migration of the Fire Island inlet, a sequence of sediments containing these structures and textures, the "inlet sequence," has been deposited under those parts of the barrier through which the Fire Island inlet has migrated. The criteria for recognizing the different units in the inlet sequence were established from the sediments in Fire Island inlet, and this sequence was then recognized in bore holes drilled on those parts of Fire Island through which the inlet is known to have migrated.

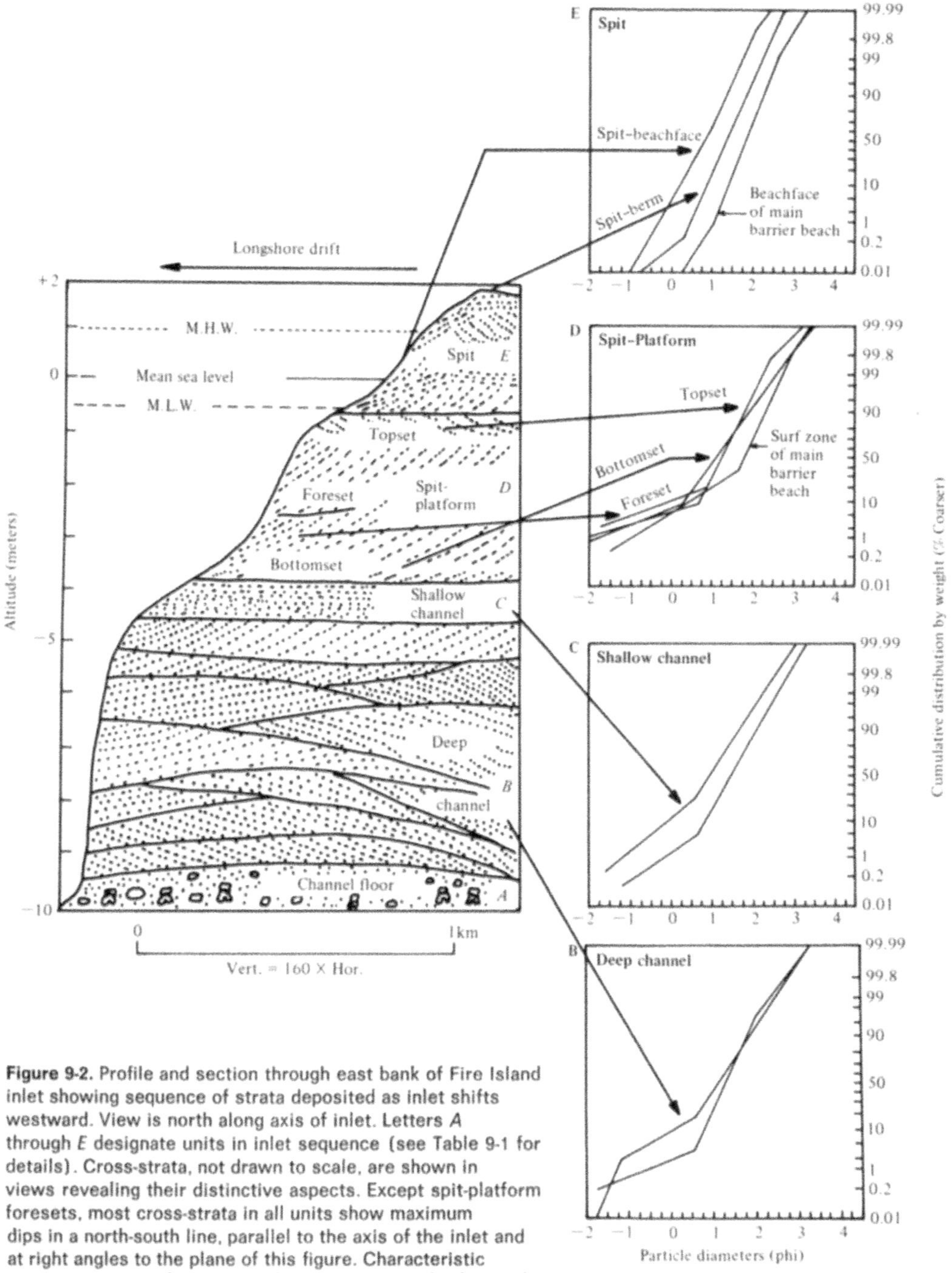

Figure 9-2. Profile and section through east bank of Fire Island inlet showing sequence of strata deposited as inlet shifts westward. View is north along axis of inlet. Letters *A* through *E* designate units in inlet sequence (see Table 9-1 for details). Cross-strata, not drawn to scale, are shown in views revealing their distinctive aspects. Except spit-platform foresets, most cross-strata in all units show maximum dips in a north-south line, parallel to the axis of the inlet and at right angles to the plane of this figure. Characteristic grain-size curves of sediments from each unit and subunit of the inlet sequence have been drawn on the right. The scale on the *y*-axis for grain-size curves is a probability scale.

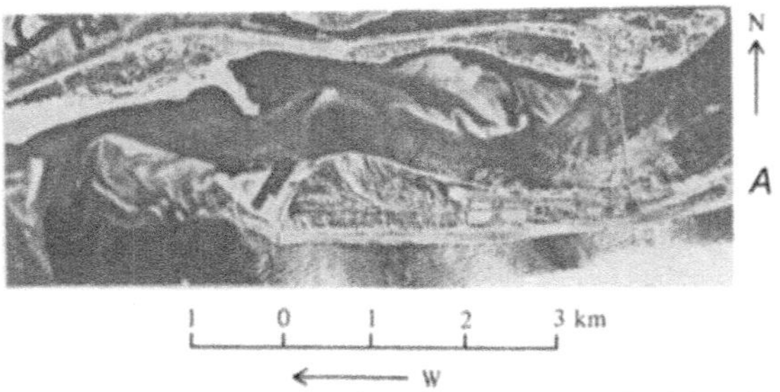

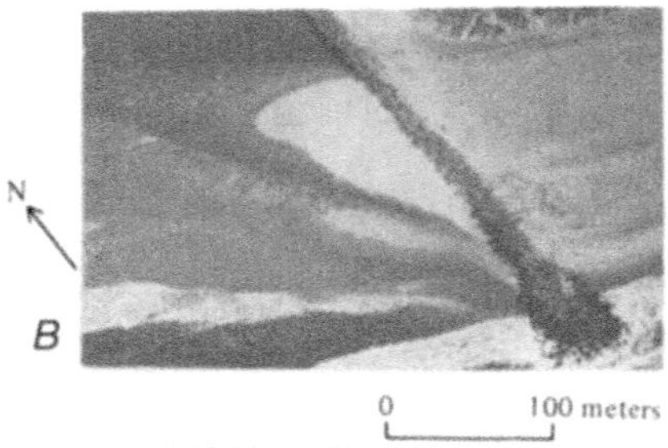

Spit:
Spit-berm
washover delta

Spit:
Spit-beam
Spit-beachface

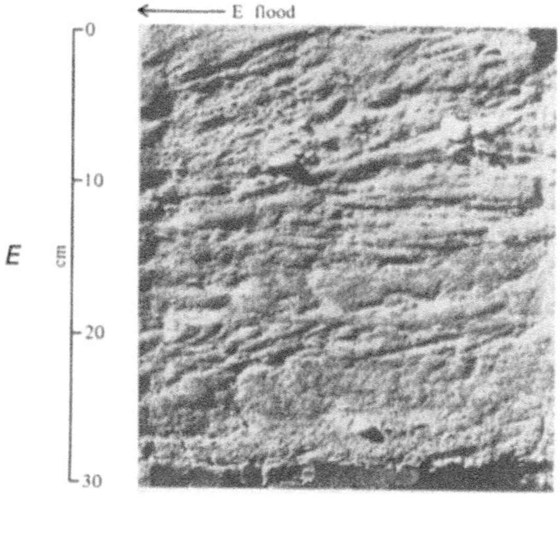

Spit-platform: topset
(depth ~ 0.6 meters)

Figure 9-3. *A.* Aerial view of Fire Island inlet on September 3, 1964. At the time the photograph was taken the eastern bank of the inlet was being dredged. *B.* Aerial view of spit-platform in Fire Island inlet on October 29, 1970. *C.* Steep landward-dipping laminae formed by washover delta dip to the right. The gentle, almost horizontal landward-dipping laminae belong to the spit-berm. The handle of shovel is 65 cm long. *D.* Steep and gentle seaward-dipping laminae of spit-beachface dip to the left; gentle landward-dipping laminae of spit-berm dip to the right. Tape measure is located almost at the spit-berm crest. *E* through *J.* Epoxy relief-peels of box or tube cores taken in different parts of the channel of Fire Island inlet at depths indicated.

Table 9–1 *Summary of criteria used in establishing the inlet sequence*

Letters corresponding to Fig. 9–2	Unit	Alt. (m)	Processes	Type of flow	Bedform	Sedimentary structures	Textures
E	**Spit:** Beachface	+2.0 to −0.6	Swash and Backwash	Upper-flow regime		Steeply and gently seaward-dipping laminae. Wide range in directions of dip.	Except bay sediments, spit sediments are similar in texture in all subenvironments.
	Berm		Swash			Heavy mineral concentrates; laminae nearly horizontal but dip gently landward.	Size-analysis curves similar to typical beachface samples.
	Bay		Settling; bioturbation			Bioturbated; thin laminae.	Grains show rounding by waves. Wind-frosted grains present. Occasional layers rich in heavy minerals.
	Washover delta		Swash during higher-than-normal sea level			Steeply landward-dipping laminae. Wide range in directions of dip.	
D	**Spit-platform:** Topset	−0.6 to ~−1.0	Pulsating unidirectional current in flood direction.	Lower-flow regime	Flood-oriented megaripples and current ripples.	Various sets of flood-oriented cross laminae.	Size-analysis curves very similar to typical surf-zone samples from open-ocean beach. Grains rounded, but less rounded than spit samples. Less percentage of wind-frosted grains than spit sediments.
	Foreset	~−1.0 to −3.0				Planar steep cross-laminae. Sets long (2 m) and thick (1 m).	
	Bottomset	−3.0 to −3.75	Ebb > flood	Lower-flow regime	Ebb-oriented megaripples.	Ebb-oriented cross-laminae.	

C	**Channel:** Shallow channel	—3.75 to —4.5	Ebb > flood	Transition or upper-flow regime	Plane-parallel beds.	Plane-parallel laminae or washed-out ripple laminae.	Size-analysis curves show a moderately sorted surface-creep population and well-sorted saltation population. No suspension population present. Grains angular, freshly broken because of intense energy. Shell fragments most common.
B	**Deep channel**	—4.5 to —10.0	Ebb > flood	Lower-flow regime	Ebb-oriented, large sand waves.	Ebb-oriented cross-laminae with gentle flood-oriented reactivation surfaces.	Size-analysis curves have three populations. Either a surface-creep, a saltation, and a suspension population, or a surface-creep and two saltation populations. Surface-creep population is larger but less well-sorted than shallow channel. Quartz grains have freshly broken surfaces. Only most resistant parts of shells present.
A	**Channel floor**	—10.0	Ebb > flood	Lower-flow regime		Lag gravel; large shells, pebbles, etc.	

An inlet sequence containing five major units has been established for Fire Island. In ascending order the units are (*A*) channel floor, characterized by a lag gravel composed of large shells, pebbles, and other coarse particles; (*B*) deep channel (—10.0 to —4.5 meters), characterized by lenticular sets of ebb-oriented "reactivation" surfaces; (*C*) shallow channel (—4.5 to —3.75 meters), characterized by plane, parallel laminae; (*D*) spit-platform (—3.75 to —0.6 meters), characterized by steeply dipping planar cross-laminae and small-scale, flood-oriented laminae near the top and ebb-oriented laminae toward the bottom; and (*E*) spit (—0.6 to +2.0 meters), characterized by steep and gentle seaward-dipping laminae and steep and gentle landward-dipping laminae in its various subenvironments.

Figure 9–2 is a schematic representation of sedimentary structures in various units of the sequence. The size-analysis curves on the right-hand side of the figure have been drawn and can be analyzed using the techniques and methods of Visher (1969). Figure 9–3*A* is an aerial view of the Fire Island inlet, and Figure 9–3*B* is an aerial view of the spit-platform associated with this inlet. Figures 9–3*C* through 9–3*J* show the sedimentary structures in relief peels of box cores and tube cores from different parts of the channel or in trenches dug on the spit.

Swift currents in the channel bottom produce a concentrate that is characteristic of the channel floor (Fig. 9–3*J*). Large sand waves, 60 to 100 meters long and 0.5 to 2.0 meters high, form and migrate within the depth range of the deep channel. These sand waves do not form and migrate in waters shallower than 4.5 meters. The thickness of the deep channel unit is therefore a function of the depth of the channel; the upper limit is always at a depth of —4.5 meters. The reactivation surfaces in Figure 9–3*I* have been described by Klein (1970) and Boersma (1969) in other areas where large sand waves are also subjected to alternating tidal flow. We do not know whether large steep planar cross-laminae exist in the large-scale sand waves beneath the depth of penetration of our box core (45 cm). At present, we think that laminae characteristic of reversing flow occur throughout these sand waves. The sediments in the shallow channel are deposited in the upper-flow regime and hence contain plane-parallel laminae (Fig. 9–3*H*).

The spit-platform (Fig. 9–3*B*) is a shallow submerged platform built by waves before a spit can form and grow subaerially (Meistrell, 1972). The spit-platform resembles a delta with characteristic topset, foreset, and bottomset surfaces; each surface has characteristic sedimentary structures (Figs. 9–3*E*, *F*, and *G*). A spit forms and grows over the topset surface of the spit-platform. The processes operating on the spit are very similar to the processes operating on the foreshore and berm of the main barrier beach, except that, because of refraction, waves approach different parts of the spit from various directions. Consequently, on a spit the landward and seaward directions may span an arc of up to 90° or more (Figs. 9–3*C* and *D*).

Table 9–1 summarizes the structural and textural criteria for recognition of various units in the sequence, see also Kumar and Sanders (1974).

References

Boersma, J. R. 1969. Internal structures of some tidal mega-ripples on a shoal in the Westerschelde Estuary, the Netherlands. *Geol. Mijnb. 48*, 409–414.

Klein, G. deV. 1970. Depositional and dispersal dynamics of intertidal sand bars. *J. Sed. Petrol. 40*, 1095–1127.

Kumar, N. and Sanders, J. E. 1974. Inlet sequence: A vertical succession of sedimentary structures and textures created by the lateral migration of tidal inlets. *Sedimentology 21*, 491–532.

Meistrell, F. J. 1972. The spit-platform concept: Laboratory observation of spit development. *In* Schwartz, M. L., ed., *Spits and Bars*. Stroudsberg, Pa., Dowden, Hutchinson and Ross, pp. 225–283.

Visher, G. S. 1969. Grain-size distributions and depositional processes. *J. Sed. Petrol. 39*, 1077–1106.

10

Sequences in Inshore Subtidal Deposits

J. H. J. Terwindt

This paper examines the sequential order of lithofacies in inshore subtidal deposits in the southwest of the Netherlands. These deposits accumulated in estuarine and tidal inlet environments. The tidal range varies between 2.0 and 3.5 meters and the maximum current velocity at 50 cm above the bottom is about 1.30 meters/second.

The several lithofacies in the area of investigation are closely associated with the intensity of the tidal currents. Indications of wave action are almost absent.

Relation Between Lithofacies and Current Intensity

Three lithofacies may be distinguished in subtidal deposits (Terwindt, 1971).

Lithofacies I. Coarse-grain sand, containing clay pebbles, shells and shell debris, peat lumps and detritus, showing large-scale cross bedding. The set thickness is 0.1 meter and more.

Lithofacies II. Medium to fine-grain sand, showing small-scale cross stratification, horizontal sand lamination, and flaser bedding. The set thickness is between 0.01 and 0.1 meters.

Lithofacies III. Fine sand and clay layers, showing lenticular and sand-clay alternate bedding. The set thickness ranges between 0.01 and 0.001 meters.

A useful parameter for the current intensity related to the formation of these lithofacies is $V_{50\ max}$: the maximum current velocity at 50 cm above the bottom during a tidal cycle.

It appears that lithofacies I may be formed if $V_{50\ max}$ is greater than 0.8 to 0.9 meter/second. Lithofacies II can be formed if $V_{50\ max}$ exceeds 0.6 meter/second. For lithofacies III $V_{50\ max}$ may vary just above and below 0.6 meter/second (Terwindt and Breusers, 1972). Although the critical values of $V_{50\ max}$ are approximate and need further confirmation, it appears that the different lithofacies are determined by relative small changes in $V_{50\ max}$.

Sequences

The value of $V_{50\ max}$ at a certain place changes in time, either suddenly or gradually. Sudden changes are reflected by sharp boundaries between the different lithofacies, with no sequential order in the succession. Both sudden and gradual changes are observed commonly in tidal deposits in the study area as well as elsewhere (De Raaf and Boersma, 1971).

The gradual changes in $V_{50\ max}$ may be separated into long-term and short-term changes. The long-term variations may be caused by shifting of channels and shoals. The short-term changes, superimposed on the long term, may be due to neap-spring tide cycles, the presence of drift currents, and so on.

A gradual long-term decrease in $V_{50\ max}$ results in a fining-upward sequence, characterized by a gradual transition of lithofacies I to III. In the same way a coarsening-upward sequence is generated when $V_{50\ max}$ gradually increases.

If $V_{50\ max}$ first decreases but afterward increases, then a fining-upward, coarsening-upward succession of lithofacies may be distinguished. A coarsening-upward, fining-upward sequence is also possible. Occasionally even fining-coarsening-fining and coarsening-fining-coarsening successions can be observed.

These gradually changing successions are bounded at their

tops by a definite break in sedimentation conditions, giving rise to an abrupt transition in the type of the lithofacies. This break is sometimes accompanied by erosion.

A short-term increase or decrease of the value $V_{50\ max}$ gives rise to a temporary change in the intensity of the sediment movement and/or sedimentation rate. A short-term decrease of $V_{50\ max}$ may result in thicker or more clayey layers alternating with sandy layers. The sedimentary effects of these short-term variations in the vertical section are normally restricted to the order of some centimeters.

Observations

Some 135 continuously cored borings, made in the tidal waters of southwest Netherlands, were analyzed in order to get a general idea of the frequency of occurrence of the several types of sequences. These borings comprise about 2,500 meters of vertical section of tidal deposits. About 600 meters or some 20 percent of the observed total did not show a sequential order in the succession of lithofacies, but instead each lithofacies has sharp upper and lower boundaries.

In the remaining 1,900 meters of vertical sections, about 500 sequences were observed. A complete fining-upward sequence was defined as follows: There should be a gradual upward transition from coarse-grain, cross-bedded megasets which may contain clay pebbles, shell debris, and peat detritus (lithofacies I), to small-scale, cross-bedded sets, or parallel-laminated sand layers, to flaser bedding (lithofacies II), to lenticular or sand-clay alternate bedding (lithofacies III). A coarsening-upward sequence should have the reverse succession. An illustrative example of the determination of the sequences is given in Figure 10–1.

Most of the observed sequences, however, are incomplete; they comprise only two of the lithofacies. Especially numerous are transitions between lithofacies II and III.

The thicknesses of the observed sequences range from 0.3 up to 5.5 meters, with a mean of about 3 meters.

About 40 percent of the observed sequences are fining-upward and some 20 percent coarsening-upward. About 20 percent are fining-upward, coarsening-upward sequences and 10 percent coarsening-fining. About 7 percent of the sequences are fining-coarsening-fining upward and about 3 percent are coarsening-fining-coarsening.

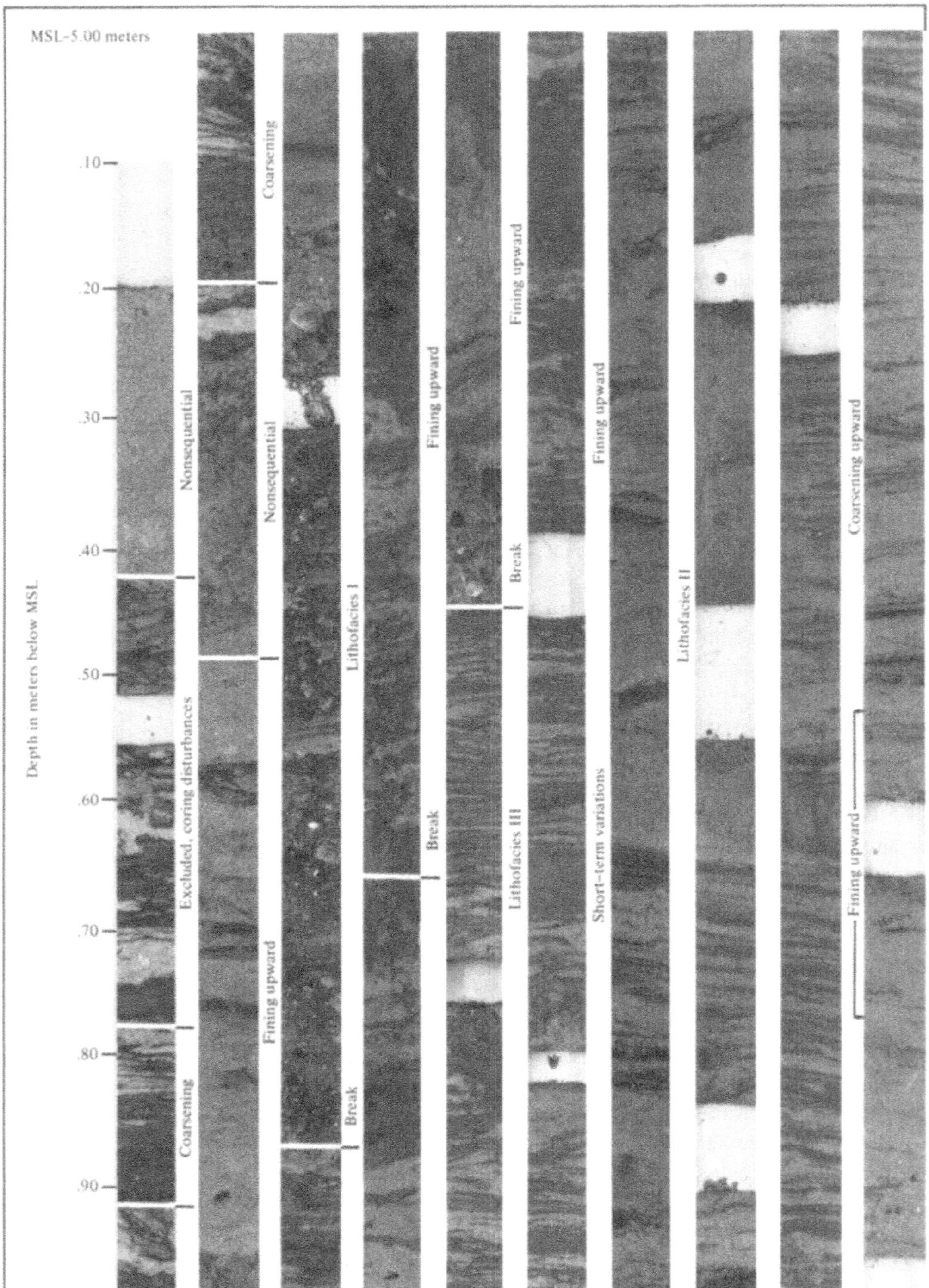
MSL-5.00 meters
Depth in meters below MSL
.10
.20
.30
.40
.50
.60
.70
.80
.90
6.00
Nonsequential
Excluded, coring disturbances
Coarsening
Coarsening
Nonsequential
Fining upward
Lithofacies I
Break
Fining upward
Break
Fining upward
Break
Lithofacies III
Fining upward
Short-term variations
Lithofacies II
Coarsening upward
Fining upward

Figure 10-1 (facing page). Example of sequential analysis.

Conclusions

The vertical succession of lithofacies in the tidal deposits of the area studied may be nonsequential (with sharp upper and lower boundaries of the lithofacies) or sequential (with gradual transitions). Most of them are fining-upward. However, coarsening-upward and other types of sequences are also observed.

References

De Raaf, J. F. M., and Boersma, J. R. 1971. Tidal deposits and their sedimentary structures. *Geol. Mijnbouw 3*, 479–504.

Terwindt, J. H. J. 1971. Litho-facies of inshore estuarine and tidal-inlet deposits. *Geol. Mijnbouw 3*, 515–526.

Terwindt, J. H. J., and Breusers, H. N. C. 1972. Experiments on the origin of flaser, lenticular and sand-clay alternating bedding. *Sedimentology 19*, 85–98.

Section II

Ancient Siliciclastic Examples

Recent tidal-flat deposits provide two of the three kinds of criteria used to identify fossil examples: diagnostic sedimentary structures and the expected vertical sequence of facies. The third kind of criterion is the regional setting and overall stratigraphic succession.

The diagnostic sedimentary structures in siliciclastic tidal flat deposits can be grouped into four categories:

1. Rapid reversals of depositing currents: herringbone cross stratification, reactivation surfaces (Klein, Papers 17, 20)
2. Small-scale alternations in slack water and strong current: flaser and lenticular bedding, clay drapes over ripples and sand waves
3. Intermittent subaerial exposure: desiccation cracks, rain prints, plant roots, evaporites, peat or coal beds, tracks of vertebrates
4. Alternating erosion and deposition: channels, scours, mud chips, winnowed sand beds and lenses

If a section of fossil marine sediment has unequivocal examples of structures from all four of these categories, or from the first three, most sedimentologists will interpret it as a tidal-flat deposit or tidalite; if only two of the categories are represented, the interpretation on structures alone is open to question.

This use of tidal flat for the environment of fossil sediments is necessarily less precise than in recent sediments. Rarely, if at all, is it possible to establish with certainty that astronomic tides were the major agents of deposition. However, an association of structures from the categories listed above does indicate, by analogy with recent examples, that deposition occurred in the zone of short-term fluctuations of sea level. Short-term fluctuations include all storm tides and seasonal changes during the year (see also the definition of intertidal in Fischer, 1964, cited in Paper 27).

The vertical sequence of Holocene tidal-flat deposits revealed by borings is generally transgressive because they accumulated during a rapid eustatic rise of sea level (Larsonneur, Paper 3; Kraft and Allen, Paper 5). Only in channels (Kumar and Sanders, Paper 9; Terwindt, Paper 10) or on the margins of a delta (Thompson, Paper 7) is the sediment supply sufficient to produce a regressive sequence.

Although the tidal-flat deposits of the North Sea and Atlantic Coast have transgressive sequences, it is possible to infer the expected regressive sequence from the pattern of subenvironments (see Thompson's summary of Klein, 1971, Paper 7; and Evans, Paper 2). The fining-upward sequence thus derived has a distinctive succession of sedimentary structures, and it is frequently cited as a criterion for tidal-flat deposition. Thompson (Paper 7) points out some of the limitations of the North Sea model, and shows that in a deltaic setting, where the tide range is similar to that of the North Sea, the combination of abundant mud-size sediment and reduced wave action gives a vertical sequence quite different from that of the North Sea model.

The first two contributions in Section II were selected to illustrate the spectrum of variation in siliciclastics; Sellwood's example has the succession of diagnostic sedimentary structures expected in the North Sea model, and it occurs between alluvial and marine sediments; on the other hand, Walker and Harms emphasize the absence of any sedimentary record of sea level fluctuations in muddy transitional sediments. The next eight contributions, arranged in stratigraphic order, are examples of the analysis of fossil sediments, each with an evaluation of the evidence for accumulation in the tidal zone. The last two summaries describe the sedimentary record in single vertical sections: an outcrop (Visher) and the core from an exploratory well (Rizzini).

11

Lower Jurassic Tidal-Flat Deposits, Bornholm, Denmark

Bruce W. Sellwood

Occurrence

The top of Gry's (1960) Lower Coal Series (Lias α-β) is well exposed in the Galgeløkke cliff section south of Rønne on the island of Bornholm. The Lower Coal Series (35 meters) consists of predominantly coarse-grain fluvial sandstones with subordinate coals overlaying the red marls of the Trias. The Coal Series is itself overlain by the ferruginous Hasle Sandstone (80 to 100 meters), which in places contains a marine fauna indicating a Lower Pliensbachian age. The sediments interpreted by Sellwood (1972a) as having a tidal-flat origin occur toward the top of the Lower Coal Series and represent a transitional facies between the fluvial sediments below and the marine sandstone above.

The author is grateful to Professors P. Allen and T. N. George and for the receipt of a Royal Society grant-in-aid.

Facies

The sequence at Galgeløkke consists of a basal sand unit that is succeeded by a predominantly flaser- and wavy-bedded unit containing thin coals (Fig. 11–1). The entire sequence constitutes a fining-upward succession, and is represented by sediments totally lacking a cement. Sections are only available in two dimensions.

Basal sand unit. This unit consists of planar-, tabular-, and wedge-shaped sets of medium- to fine-grain sands. Many of the individual sets are draped by thin clay laminae. Foreset dips are generally toward the south, although northerly components may be represented, especially where herringbone cross-bedding is developed. Usually, however, bipolarity of current flow is not indicated by the cross-bedding in immediately successive beds, but opposition in foreset dips may be found in units several beds apart. Much of the sediment is either flat-bedded or consists of very low-angle cross-lamination, but numerous sets of ripple-laminated sands also occur. Upper-flow-regime conditions were dominant during basal sand deposition but lower-flow-regime conditions produced symmetrical and asymmetrical ripples also.

The unit also contains a number of megaripples, each of which displays a well-developed reactivation surface (Klein, 1970). These megaripples (Fig. 11–2) are draped externally by clays, and *Skolithos* burrows sometimes descend from the reactivated surfaces, which thus remained stable for some time after their development. Foreset laminae within the megaripples

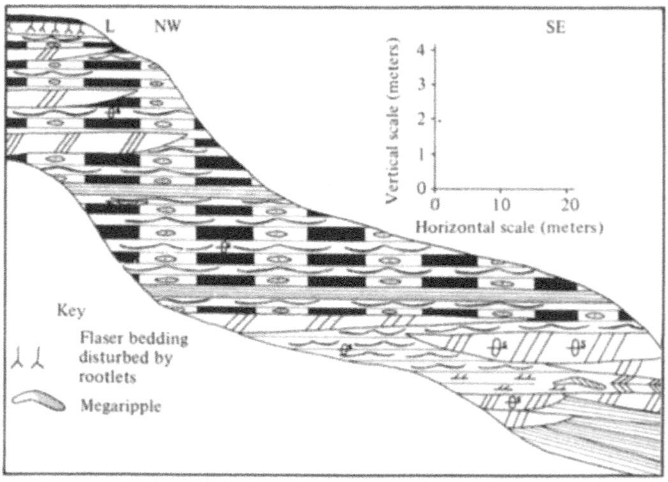

Figure 11-1. Section at Galgelokke illustrating horizontal and vertical nature of facies distributions.

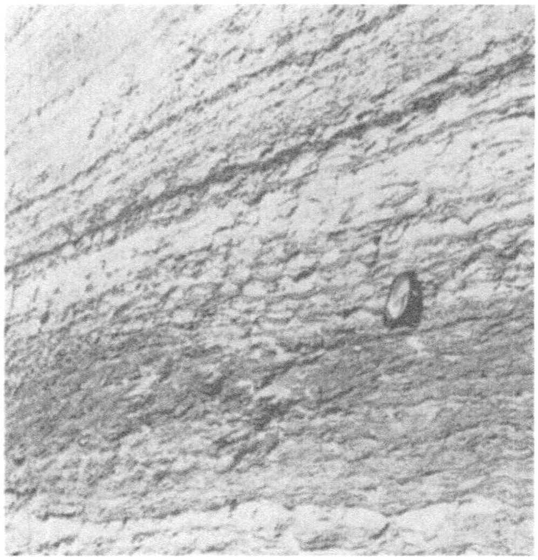

Figure 11-2 (left). Megaripple with well-developed reactivation surface. Reactivation surface and some foresets are clay-draped and penetrated by burrows. Basal sand unit; knife 15 cm.

Figure 11-3 (right). General view of wavy and flaser bedding showing lateral continuity of clay and rippled-sand laminae. Flaser-bedded unit.

are often draped with clay, and vertical burrows may descend from individual lamination surfaces. This again reflects bed stabilization between phases of ripple accretion. These stable surfaces may only reflect phases lasting periods of a few hours or days (cf. Howard and Elders, 1970).

Channels up to 1 meter deep cut through the sands, and are filled with coarse- to medium-grade sands and clay drapes. The channel bases are marked by erosion surfaces overlain by clay-clast conglomerates. These are succeeded by large- or small-scale trough cross-bedded sands sometimes associated with ripple-drift lamination. These channel-floor sediments pass laterally and vertically into clay-draped lateral-accretion sets up to 5 cm thick composed of ripple-laminated sand.

Most of the channels contain undisturbed lateral-accretion sets, but some exhibit simple tubular burrows at right angles to the gently inclined laminae.

Cross-bed laminations within the sequence are well defined by the presence of abundant comminuted plant debris.

FLASER-BEDDED UNIT AND COALS. Above the basal sands, the clay content of the sediment increases (Figs. 11–1, 11–3, and 11–4) and much of the section consists of the flaser, wavy, and lenticular bedding types of Reineck and Wunderlich (1968). The sand layers are mostly current-rippled, although many units of symmetrical ripples also occur.

Bioturbation occurs at some levels, but mostly laminations are undisturbed. A striking feature of the unit is the lateral con-

Figure 11-4 (left). Ripple-laminated sands with well-developed clay drapes interbedded with lenticular- and flaser-bedded sediments. Flaser-bedded unit. Lens cap 3 cm.

Figure 11-5 (right). General view of part of flaser-bedded unit, including a section through a megaripple (*arrowed*) which also displays a well-developed clay-draped reactivation surface. Knife 15 cm.

tinuity of individual clay laminae which may be traced for almost 100 meters. With the exception of agglutinating Foraminifera, body fossils are absent, possibly resulting from reduced salinity conditions at the time of deposition.

Channel fills up to 1 meter thick cut through these flaser-bedded sediments (Fig. 11–1), and they exhibit a similar range of structures to those from the basal sand unit.

Megaripples are also included within the lower part of the flaser-bedded unit (Fig. 11–5), but they never achieve the dimensions of those from the basal sands, seldom exceeding 300 mm in ripple length and 100 mm in height. Here, too, the ripple forms have been rounded and eroded. These reactivation surfaces are mostly draped with clay (Fig. 11–5) and, as with those from the sand unit, burrows are sometimes seen descending from the rounded tops of the ripples.

As well as being dissected by channels, the flaser-bedded sediments are also interrupted by laterally extensive beds of flat-laminated fine- to medium-grade sand. These beds are up to 0.5 meter thick, and their lower contacts are mostly nonerosive, particularly where they rest upon clay-draped ripple surfaces. The tops of the beds often show a transition to oscillation ripples and these ripple forms, in turn, are draped with clay. Seldom are these beds disturbed by bioturbation, but rare escape burrows may be present.

Trough-shaped scours up to 1.1 meters deep also cut through the flaser-bedded sediments, but, unlike the channels, these scours are filled by parallel drapings of plant-rich clays interbedded with ripple-laminated sands. Water-escape structures are common at some levels (Fig. 11–6).

Sellwood (1972a) attributed an angular unconformity of about 2° within the unit to tectonic processes. However, it is more probable that the dipping rippled sands with their clay drapings represent the flaser- and wavy-bedded filling of a large abandoned channel.

Toward the top of the unit (Fig. 11–1) several 300 mm thick coal seams occur, each resting upon a rootlet bed. The rootlets penetrated and disturbed the flaser-bedded sediments to depths of 1 meter and often to the extent that the original lenticular and flaser bedding has been destroyed.

Several coals are present, each separated by 1 to 2 meters of lenticular- and flaser-bedded sediment. The topmost coal is overlain by 2 to 4 meters of similar material before the sequence is cut out by the basal conglomeratic ironstone of the overlying marine sands.

Evidence for Tidal Origin

The entire sequence at Galgeløkke (Fig. 11–1) shows a fining-upward succession capped by coals, and bears a striking similarity to the expected sequence from a modern regressive tidal flat in which sand flats pass upward through mixed flats, mud flats, and supratidal salt marsh deposits (cf. Evans, 1965; pp. 13–20, above; Reineck, 1972; pp. 5–12, above; Straaten,

Figure 11-6 (left). Water-escape structures. Flaser-bedded unit; knife, 15 cm.

Figure 11-7 (right). Slightly bioturbated flaser-bedded sediments overlaying flat-laminated sand. Note escape burrow in the sand and development of clay-draped current ripples at top of sand bed.

1954). The entire sedimentary assemblage thus provides the most convincing evidence of tidal influence during deposition.

Of the individual structures, those most characteristic of modern tidal flats are the extensive wavy-, lenticular-, and flaser-bedded units; herringbone cross stratification; and megaripples with reactivation surfaces (Klein, 1971). Bipolarity of current flow and the development of reactivation surfaces probably reflects successive ebb- and flood-dominated current episodes (Klein, 1970). Examples of some of the typical sedimentary structures are given in Figs. 11–7 to 11–12.

Klein (1971) has reviewed in detail the relative merits of these structures as tidal criteria. At the present time, although individual structures should not be considered diagnostic tidal criteria on their own, in the context of the sequence described the assemblage is believed to conform well with a tidal-flat model.

The coals represent the emergent salt marsh portions of the sequence; the wavy-, lenticular-, and flaser-bedded sediments upon which they rest are believed to represent a mud-flat and mixed-intertidal-flat assemblage. These flats were dissected by channels also probably of tidal origin indicated by the bipolar cross-bedding of their fills and the extensive clay drapings on their lateral accretion sets. Deposition of these latter features would have been favored by slack-water episodes under tidal conditions (cf. Reineck and Wunderlich, 1968).

The units of flat-bedded sand included within the flaser-bedded unit are believed to have been deposited by storm in-

Figure 11-8 (left). Opposed and clay-draped foresets within megaripples developed toward the top of the basal sand unit. Section 5 meters high.

Figure 11-9 (right). Scour with parallel-laminated filling. Basal sand unit. Section 4.5 meters high.

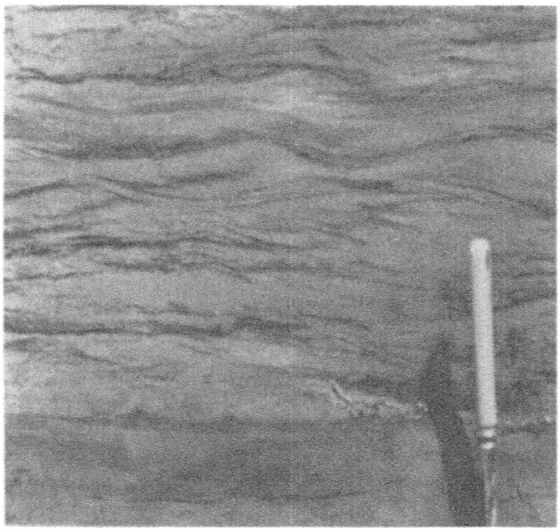

Figure 11-10 (left). Ripple-drift laminations. Ripple forms draped with thin clay drapes. Channel fill from flaser-bedded unit.

Figure 11-11 (right). Ripple-laminated sands with complete and ruptured clay drapes. Flaser-bedded unit.

undation and the draped scours to be sections through ripples. In The Wash, water-escape structures are very common, especially in the vicinity of channels (Evans, 1965; pp. 13–20, above).

The basal sand unit reflects deposition under more turbulent conditions than those of the flaser-bedded unit. It was probably deposited in a lower sand flat or upper shoreface environment which was also cut by tidal channels. Pulsatory current flow typical of tidal areas is strongly indicated by the development of clay drapes on foreset laminae within channels and megaripples. It seems that clay-clast conglomerates can best be formed by the erosion of partially desiccated clays. Although sun cracks

Figure 11-12. Several megaripples with clay-draped reactivation surfaces from near the top of the basal sand unit.

are not observable in this sequence, the presence of these clasts may signify the effects of exposure during deposition.

Discussion

The sequence occurs as a transitional facies between marine and nonmarine strata—in precisely the stratigraphic position where one would expect to observe signs of tidal influence.

Strongly tidal areas usually occur under marine conditions but indications of a normal marine environment are not present in the example cited here. The burrows are nondiagnostic and faunas are lacking. This, along with the abundance of land-derived plant material, probably indicates deposition under low-salinity (?estuarine) conditions. Without diagnostic structures indicating exposure during deposition, an "intertidal" interpretation should be regarded with some caution, and from the sedimentologic point of view, there is still no agreement on what structures result wholly from diurnal tidal processes.

During the Early Jurassic much of northern Europe was covered by an epeiric sea (Sellwood, 1972b), and, according to Shaw (1964) and Irwin (1965), such a sea should be tideless or at least with only minimal tidal influence. Facies and faunal distributions would therefore be expected to be simple. These concepts are not supported for the European Liassic basin if a tidal interpretation for the Bornholm sediments is accepted, and the recognition of tidality in an epeiric sea throws into question the existing theoretical concepts of sedimentation in other epeiric basins.

The geologic evidence of tidal influence during deposition is less equivocal in intertidal situations where bipolarity of currents and slack-water/strong-current episodes are associated with all the other well-established criteria for subaerial exposure. These indications may also be enhanced biologically by the recognition of diurnal and monthly periodicity of shell growth in the faunal component. In open neritic environments, on the other hand, the effect of tides is often masked by predominantly wave-induced processes (McCave, 1971), and the faunas give less well-marked diurnal and monthly responses, which may be circadian or related only to the lunar presence rather than to a direct tidal control.

References

Evans, G. 1965. Intertidal flat sediments and their environments of deposition in the Wash. *Quart. J. Geol. Soc. London 121*, 209–245.

GRY, H. 1960. Geology of Bornholm: Guide to excursion nos. A45 and C40. *In* Sorgenfrei, T., ed., Intern. Geol. Congr. 21st. Copenhagen, Rept. Sess., Norden, 3–16.

HOWARD, J. D., and ELDERS, C. A. 1970. Burrowing patterns in haustoriid amphipods from Sapelo Island, Georgia. *In* Crimes, T. P., and Harper, J. C., eds., *Trace fossils. Geol. J. 3*, 243–262.

IRWIN, M. L. 1965. General theory of epeiric clear water sedimentation. *Bull. Am. Assoc. Petrol. Geol. 49*, 445–459.

KLEIN, G. DEV. 1970. Tidal origin of a Precambrian quartzite—The lower fine grained quartzite (Middle Dalradian) of Islay, Scotland. *J. Sed. Petrol. 40*, 973–985.

——— 1971. A sedimentary model for determining paleotidal range. *Geol. Soc. Am. Bull. 82*, 2585–2592.

MCCAVE, I. N. 1971. Wave effectiveness at the sea bed and its relationship to bed forms and deposition of mud. *J. Sed. Petrol. 41*, 89–96.

REINECK, H.-E. 1972. Tidal flats. *In* Rigby, D. K., and Hamblin, W. K., eds., Recognition of ancient sedimentary environments. *S.E.P.M. Spec. Publ. 16*, 146–159.

REINECK, H.-E. and WUNDERLICH, F. 1968. Classification and origin of flaser and lenticular bedding. *Sedimentology 11*, 99–104.

SELLWOOD, B. W. 1972a. Tidal-flat sedimentation in the Lower Jurassic of Bornholm, Denmark. *Palaeogeog. Palaeoclimatol. Palaeoecol. 11*, 93–106.

——— 1972b. Regional environmental changes across a Lower Jurassic stage-boundary in Britain. *Palaeontology 15*, 125–157.

SHAW, A. B. 1964. *Time in Stratigraphy*. New York, McGraw-Hill, 365 pp.

STRAATEN, L. M. J. U. VAN. 1954. Sedimentology of Recent tidal flat deposits and the Psammites du Condroz (Devonian). *Geol. Mijn. 16*, 25–47.

12

Shorelines of Weak Tidal Activity: Upper Devonian Catskill Formation, Central Pennsylvania

Roger G. Walker and John C. Harms

A common expectation among stratigraphers is to find tidal deposits where marine strata merge with nonmarine strata, either in laterally or vertically exposed sequences. This expectation would seem especially well founded where the marine to nonmarine transition occurs in mud-rich sediments. However, we see no evidence of significant thicknesses of intertidal deposits in a part of the Upper Devonian Catskill Formation, which is dominantly muddy and contains more than 20 marine-nonmarine regressive sequences (Walker and Harms, 1971). Although the regional setting would appear favorable for extensive tidal-flat development (and these sediments have been interpreted as tidal deposits) (Allen and Friend, 1968), no features of these sediments suggest periodic flood and ebb of tides.

We are indebted to the following for their assistance during this study: D. M. Hoskins and his colleagues at the Pennsylvania Geological Survey; R. John Knight, C. H. Carter, G. V. Middleton, and R. Goldring. For funds, Walker is indebted to NATO and the National Research Council of Canada. Permission to publish has been given by the Marathon Oil Company.

We believe that a muddy shoreline prograded rapidly into a sea of low tidal range, and thus any intertidal fringe was restricted in area and migrated too rapidly to form a significantly thick record.

Two aspects of these rocks should be stressed: first, the sedimentary features that define the regressive sequences; and second, the predicted tidally induced features that appear to be consistently absent.

It is not our intention to reiterate work already published. Full, illustrated descriptions have been given by Walker and Harms (1971) and Walker (1971), and a field guidebook with full measured sections has recently been published (Walker, 1972).

Occurrence

The lower 600 meters of the Catskill Formation (the Irish Valley Member) contains a series of repeated marine-nonmarine lithofacies sequences (termed "motifs" by Walker and Harms, 1971), each of which contains low tidal activity shoreline deposits. There are about 20 to 25 repeated motifs in the Susquehanna Valley area, ranging in thickness from 4 to 45 meters, and averaging about 25 meters. They overlie turbidities and slope shales (Trimmers Rock Formation), and pass upward into nonmarine alluvial fining-upward cyclothems (Walker and Harms, 1971; Walker, 1971, 1972). The motifs extend along the depositional strike for at least 50 mi, and each one may have prograded as much as 20 mi into the basin.

Lithofacies Sequences in the Motifs

Individual examples of the Irish Valley motifs vary in thickness, facies type, and lithologic type and proportion. All motifs, however, contain elements that reflect, by their character and organization, related events and sedimentation processes. Consequently, an idealized motif can be described here (Fig. 12–1), and readers are referred to completed measured sections (Walker, 1972) for details of individual motifs. The thicknesses given below refer to Figure 12–1, but motif thicknesses vary from 4 to 45 meters.

There are five parts:

1. 0 to 0.5 meters. A basal bioturbate sandstone with scattered quartz granules, brachiopod and crinoid fragments, and some phosphatic nodules and bone fragments. The lower surface of the sand is commonly sharp and planar, but in places is burrowed and irregular. The maximum thickness is about 50

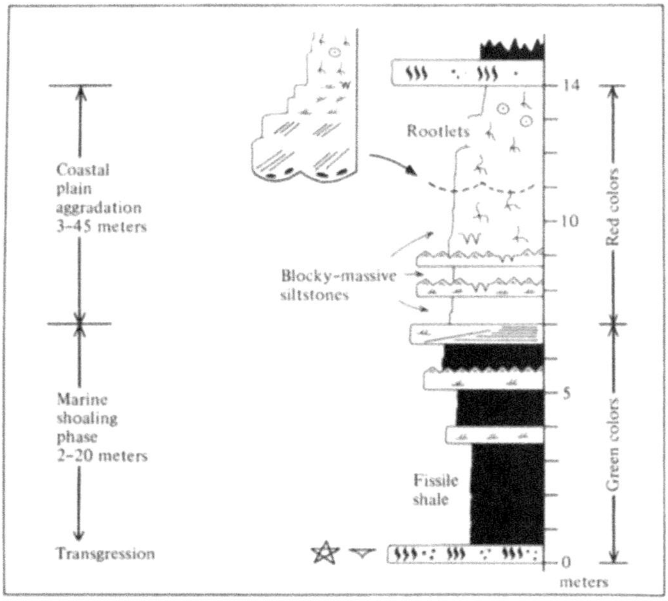

Figure 12-1. An idealized Irish Valley motif, here shown 14 meters thick. In the Susquehanna Valley area, individual motifs range from 4 to 45 meters thick. The five main parts of the motif are discussed in the text. In less than 10 percent of the motifs observed, an alluvial fining-upward sequence was observed, shown diagrammatically in the top center of the figure.

cm. All sedimentary structures have been destroyed by the bioturbation. This sand is followed rather abruptly by

2. *0.5 to 3.5 meters.* Olive green fissile shales, with scattered brachiopod and crinoid fragments. Rarely, this unit may form the base of the motif where the bioturbate sand is absent. There are no siltstone or sandstone laminae in these fissile shales. They grade upward into

3. *3.5 to 7.0 meters.* Green siltstones and mudstones, with very fine grain sandstone beds between 10 and 50 cm thick. The sandstones show low-angle cross-bedding, horizontal lamination, or ripple cross-lamination. A marine fauna is rare, but symmetrical ripples with 10 to 15 cm wavelengths are common on the top of sandstones. Burrows and trails are common. Conspicuously absent are flaser and lenticular bedding, as well as scouring or channeling. This unit grades upward into

4. *7.0 to 12.0 meters.* Drab red siltstones characterized by rootlet traces and desiccation cracks. Thin sandstone beds (2 to 10 cm thick) commonly have symmetrical ripples of 2 to 10 cm wavelength, and vertical and horizontal trails and burrows are common. This unit grades upward into

5. *12.0 to 14.0 meters.* Massive, blocky, red mudstones characterized by rootlet traces, desiccation cracks, and tan calcareous concretions. Calcite is disseminated throughout the mudstones. Sandstones are rare, unless occurring in a fining-

upward alluvial facies sequence. The sandstones in such sequences are 2 to 9 meters thick, have an intraformational "lag" conglomerate at the base, and are characterized by parallel lamination, cross-bedding, and, at the top, ripple cross-lamination.

"Tidal" Features Notable by Their Absence

Despite the dominantly silty and muddy nature of the motifs, the flaser and lenticular bedding so characteristic of the North Sea tidal flats (Straaten, 1954; Reineck, 1960) was never observed. Small channels, with low-angle muddy and silty lateral accretion surfaces, also characteristic of the lower parts of tidal flats, were never observed. In the rare sand bodies thicker than about 1 meter, there were no reactivation surfaces, or features indicating tidal emergence runoff (Klein, 1970). Despite the prograding nature of the shoreline in these motifs, the tidal "fining-upward" facies sequence (Klein, 1971) was never developed. Our interpretation of the motifs is therefore based upon the idea of weak, low flux tidal activity.

Basic Interpretation of the Motifs

The motifs represent prograding shorelines. The marine fissile shales (0.5 to 3.5 meters) represent offshore clay deposition, and the drab red siltstones (7.0 to 12.0 meters), with rootlets and desiccation cracks, represent vegetated mud flats at the distal margin of an alluvial plain. The portion from 3.5 to 7.0 meters, by its position in the sequence, must be nearshore marine/coastal margin. However, winnowed sand bodies are conspicuously absent, and there is no evidence of offshore bars or barriers, nor of beaches, nor of lower tidal-flat sheet or channel sands. The uppermost unit (12.0 to 14.0 meters) is also interpreted as distal floodplain. The tan calcareous concretions are calichelike, and the fining-upward sandstones probably represent meandering channels draining the alluvial plain.

After maximum progradation, cut off of sediment supply, coupled with subsidence, resulted in marine transgression and initiation of a new motif. The bioturbate sand (0 to 0.5 meter) contains quartz granules winnowed from the underlying motif. The very slow deposition during transgression results in total bioturbation, and as fully marine conditions are reestablished, brachiopods, and crinoids reappear.

In essence, therefore, each motif represents a prograding, low-energy, muddy shoreline, where tidal and other longshore or semipermanent currents are incapable of winnowing sediment into clean sand bodies during progradation. The closest modern equivalent is the prograding muddy shoreline of southwestern Louisiana (Beall, 1968), and detailed comparisons have been made by Walker and Harms (1971). The extent of the motifs and lack of associated sand bodies probably eliminate an estuarine or bay environment.

Tidal Range

Three lines of evidence suggest a low tidal range:

1. *Absence of channels* (excluding the fining-upward alluvial sequences). In modern areas of moderate to high tidal range, water is carried off the inundated tidal flats in a network of tidal channels. In the Irish Valley motifs, the absence of such channels in the zone of marine to nonmarine transition suggests that large quantities of water were *not* carried off tidal flats following each high tide, implying a low tidal range.

2. *Close vertical association of marine and nonmarine elements.* In some of the motifs, rootlets and desiccation cracks occur less than 2 meters stratigraphically above marine faunas. The tidal range is unlikely to have exceeded 2 meters; otherwise, brachiopods would have been exposed at low tide and the plants inundated at high tide. Even allowing for compaction and subsidence during progradation, approximately 2 meters can be proposed for the *maximum* possible tidal range. In this critical transitional zone, there are no small-scale directional or textural alternations that would suggest ebb and flood, nor are there any overall textural gradients that would suggest vertical succession of a lower, slightly coarser tidal flat below an upper slightly finer tidal flat.

3. *Absence of winnowed sand bodies.* The nearshore and shoreline area can be readily identified in the motifs, as it must occur between the green marine facies and the red siltstones with rootlets and desiccation cracks. However, clean winnowed sand bodies thicker than 50 cm are absent in over 90 percent of the motifs studied, implying no formation or preservation during shoreline progradation of beaches and beach ridges, cheniers, surf zone sandbars, or barrier island complexes (it is not clear in Allen and Friend's (1968) publication why barrier island-lagoon complexes were suggested for this part of the section). The winnowing process is primarily controlled by wave action and the rate of supply of mud. If mud is supplied faster than the waves can winnow sand, a muddy shoreline will prograde. However, the problem remaining in the context of this volume

is the importance of tidal currents and tidal range, rather than wave action alone. A high tidal range, and hence a higher tidal flux, should result in more powerful currents in the nearshore area. These, in turn, should result in more channeling and winnowing, perhaps more than is recorded in the motifs. The absence of winnowed sand bodies can therefore be related primarily to a high rate of mud input and low activity, and, secondarily, to low tidal flux.

References

ALLEN, J. R. L., and FRIEND, P. F. 1968. Deposition of the Catskill Facies, Appalachian region: With notes on some other Old Red Sandstone basins. *In* Klein, G. deV., ed., Late Paleozoic and Mesozoic continental sedimentation, northeastern North America. *Geol. Soc. Am. Special Paper 106*, 21–74.

BEALL, A. O. 1968. Sedimentary processes operative along the western Louisiana shoreline. *J. Sed. Petrol. 38*, 869–877.

KLEIN, G. DEV. 1970. Depositional and dispersal dynamics of intertidal sand bars. *J. Sed. Petrol. 40*, 1095–1127.

——— 1971. A sedimentary model for determining paleotidal range. *Geol. Soc. Am. Bull. 82*, 2585–2592.

REINECK, H.-E. 1960. Über die Entstehung von Linsen- und Flaserchichten. *Abhandl. Deut. Akad. Wiss. Berlin 1*(3), 369–374.

STRAATEN, L. M. J. U. VAN. 1954. Composition and structure of recent marine sediments in the Netherlands. *Leidse Geol. Mededel. 19*, 1–110.

WALKER, R. G. 1971. Nondeltaic depositional environments in the Catskill clastic wedge (Upper Devonian) of central Pennsylvania. *Geol. Soc. Am. Bull. 82*, 1305–1326.

——— 1972. Upper Devonian marine-nonmarine transition, Southern Pennsylvania. *Pennsylvania Geol. Surv. Bull. G.62*, 25 pp.

WALKER, R. G., and HARMS, J. C. 1971. The "Catskill Delta": A prograding muddy shoreline in central Pennsylvania. *J. Geol. 79*, 381–399.

13

Miocene-Pliocene Beach and Tidal Deposits, Southern New Jersey

Charles H. Carter

Occurrence

The Cohansey Sand, a Miocene-Pliocene quartz arenite, underlies more than two-thirds of the New Jersey coastal plain, an area greater than 5,000 km^2. This formation is underlain conformably throughout the coastal plain by the marine Miocene Kirkwood Formation, and is overlain unconformably, in various areas, by continental Pleistocene formations.

Facies and Facies Sequences

Two facies sequences were identified in the Cohansey Sand. Sequence A (Fig. 13–1) is characterized by conformable, non-channeled facies contacts and by the laminated sand facies.

This paper is part of a Ph.D. dissertation done at The Johns Hopkins University, Baltimore, Maryland. My advisers were Professors F. J. Pettijohn and C. B. Hunt. The University and the United States Geological Survey gave me financial support for the study.

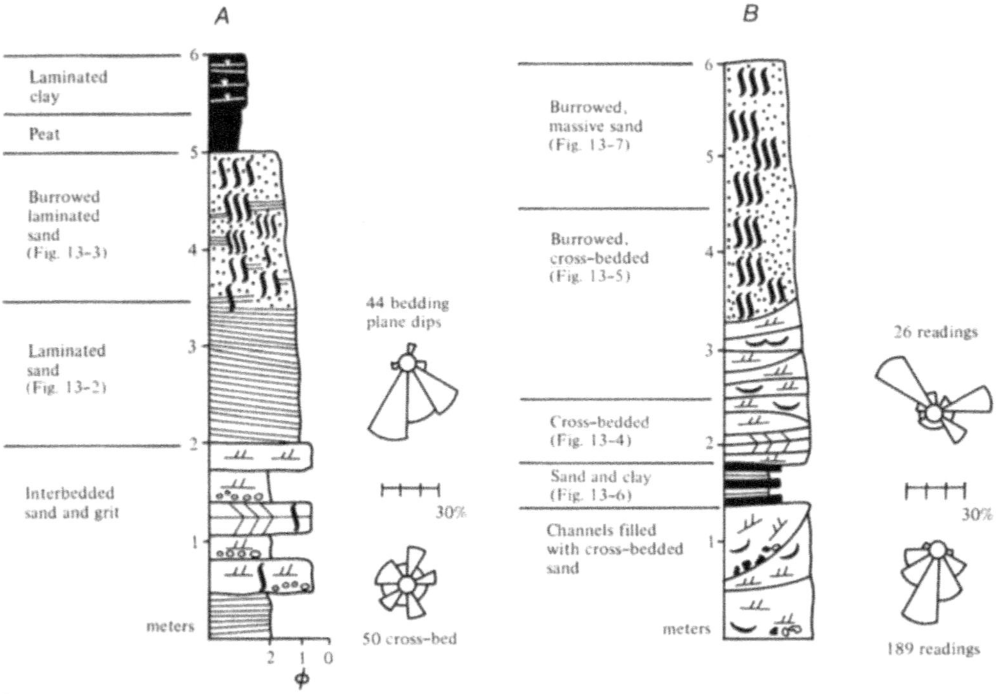

Figure 13-1. Generalized sequences.

Sequence B (Fig. 13–1), on the other hand, is characterized by channels, *Ophiomorpha* burrows, and the burrowed, cross-bedded sand facies.

Sequence A

In sequence A the interbedded sand and grit facies is characterized by multidirectional trough sets. Large burrows similar in size, shape, and orientation to *Ophiomorpha*, but without the clay-rich walls, truncate the bedding here and there.

The laminated sand facies is characterized by gently dipping laminations composed of well-sorted sand. The individual laminations are traceable for as much as 10 meters; commonly, they dip 5° to 10° to the south-southeast. In places, randomly oriented burrows, generally about 0.5 to 1 cm in diameter and less than 15 cm long, identified by heavy mineral accumulations, disrupt the bedding.

The burrowed, laminated sand facies is characterized by burrows, remnant stratification, and abundant heavy minerals. Even where intensive burrowing has destroyed the primary stratification, horizontal heavy-mineral-rich zones up to 30 cm thick indicate nearly horizontal bedding. Carbonaceous fragments are locally present, and a fine, dark gray organic material that adheres to the sediment is present near the top of this facies. Vertical changes within the facies include an upward increase in burrowing and organic debris, whereas there is an upward decrease in heavy minerals and primary stratification.

The compact, brownish-black peat contains many roots (less than 1 cm in diameter) and leaf fragments. Thin lenses and laminations of sand, as well as burrows, are common at the base and the top of the facies.

The laminated clay facies is multicolored (gray, red, and yellow) and thinly laminated. It contains intercalated silt and sand lenses, peat fragments, and irregular, silt-filled burrows (less than 1 cm in diameter).

Sequence B

The cross-bedded sand facies is characterized by tabular and trough sets. Channels up to 20 meters wide, 5 meters deep, and 200 meters long, filled with cross-bedded sand or interbedded sand and clay, are common. Clay clasts and quartz pebbles are common at the base of the channels. Both the tabular and trough sets occur in groups, not as solitary sets. A few cross-bedded layers contain laminated clay lenses (flasers) that are parallel to the sand laminations or the basal surface of the cross bed. There are a few large (about 35 cm wide by 60 cm high) pouch-shaped burrows devoid of primary stratification, as well as smaller, V-shaped burrows with cone-in-cone laminations. The V-shaped burrows are similar to those made by modern anemones.

Figure 13-2. Laminated sand facies.

Figure 13-3. Burrowed laminated sand facies.

Ophiomorpha burrows and remnant trough sets characterize the burrowed, cross-bedded sand facies. The *Ophiomorpha* burrows are vertical or nearly so, and generally consist of a simple linear form (although some branch and/or curve), up to 1 meter in length and 2 to 5 cm in diameter. V-shaped anemone burrows similar to those in the cross-bedded sand facies are also present. The trough sets are commonly found in channels similar to those in the cross-bedded sand facies. Flasers parallel the laminations in some troughs and cross-bedding measurements are bimodal.

Thinly laminated, multicolored clays and burrowed, cross-bedded sands characterize the interbedded sand and clay facies. This facies is confined to elongate, crescent-shaped bodies (con-

Figure 13-4. Channel in cross-bedded sand facies.

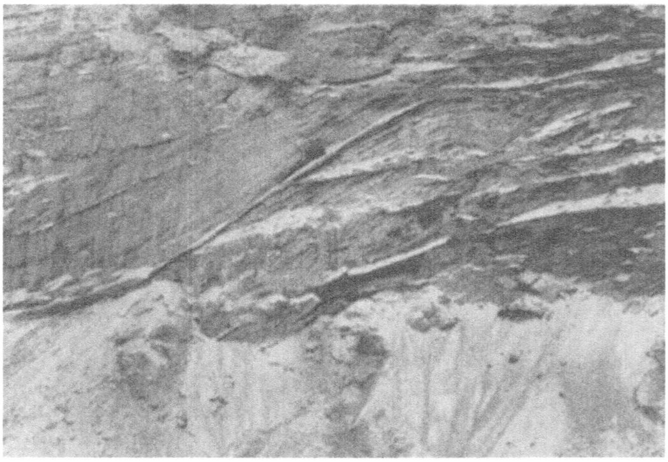

cave side up) up to 5 meters thick and 300 meters long that fill channels cut into the cross-bedded and burrowed, cross-bedded sand facies. The clay beds have a mean thickness of about 10 cm and a range of 2 to 30 cm. The sand beds range in thickness from 2 cm to 1.5 meters. They contain a few solitary tabular and trough sets and groups of small (less than 5 cm thick) trough sets. There are anemone burrows in the sand beds.

The burrowed, massive sand facies is characterized by an intensely burrowed, multicolored (red, brown, and gray), fine-grain sand. *Ophiomorpha* burrows are abundant at the base of a section but decrease in abundance upward. There are a few, small (no more than 10 cm long and 0.5 cm thick), light gray clay lenses near the top of this facies.

Discussion

Sequence A

A BARRIER ISLAND DEPOSIT. The interbedded sand and grit facies is interpreted as a surf zone deposit. "Near the outer portion of the surf zone the bed form is planar (outer planar facies), but, in the inner portion of the zone, an area of large scale bed roughness (inner rough facies) commonly is present. . . . Structures in the inner rough zone produce medium scale foresets that mostly dip directly or obliquely seaward although landward dipping foresets also occur." (Clifton et al., 1971, p. 651). The sand in the inner rough zone (p. 657) ". . . generally is relatively coarse and quite loosely packed."

The laminated sand facies is interpreted as a swash zone (beach foreshore) deposit. This zone is characterized by a

Figure 13-5. Burrowed, cross-bedded sand facies.

seaward-dipping, planar surface. Seaward-dipping laminations (which may exhibit inverse grading) are produced by sheet flow (upper-flow regime) in this zone (Clifton, et al., 1971).

The burrowed, laminated sand facies is interpreted as a backshore-lower dune deposit. The backshore is covered by water only during very high tides and/or storms, and its stratification consists mostly of landward-dipping laminations (McKee, 1957, p. 1707). Moreover, the backshore contains high concentrations of heavy minerals. I have observed horizontally laminated sands in the basal meter of South Carolina beach dunes; in places, roots and burrows have completely destroyed this stratification.

The peat facies is interpreted as a salt water marsh deposit. This interpretation is consistent with an analysis of the peat: . . . pollen and spore of this character are found in lagoonal sequences (back-barrier sequences) (Owens, personal communication, 1971).

The laminated clay facies is interpreted as an outer marsh deposit. For example, salt marshes in The Wash, England, are characterized by well-laminated silty clays and clayey silts, both containing small amounts of sand (Evans, 1965).

A HOLOCENE MODEL FOR SEQUENCE A. A probable Holocene model for sequence A of the Cohansey is present in the North Sea region (Holland), where cores of a Holocene sand contain a sequence similar to sequence A (Straaten, 1965). This sequence, which van Straaten interprets as a regressive barrier island complex, ranges from about 4 to 8 meters in thickness

Figure 13-6. Interbedded sand and clay facies.

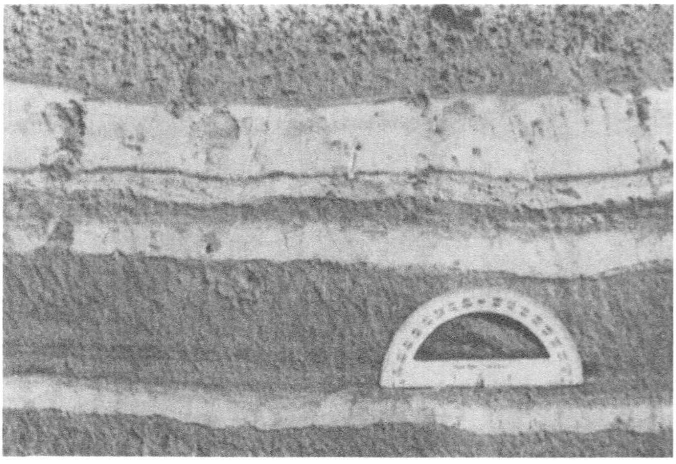

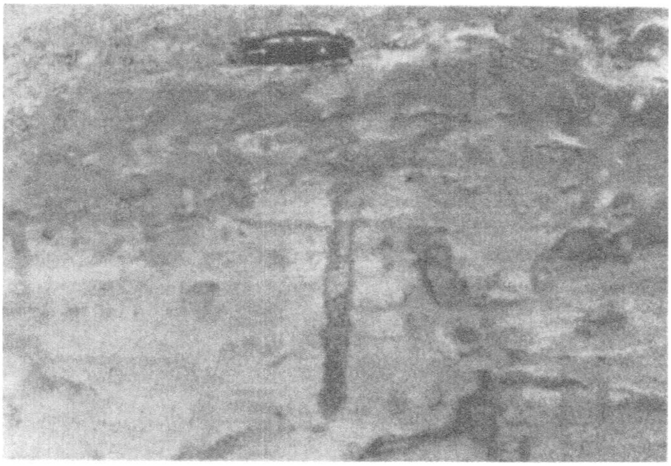

Figure 13-7. Burrowed massive sand facies.

and commonly consists from the base up of: (1) ". . . wildly cross-laminated sands . . ." (p. 63); (2) horizontally laminated sands; (3) soil-sand, disrupted by plant roots and burrowing land animals; and (4) cross-bedded sands or peat. This North Sea sequence resembles the interval between the interbedded sand and grit and the peat in sequence A of the Cohansey. The facies making up sequence A, then, appear to represent discrete environments within the barrier island complex; the sequence was probably produced by a seaward migration of the adjacent environments in this complex.

Sequence B

A TIDAL DEPOSIT. The cross-bedded sand facies is interpreted as a tidal channel deposit. Large-scale cross-beds produced by ripples and point bars are common in this environment.

The burrowed, cross-bedded sand facies is interpreted as a lower tidal deposit. Mud flasers in cross-beds were observed by Straaten (1954) and Reineck (1967) in this environment. They also observed an increase in burrowing and fewer cross-beds than in the tidal channel deposits.

The interbedded sand and clay facies is interpreted as an abandoned tidal channel fill deposit. Abandoned Holocene tidal channels in the mid-Atlantic region, U.S.A., are being filled with mud and ". . . if preserved in the geologic record the sandy walls would stand out in strong contrast to the black, shale-filled channel" (Harrison, 1971, p. 180).

The burrowed massive sand facies is interpreted as an upper tidal deposit (the zone between mean sea level and mean high water). Intense burrowing and a scarcity of cross-bedding are important features of this environment (Straaten, 1954; Reineck, 1967).

A HOLOCENE MODEL FOR SEQUENCE B. A probable Holocene model for sequence B of the Cohansey is present in the North Sea region (Holland) where cores of a Holocene deposit in the Wadden Sea contain a sequence similar to sequence B (Evans, 1970). This sequence, which Evans interprets as a prograding tidal-channel–tidal-flat complex, is about 8 meters thick and consists from the base up of (1) cross-bedded sand with some mud laminations, shell gravel, and peat and clay pebbles; (2) laminated sands, sandy muds and muds, intense burrowing near top; (3) laminated silty clays with plant roots; and (4) peat. The major part of this North Sea sequence resembles sequence B of the Cohansey.

The facies making up sequence B, then, appear to represent discrete environments within a tidal-channel–tidal-flat complex; the sequence was probably produced largely by seaward migration of adjacent environments in this complex.

References

CLIFTON, H. E., HUNTER, R. E., and PHILLIPS, R. L. 1971. Depositional structures and processes in the non-barred high-energy nearshore. *J. Sed. Petrol. 41*, 651–670.

EVANS, G. 1965. Intertidal flat sediments and their environments of deposition in the Wash. *Geol. Soc. London Quart. J. 121*, 209–245.

——— 1970. Coastal and nearshore sedimentation; a comparison of clastic and carbonate deposition. *Geol. Assoc. (London) Proc. 81*, 3, 493–508.

HARRISON, S. C. 1971. The sediments and sedimentary processes of the Holocene tidal flat complex, Delmarva Peninsula, Virginia. Ph.D. dissertation, Johns Hopkins Univ., Baltimore, 202 pp.

MCKEE, E. D. 1957. Primary structures in some Recent sediments. *Am. Assoc. Petrol. Geol. Bull. 41*, 1704–1747.

REINECK, H.E. 1967. Layered sediments of tidal flats, beaches, and shelf bottoms of the North Sea. *In* Lauff, G., ed., *Estuaries*, Am. Assoc. Advan. Sci. Publ. 83, 191–206.

STRAATEN, L. M. J. U. VAN. 1954. Composition and structure of Recent marine sediments in the Netherlands. *Leidse Geol. Mededel. 19*, 1–110.

——— 1965. Coastal barrier deposits in south and north Holland in particular in the areas around Scheveningen and Ijmuiden. *Mededel. Geol. Stichting 17*, 41–75.

14

Tidal Sand Flat Deposits in Lower Cretaceous Dakota Group Near Denver, Colorado

David B. MacKenzie

Occurrence

West of Denver, on the east side of the Dakota hogback, large bedding plane surfaces dipping 35° to 40° east expose the upper 30 meters of the 100-meter thick Lower Cretaceous Dakota Group (Fig. 14–1). The Dakota Group of this area is primarily a shoal-water deltaic assemblage within which many different subenvironments are recognized (MacKenzie, 1971, 1972; Weimer and Land, 1972).

Facies

The section (Fig. 14–2) is made up of well-sorted fine- to very fine-grained quartzose sandstones (Table 14–1) with less than 10 percent interbedded shales. The lower half consists

Discussion with those familiar with Holocene tidal flats—H. E. Reineck and J. D. Howard, in particular—has been very helpful. Observations made by my colleague J. C. Harms, as well as his helpful review of this manuscript, are gratefully acknowledged.

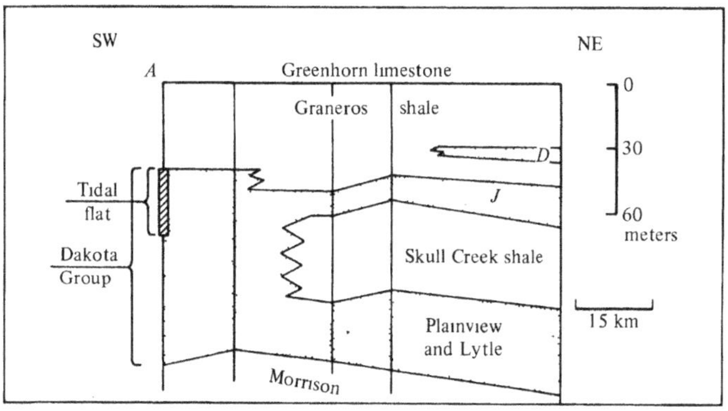

Figure 14-1. Southwest-northeast stratigraphic cross section showing position of tidal-flat deposits. *A* is The Alameda Avenue section located about 10 miles west-southwest of the state capitol building in Denver.

mostly of a fining-upward sequence of fine-grained, cross-stratified sandstone; the upper half consists of very fine-grained tabular-bedded sandstones, locally incised by sand- or partially mud-filled channels.

Ripple-modified tabular sets of cross-strata. A 9-meter thick unit in the lower-middle part of the section (*F* in Fig. 14–2) is a fine-to medium-grained sandstone with tabular sets of cross strata 0.1 to 0.6 meters thick. Carbonized wood fragments are abundant throughout. The cross-laminae, which commonly consist of alternating fine- to medium-grained sandstones, are concave up and have maximum dips (restored) of 25°. They dip uniformly north. In plan view, the traces of the

Table 14–1 *Grain size analyses*

Bed	Comments	Median ϕ	Median μm	5th percentile ϕ	Sorting (Folk and Ward), ϕ
U		3.25	105	2.56	2.01
B		3.93	66	3.08	3.20
N	Rippled pocket	3.51	88	2.85	1.45
	Unrippled surface	3.36	97	2.55	2.64
X		4.28	52	3.18	2.84
R		3.68	78	2.68	2.06
D		2.78	146	2.42	1.34
Z		3.45	92	2.63	2.06
P	ss above scour	2.63	162	2.24	0.51
	ss below scour	3.08	119	2.52	1.19
F		2.15	226	1.85	0.56
G		3.10	117	2.58	0.97

cross-laminae on boundaries between sets are remarkably straight—in some places for at least 6 meters

A distinctive feature is the presence of long-crested, asymmetric ripples on the boundaries between sets of cross-strata (Fig. 14–3*D*). The ripple crests trend at a large angle to the strike of the cross-strata; the crests trend 009° and are generally asymmetric toward the west. The ripples have wavelengths of 4 to 8 cm and amplitudes of 2 to 5 mm. Shorter wavelengths are commoner toward the top of the unit. Invertebrate tracks and trails are present locally on rippled surfaces in the lower part. Either low amplitude symmetric ripples or straie are also present locally on the surfaces of cross-strata.

Clearly, the dominant mode of sedimentary transport was unidirectional toward the north; equally clearly, the body of water was shallow enough and broad enough for closely spaced ripples to form intermittently on the bed as a result of waves moving transverse to average current direction. Repeated alternations of currents and wavy slack water are consistent with a

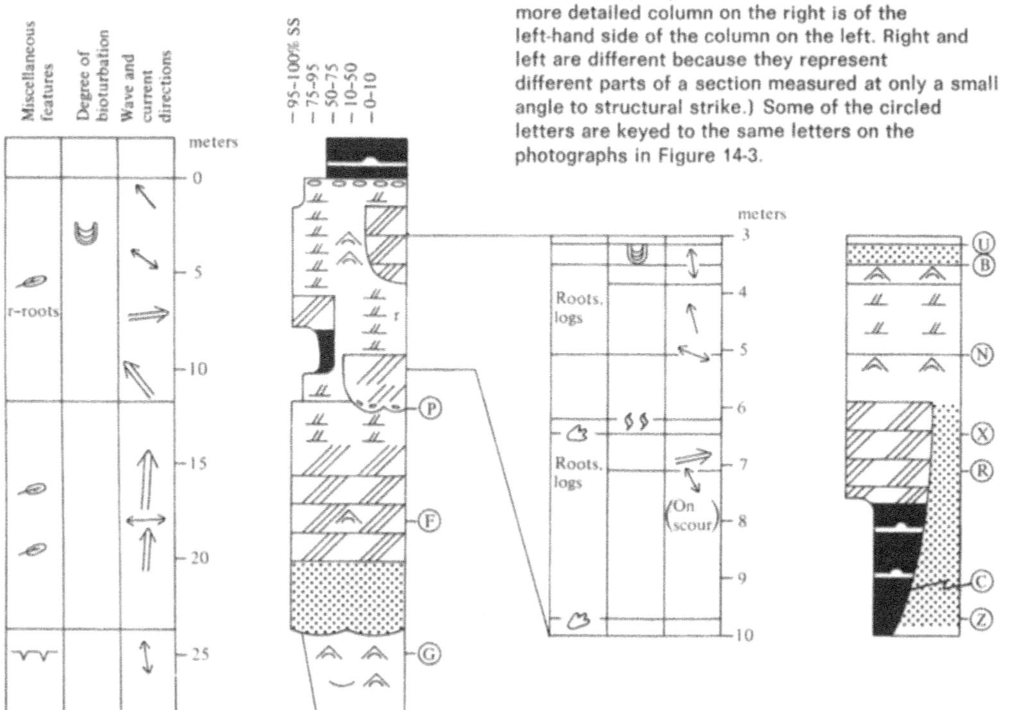

Figure 14-2. Columnar section of upper 30 meters of Dakota Group exposed at Alameda Avenue. (The more detailed column on the right is of the left-hand side of the column on the left. Right and left are different because they represent different parts of a section measured at only a small angle to structural strike.) Some of the circled letters are keyed to the same letters on the photographs in Figure 14-3.

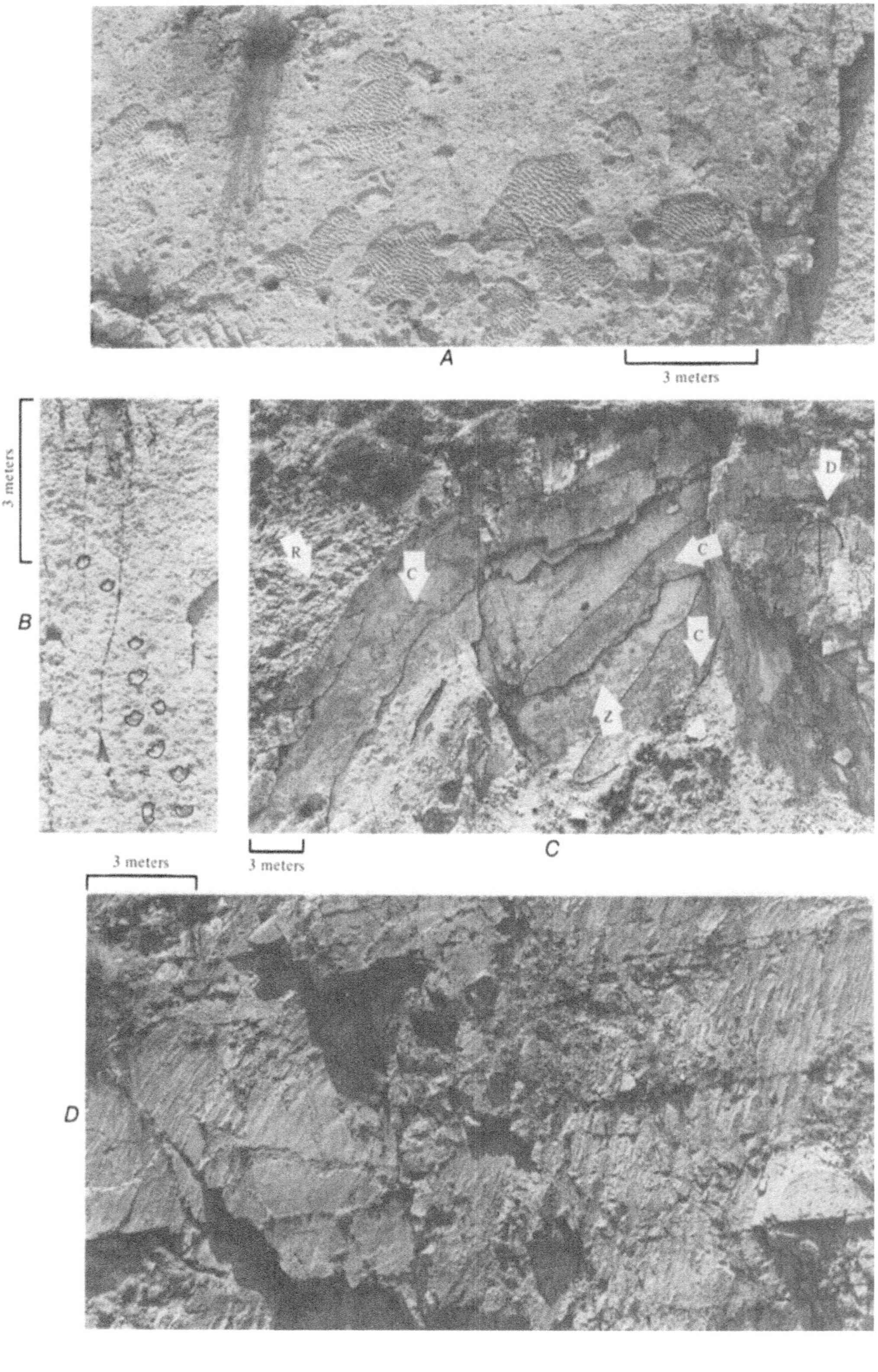
A
3 meters
B
3 meters
R
C
C
C
D
Z
C
3 meters
3 meters
D

Figure 14-3 (opposite). *A.* A surface with long-crested ripples at position *N* in Fig. 14.2. Width of photograph is 25 meters. *B.* A surface with dinosaur footprints at position *X* in Fig. 14.2. Width of photograph is 4 meters. *C.* Scour beneath mud-filled channel in the lower half of the right-hand column of Fig. 14-2. Width of photograph is 35 meters. *D.* A surface showing ripple-modified sets of cross strata at position *F* in Fig. 14-2. Width of photograph is 22 meters.

tidal environment. The preservation of wave ripples beneath advancing sand waves may require some sort of stabilization of the sand—perhaps binding by algal mat.

CHANNEL, PARTIALLY MUD-FILLED. From about 6 to 9 meters stratigraphically beneath the top of the Dakota Group (*C* in Fig. 14–2) is an erosional scour at least 60 meters broad, overlain first by shale and siltstone and then by sandstone. Preserved for observation are three fragments (each marked *C* in Fig. 14–3*C*) of a once continuous curved surface that truncates at least 2.5 meters of underlying beds. The upper part of the surface has ripples with long crests subparallel to the dip. In Figure 14–3*C* the scour can be seen to truncate the surface labeled *R*, characterized by rootlets and by imprints of driftwood in a very fine-grained sandstone. The surface of truncation is overlain by 2 meters of medium gray silty shale with interlaminated siltstone and then by 2 meters of fine-grained, trough cross-stratified sandstone. The axes of the troughs dip eastward (080°) (Fig. 14–3*C*).

RIPPLES. Some bedding plane surfaces have remnants of current ripples; others have conspicuous, long-crested, symmetrical, wave-generated ripples (Fig. 14–3*A*). The long-crested, symmetrical ripples on the very fine-grained sandstone at *B* have a wavelength of 5 cm, an amplitude of 5 mm, and an orientation of 082°. Those on the surface of the very fine-grained sandstone at *N* occur in irregular pockets on an otherwise unrippled surface. They much resemble features observed in intertidal zones at Sapelo Island, Georgia, and in the North Sea (H. E. Reineck, personal communication). They have an average wavelength of 12 cm, an amplitude of 1.5 cm, and an orientation of about 020°.

At the edges of the irregular patches at *N*, the ripples rise to the level of the unrippled surface. So the patches are not the result of recent selective removal of an overlying layer. How they do originate is not clear. One interpretation is that the unrippled part was flattened by swash action and then stabilized by algal mat. It is only where this algal mat was somehow ripped up that ripples were able to form. Although algal mat is common in the upper parts of some intertidal sand flats, once formed it is probably too tough to be breached later by relatively gentle wave motion in shallow water. Had the rippled depressions on surface *N* been formed by ripping up of algal mat, one might expect associated features indicating a more vigorous environment.

An alternative interpretation is that ripples are preserved only in gentle depressions, those once present on the slightly higher and much larger surface having been destroyed as the water shoaled and swash or shallow sheet flow flattened the bed. In either interpretation, short-period water depth changes are indicated.

MUD CRACKS. The rippled, very fine- grained sandstone, indicated by *G*, is mud cracked locally. The mud cracks are up to 3 mm wide and several centimeters long. The thin yellow film that cracked was deposited as a mud in the hollows between ripple crests. The sequence of mud deposition (slack water), mud cracks (emergence), and overlying rippled sand (resubmergence) is suggestive of short-period fluctuations in water level.

NESTED U-SHAPED BURROWS. Although tracks and trails are common on many of the bedding planes in the upper third of the section, bioturbation is conspicuous in only one bed (*U* in Fig. 14–2). It is a 34-cm thick, very fine-grained sandstone. Seen in vertical section, each burrow has the shape of many *U*s nested together (Fig. 14–4). A typical one is 4 cm wide and 18 cm deep. Their presence is often revealed only by closely spaced vertical planes of weakness along which the sandstone of the *U* bed breaks.

Burrows similar to these have been described earlier from the Dakota Group (Howell, 1957) and from the Devonian Baggy Beds in Devonshire, England: *Diplocraterion yoyo* (Goldring, 1962). The organisms that made this burrow cannot be identified. But a crustacean *Corophium* does make this kind of burrow in the upper parts of tidal flats of Jade Bay, Germany (Reineck, 1967), and some species of clams may make a similar burrow also.

PLANT ROOTLETS. Tiny outlines of plant rootlets are visible in at least two beds (Fig. 14–2). Sometimes the carbonaceous

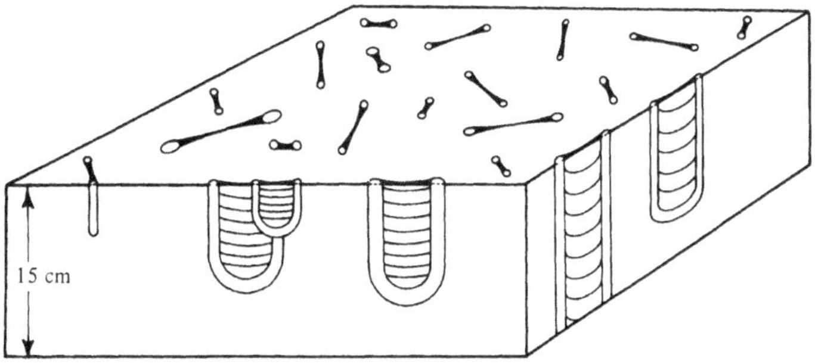

Figure 14-4. Block diagram showing nested U-shaped burrows in bed *U* of Fig. 14-2.

filaments are preserved (especially on freshly broken surfaces); sometimes they have been oxidized so that only the molds around them remain. Their presence in a bed means that before the bed was covered by other sediments, plants became rooted in it. The waters covering the surface of the bed must have been very shallow, or the bed may have been emergent. The size of the rootlets (1 to 2 mm in diameter and up to 8 cm long) suggests rushes or small bushes.

DINOSAUR FOOTPRINTS. Dinosaur footprints are found on beds *X* and *Z*. Bed *X* is a coarse siltstone with at least three different sets of footprints. Selected individual prints are outlined in black in Figure 14–3*B*. The footprints are three-toed and either 55 cm wide and long or 40 cm wide and long. Most of the prints are indistinct, but deep. This suggests that although the tracks were formed under water, most of the animals' weight was borne by the bottom; hence the water must have been quite shallow.

SEDIMENTARY TRANSPORT DIRECTIONS. As shown to the left of the columnar section (Fig. 14–2), the orientations of the primary sedimentary structures are variable, particularly the long-crested ripples. However, the cross-strata of the sandstones generally show a north component of sedimentary transport. The more marine facies of the Dakota Group also lie to the north and east.

Examples of local upward alternation in direction of sedimentary transport—herringbone cross-stratification (Klein, 1971)—are rare. A possible example occurs at the base of the sandstone channel fill at *P*.

Comparison with Holocene Deposits

Observation of modern tidal flats and their subtidal equivalents suggests that progradation would result in a 3- to 10-meter thick, fining-upward sequence (MacKenzie, 1972). The fills of migrating ebb gullies would be selectively preserved in the lower parts; the deposits of the higher parts of the tidal flat of one cycle would tend to be selectively removed by erosion associated with succeeding cycles.

Comparison of the Alameda Avenue deposits with this Holocene model points up one obvious disparity—thickness. Either the 30-meter thick section contains several cycles, or subsidence was sufficiently pronounced for the thickness of one cycle to be greatly increased. The presence of mud cracks in the lower part of the section suggests that there are at least two and probably more cycles present.

The cross-stratified sandstones of the lower part of the Alameda Avenue sequence could well represent the fills of migrating ebb gullies. Many of the features associated with the higher parts of Holocene tidal flats are, in fact, found only in the upper 10 meters of the Alameda Avenue section: root beds, intense bioturbation, partially mud-filled channel, vertebrate tracks. Not found at Alameda Avenue but reported from Holocene tidal flats are "longitudinal oblique beds" as parts of channel fills, features attributable to "emergence runoff" (Klein, 1971), flaser bedding, or algal mats.

Any alternative to the tidal sand flat interpretation presented here should account for the following key features:

1. Overall fining-upward sequence
2. Presence of broad surfaces with abundant tracks and trails, interbedded with root beds indicating emergence
3. Widespread occurrence of wave-generated ripples
4. Presence of fairly numerous channels filled with sand deposited by currents flowing generally in a seaward direction.

To explain this association one might invoke other parts of a shoal water delta where tidal influences were minimal or nonexistent. In that case, the local scours would be interpreted as deltaic distributary channels; the broad, thin beds with abundant tracks and trails as interdistributary bay deposits; and the root beds as deposits of interdistributary swamps. One argument against such an interpretation is the scarcity of muds; present shoal water deltas of the world contain a much higher proportion of mud than does the upper 30 meters of the Dakota Group at Alameda Avenue.

Many features of these rocks indicate deposition in a marginal marine environment in which water level was undergoing short-period fluctuation. Neither period nor range is known. Of present-day processes, tidal movement, whether wind-driven or astronomic in origin, is the preferred explanation.

References

GOLDRING, R. 1962. The trace fossils of the Baggy Beds (Upper Devonian) of North Devon, England. *Palaontol. Z. 36*, 232–251.

HOWELL, B. F. 1957. New Cretaceous scoleciform annelid from Colorado. *J. Paleontol. Soc. India 2*, 149–152, pl. 16.

KLEIN, G. DEV. 1971. A sedimentary model for determining paleotidal range. *Bull. Geol. Soc. Am. 82*, 2585–2592.

MACKENZIE, D. B. 1971. Post-Lytle Dakota Group on west flank of Denver basin, Colorado. *Mountain Geol. 8*, 91–131.

——— 1972. Tidal sand flat deposits in Lower Cretaceous Dakota Group near Denver, Colorado. *Mountain Geol. 9*, 269–277.

REINECK, H.-E. 1967. Layered sediments of tidal flats, beaches, and shelf bottoms of the North Sea. *In* Lauff, G. H., ed., *Estuaries.* Am. Assoc. Advan. Sci., pp. 191–206.

WEIMER, R. J., and LAND, C. B., JR. 1972. Field guide to Dakota Group (Cretaceous) stratigraphy, Golden-Morrison area, Colorado. *Mountain Geol. 9*, 241–267.

15

Tidal Origin of Parts of the Karheen Formation (Lower Devonian), Southeastern Alaska

A. Thomas Ovenshine

Occurrence

Two types of deposits of intertidal or near-tidal origin are presently recognized in the Lower Devonian Karheen Formation of the northwest coastal area of Prince of Wales Island in the Alexander Archipelago of Alaska. One is lithic sandstone-red mudstone cycles from 1 to 3 meters thick; the second is about 150 meters of alternating red and green stromatolitic, calcareous mudstone with ubiquitous mud cracks.

The Karheen Formation is a clastic wedge that over a distance of 65 mi thins and fines northward from boulder conglomerate in the Cruz Islands to sandstone and argillite in Tonowek Narrows and calcareous mudstone on Kosciusko Island. In the south, where conglomerate predominates, the Karheen unconformably overlies Ordovician and Lower Silurian rocks of the Descon Formation; more than 10,000 ft of Middle and Upper Silurian limestone are absent across the unconformity. Northward, in the areas where sandstone and mudstone predominate, the Karheen conformably overlies the Middle and Upper Silurian Heceta Limestone. The Karheen Formation is overlain by the Middle or Upper Devonian Wadleigh Limestone, a shallow water biothermal limestone.

Marine fossils occur sparingly in the Karheen Formation, and are typically restricted to rare limestone beds intercalated in the sandstone and conglomerate. Most of the formation is probably marine in origin, although the sedimentary structures of some of the coarsest conglomerates suggest subaerial deposition.

Facies

Staney Island

Sandstone-mudstone cycles of possible tidal origin are well developed on islets near Staney Island (55°48.6′ N, 133°10.7′ W) in Karheen Passage. As illustrated in Figure 15–1, the cycles consist of four subunits:

1. Trough cross-stratified sandstone. The basal unit of the cycle is 0.5 to 2 meters thick and consists of poorly to moderately sorted medium-grain, lithic-feldspathic subgraywacke. The unit disconformably overlies red mudstone of the highest part of the previous cycle with up to 15 cm of relief on the erosional surface. Slumped cross-stratification and mudstone intraclasts may occur in the sandstone just above the erosion surface. Trough cross-stratification greatly predominates over rare planar cross stratification. The largest values of both average and maximum grain size occur in this subunit of the cycle.

2. Lower red mudstone. Hematitic, homogeneous, or wispy-bedded sandy mudstone succeeds the trough cross-stratified unit. The lower contact of the unit is gradational because of intercalation and/or bioturbation, but the upper is sharp. Flaser bedding, ripple marks, thin parallel-laminated sandstone layers, and parting lineation were the primary depositional structures of this unit. Typically, however, they have been obliterated by pervasive bioturbation (Fig. 15–2*A*).

3. Parallel-laminated fine sandstone. Several 5- to 50-cm beds of fine-grained, moderately well-sorted, parallel-laminated, lithic-feldspathic subgraywacke usually occur between the lower and upper red mudstone units, although in some cycles the unit is absent. Subordinate red mudstone and siltstone may also occur within the parallel-laminated unit. Ripple marks, small-scale cross-stratification, and parting lineation are common sedimentary structures. Bioturbation is not prevalent.

4. Upper red mudstone. This pervasively bioturbated unit

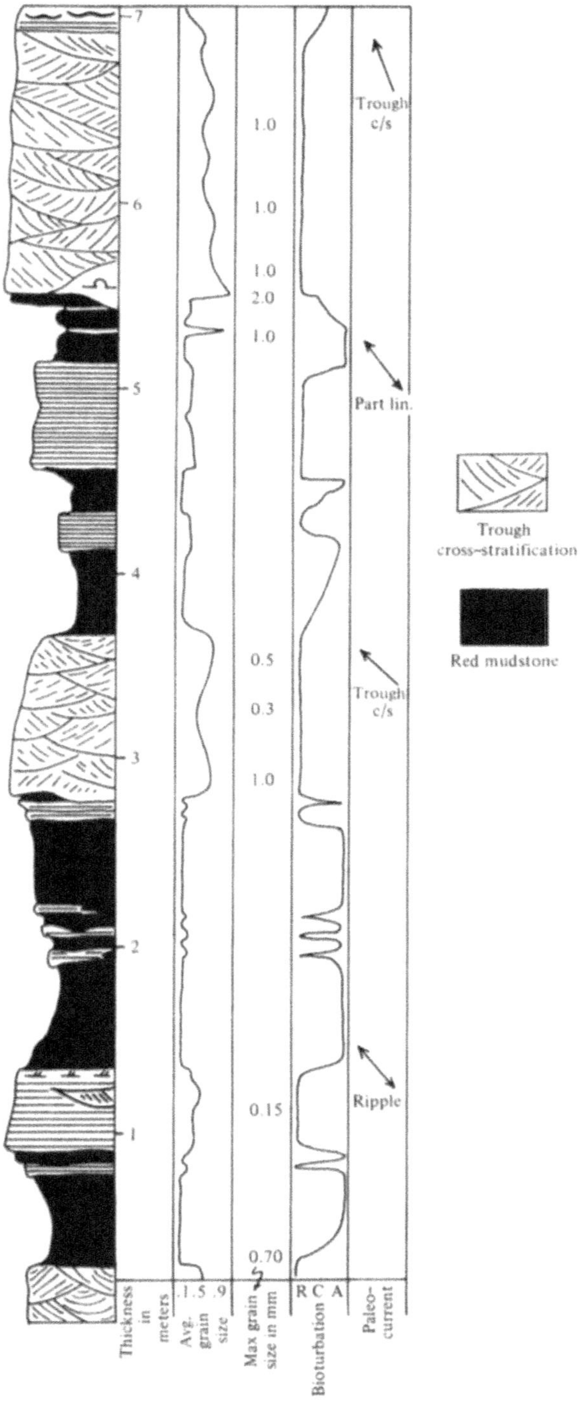

Figure 15-1. Sandstone-red mudstone cycles from the lower part of the Karheen Formation in the Staney Island area (55° 48.6′N, 133° 10.7′W), southeastern Alaska.

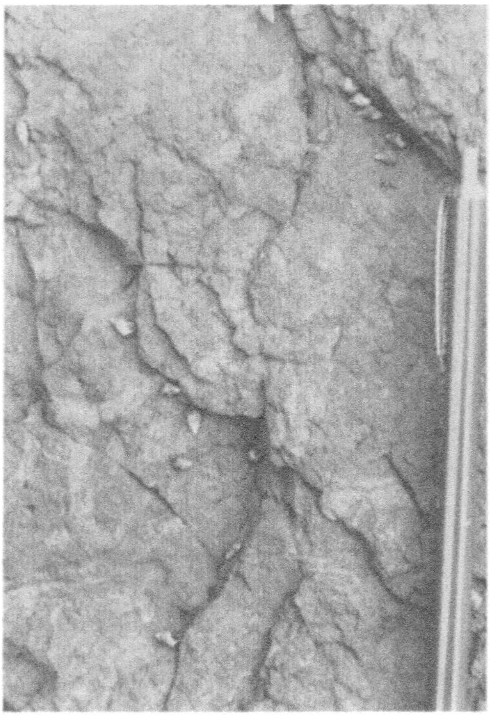
A

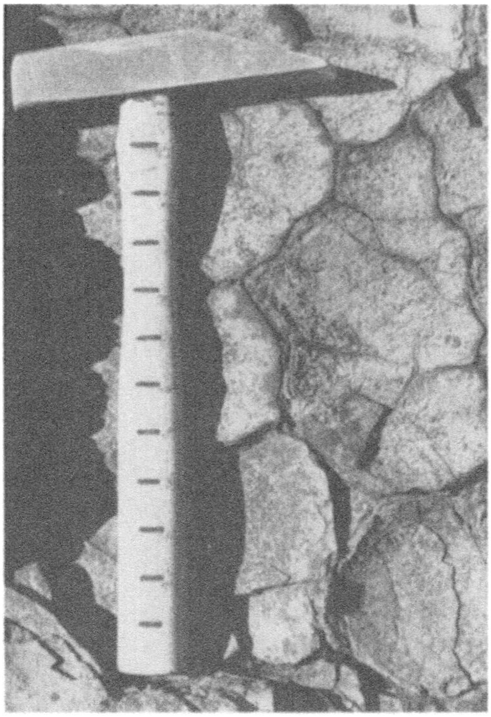
B

Figure 15-2. Sedimentary structures of the Karheen Formation *A*. Bedding plane view of feeding trails in poorly sorted sandstone of the upper Karheen Formation; Staney Island (55° 48.6′N, 133° 10.7′W), southeastern Alaska. The gastropods are contemporary. *B*. Bedding plane view of mud cracks in laminated calcareous red mudstone, Kosciusko Island (55° 55.4′N, 133° 39.7′W), southeastern Alaska. *C*. Congruent stacking of mud cracks in red calcareous mudstone, Kosciusko Island (55° 55.4′N, 133° 39.7′W), southeastern Alaska.

C

is identical to the lower red mudstone except near the top where it may contain 1- to 5-cm beds and lenses of medium to coarse subgraywacke. These are interpreted as precursors of the basal unit of the next cycle.

Kosciusko Island

Intertidal deposits of calcareous mudstone occur on the shoreline of Kosciusko Island (55°55.4′ N, 133°39.7′ W), west of the entrance to Edna Bay. About 150 meters of beds are exposed, consisting of rhythmically alternating zones of red hematitic calcareous mudstone (1 to 3 meters) and pale green calcareous mudstone (0.25 to 1 meter). Small ostracods, gastropods, and, rarely, brachiopods occur sparingly in the green zones but have not been found in the red zones. In hand specimens or thin section the mudstone is millimeter laminated, consisting of alternating layers of gray magnesium limestone and red or green mudstone. Groups of laminations of similar thickness and lithology form beds 0.5 to 10 cm thick, averaging 4 to 5 cm.

Primary sedimentary features of the calcareous mudstones are mud cracks, ripple marks, parting lineation, mudchip breccias, bioturbation, and rare "xenoclasts." Mud cracks, which are most abundant in the red zones (Fig. 15–2*B*), range in diameter from 5 to 30 cm and average about 5 cm. Although isolated laminations and single beds of mud cracks occur, most outcrops exhibit congruent stacking of the concave-upward polygons (Fig. 15–2*C*) resulting from persistence of the same mud-crack pattern through numerous episodes of deposition.

Penecontemporaneously eroded and transported mudchip breccia in layers a few centimeters thick is common, and may comprise one-fourth the total section. Bioturbation, in the form of isolated feeding trails, biogenic mudchip breccias, and homogenized beds, is widely distributed but not a predominant feature. The isolated xenoclasts are small to large pebbles of graywacke, andesite, greenstone, and hornfels that occur sporadically and sparsely in the calcareous mudstone in densities on the order of one class per 100 cu meters of mudstone. Most clasts are spherical, well rounded, and highly polished, and are of unknown origin and significance.

Discussion and Interpretation

The millimeter-laminated red and green mud-cracked beds of Kosciusko Island must be of intertidal origin because of the co-occurrence of invertebrates (= marine deposition) with mud cracks and redbeds (= subaerial interludes). The magnesium content of the limestone, the mudchip breccia, and stromatolitelike lamination are compatible with this interpretation. Although the origin and significance of the rhythmic red and green zones have not been studied, perhaps the green zones reflect submergence during all but extreme low water.

The interpretation of the sandstone-red mudstone cycles of Staney Island is less certain. As shown in the following tabulation, there is some correspondence between the Staney Island sandstone-red mudstone cycles and the array of sedimentary environments described by Evans (1965) of the intertidal flats of The Wash in England.

The Wash (Evans, 1965)	*Staney Island* (this paper)
Salt marsh	Disconformity?
Higher mud flats	Upper red mudstone
Arenicola sand flats	Parallel-laminated fine sandstone
Lower mud flats	Lower red mudstone
Lower sand flats	Trough cross-stratified sandstone

At Staney Island, however, the distribution of bioturbation in the sedimentary environments (Fig. 15–1) does not correspond closely to the distribution observed by Evans (1965, p. 229). Whereas at Staney Island the lower and upper mudstones are the loci of intense organic reworking, at The Wash intensive bioturbation occurs in the *Arenicola* sand flats that occupy the middle of the cycle.

Further uncertainty in the interpretation of the cycles at Staney Island results from their general similarity to the fining-upward cycles (Allen, 1965) that are usually ascribed to the combined effects of progradation and channel migration in settings ranging from fluviatile to nearshore-marine. The Staney Island cycles, however, differ from those illustrated by Allen (1965) in containing a medial, parallel-laminated sandstone.

The sandstone-red mudstone cycles were deposited in a marine environment because correlative strata to the south (toward the sediment source), west, and north contain marine fossils. But it is unknown whether the cycles have resulted from simple seaward progradation of intertidal zones or from a complex interaction between progradation and lateral migration of intertidal or subtidal channelways.

References

ALLEN, J. R. L. 1965. Fining-upwards cycles in alluvial successions. *Geol. J. 4*, 229–246.

EVANS, G. 1965. Intertidal flat sediments and their environments of deposition in the Wash. *Quart. J. Geol. Soc. London 121*, 209–245.

16

Clastic Coastal Environments in Ordovician Molasse, Central Appalachians

Allan M. Thompson

Occurrence

Marginal-marine rocks of inferred tide-related origin occur in the lower Bald Eagle Formation (Upper Ordovician) of the Central Appalachian miogeosynclinal sequence. The marginal-marine sequence averages 65 meters thick, and separates underlying delta-slope shales and turbidities (Reedsville Formation) from overlying fluvial sandstones and conglomerates of the upper Bald Eagle (Thompson, 1968). The entire lithofacies sequence comprises the lower parts of the Taconic clastic wedge, which prograded northwestward during Late Ordovician time as indicated by both marine and continental facies changes and paleocurrents. The resultant northwest-sloping paleoslope provided the framework for the origin and geometry of the marginal-marine rocks, which developed along the margins of an advancing delta front.

Facies

The marginal-marine rocks contain three major facies, one dominantly fine grained and the others coarse grained (see Fig. 16–1 for a composite vertical section). The sequence may be best understood as a variable intermixing of the three facies in intervals from 20 cm to 6 meters thick.

FACIES A. Facies A contains thinly interbedded lithic wacke to arenite, siltstone, and mudstone. Beds are typically very thin, and range from 0.01 to 50 cm or more, averaging 3 to 5 cm (Fig. 16–1, column A). Individual beds are flatly lenticular, and often approach parallelism (Fig. 16–2); lenses are frequently small, and both mud and sand flasers are common.

Ripple-generated primary structures characterize the thinly interbedded sandstones, siltstones, and mudstones. Such structures include pervasive ripple-bedding (Fig. 16–3), preserved ripples and ripple trains (Fig. 16–4), mud drapes on sand ripples (Fig. 16–4), flaser bedding, ladder ripples and others. The mudstones and siltstones often contain mud cracks and tracks of epifaunal browsing taxa.

The thicker sandstones are usually cross-bedded. Single sets of planar, steep-foreset cross-strata are common (Fig. 16–5), and are often intercalated with ripple-bedded siltstones; foresets are usually inclined to the northwest. Cosets of small-scale

Figure 16-1. Schematic columnar section of marginal-marine sequence (left) and representations of the three facies.

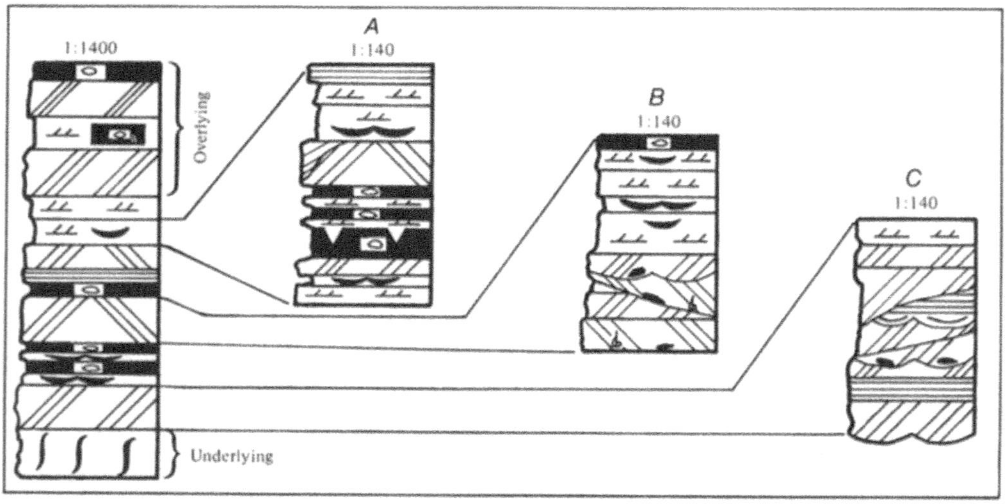

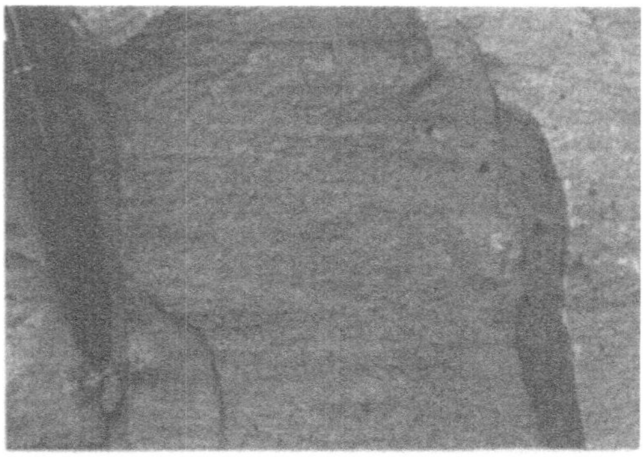

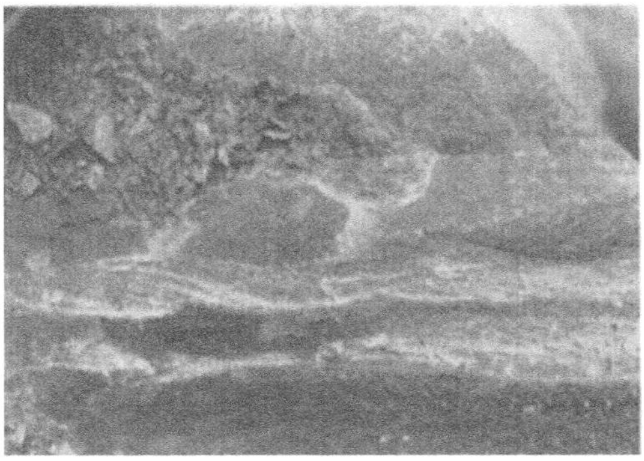

Figure 16-2. Facies A. Thinly interbedded siltstone and mudstone at left. Solitary, planar cross-strata in center and right.

Figure 16-3. Close-up view of ripple-bedding in part of a 1.5-meter sandstone of facies A. Visible part of pen is 9 cm long.

Figure 16-4. Current ripples and mud drapes in facies A. Thickest part of upper ripple train is 1.5 cm. Black spots are road tar.

Figure 16-5. Planar cross-bedded sandstone in facies A. Note envelope of mudstone. Pen is 14 cm long. Black spots are road tar.

trough cross-strata are up to 50 cm thick, and show bipolar orientation and herringbone pattern. Several sets show smoothly rounded upper surfaces, which resemble reactivation surfaces (Klein, 1970).

Facies A rocks are sparingly fossiliferous, with rare lingulid brachiopods. Isolated, vertical burrows are sparingly present in all lithologies (Fig. 16–6).

Collectively, these data indicate deposition from currents of reversing flow direction in very shallow water. Abundant ripples, flaser bedding, bipolar cross-strata, reactivation surfaces, and mud cracks support this interpretation. Moreover, the reversing current flow was located in a coastal, possibly inter-

Figure 16-6 (left). Vertical burrow structures in ripple-bedded to wavy-bedded sandstone of facies A. Note erosional contact with underlying mudstone. Pen is 14 cm long.

Figure 16-7 (right). Cross-bedded sandstones of facies B, gradationally overlain by mudstones and siltstones of facies A.

tidal, zone. The sedimentary structures and faunal relationships strongly resemble those documented from modern intertidal environments (Klein, 1970; Wünderlich, 1970; Evans, 1965; Straaten, 1961). These data suggest that facies A sediments originated by deposition from relatively low-energy tidal currents on intertidal to possibly supratidal flats. Bedding features indicate that low-velocity traction currents existed during sand deposition. Bedforms of at least two scales migrated across the flats: large dunes generated solitary sets of planar cross-strata, while smaller ripples were moved in interdune areas and across muddier flats. The thin sandstone-mudstone interbedding resembles tidal bedding described by Wünderlich (1970), and suggests rapidly alternating periods of traction-current and suspension deposition. Mud-draped current ripples indicate the abruptness of current velocity fluctuations. Desiccation cracks were developed in sediments exposed for considerable periods of time. Such features support the interpretation of alternating periods of flowing and slack water, and intermittent exposure, generated in tidal systems.

FACIES B. Facies B consists of quartz wacke and lithic wacke, both with considerable matrix, and minor thin siltstone and mudstone beds (Fig. 16–7; Fig. 16–1, column B). The sandstones are large-scale trough cross-bedded to ripple-bedded, and contain channel cuts and fills (Fig. 16–7) and ripples on bedding planes. Planar-base cross-strata are rare. Foreset inclinations of trough sets are bipolar, and indicate general northwest and southeast transport directions. Herringbone cross-stratification is occasionally developed, while reactivation surfaces are rare to absent. Ripple-bedded sandstones reach 50 cm thick, and generally lie above cross-bedded zones.

Figure 16-8 (left). Contact of facies A and overlying facies B. Person points to basal, southeast-dipping (originally), trough cross strata of facies B; succeeding cross strata dip gently in opposite direction.

Figure 16-9 (right). Facies C. Note light, basal, thin-bedded zone, and upper cross bedding and channeling. Dark strata at upper right are facies A.

Facies B sandstones occur in discrete intervals up to 3 meters thick. Basal trough sets are large scale, and contain southeast-facing foresets (Fig. 16–8). Sparse lag conglomerates contain abundant mud clasts and rare marine skeletal debris. Higher sets are small-scale trough cross-bedded and ripple-bedded, and frequently grade up into facies A sequences (Fig. 16–7).

Facies B is interpreted to represent traction-current deposition in relatively high-energy areas of reversing current flow, with decreasing energy higher in the intervals. The internal organization of facies B sequences and their gradation into overlying facies A sequences resemble those of point-bar deposits of meandering streams (Allen, 1964; Klein, 1971) and indicate steadily declining current velocities. Reversing current directions suggest that the streams were affected by coastal rise and fall of water, and thus traversed intertidal flats. Flooding of the flats themselves occurred by overflow from these creeks; thus deposits on the flats (facies A) and those in the creeks (facies B) are gradational. Mud lenses and drapes were generated by mud deposited from slack water at high- and low-water stands.

Facies C. Facies C consists of fine- to medium-grained, clean, well-sorted quartz arenite and lithic arenite, with no interbedded siltstone or mudstone (Fig. 16–1, column C). Many rocks are thin bedded to laminated, with primary current lineations and current crescents on bedding planes. Thin-bedded zones reach 1 meter thick, and are erosionally interbedded with cross-bedded units (Fig. 16–9).

Sets of trough cross-strata reach 75 cm thick and cosets reach 2 meters. Foresets are inclined at variable angles in the range 5° to 20°, and are usually tangential to set bases. Several angles of inclination may occur in a single set (Fig. 16–9). Foresets dip consistently to the northwest, with some scatter. Interset scours are common, and assume low angles of inclination; sediments filling the scours are of the same grain size as those scoured (Fig. 16–10). Herringbone cross-strata and reactivation surfaces were not observed in this facies, and planar cross-strata are very rare.

Facies C rocks occur as discrete intervals up to 4 meters thick, and are bounded below and above by scour surfaces (Figs. 16–9 and 16–10). Basal portions of such intervals are slightly coarser grained than higher intervals, and usually contain any thin bedding and lamination present. Large-scale cross-bedding and then small-scale cross-bedding and ripple-bedding characterize successively higher intervals. Intraformational mud clasts commonly occur in the lower portions; fossil debris is notably absent.

Facies C sandstones are interpreted to represent traction-current deposition from relatively high energy currents of essentially unidirectional flow. The absence of marine fossils, evidence of high current velocity (flat bedding and laminations), unimodal current directions, and absence of herringbone cross-strata and reactivation surfaces all indicate that current flow was not reversing but was consistently to the northwest.

The total absence of mud from this facies suggests that, although available for deposition in other nearby environments, mud was either not available for deposition here owing to absence in the suspension load, or was not deposited because of consistently high current velocities. A lack of mud in the suspension load is the more reasonable alternative, because if mud were present at least some would have been deposited with the sand. These restrictions suggest that the clean sandstones represent local distributary channels on the delta front, which

Figure 16-10 (left). Facies C sandstone sequence, bounded above and below by facies A sequences. Channel structure (containing hammer) and cross-strata indicate northwest-flowing currents.

Figure 16-11 (right). Fossiliferous siltstone and thin coquinoid beds (dark) of underlying rocks. Note sharp, erosional bases of coquinas and irregular upper surfaces. Pen is 14 cm long.

flowed northwestward down the paleoslope. If reversing flow occurred in these channels, it was strongly ebb-dominated, and is not reflected in the primary structures. The reversing-flow environments of facies A and B probably occupied interdistributary areas between channels.

Adjacent Rocks

UNDERLYING ROCKS. Rocks beneath those described here are interbedded fine sandstones, siltstones, and mudstones—all fossiliferous with marine taxa. A basal zone of thinly interbedded siltstone and shale is overlain by a 3- to 6-meter thick zone of well-sorted, parallel-laminated sandstone containing primary current lineations and current crescents. Above this are 3 to 15 meters of intensely bioturbated sandstone and siltstone containing thin, discontinuous, size-graded beds of small-mollusk coquina which carry the same taxa as the surrounding sediment (Fig. 16–11). Lingulid brachiopods commonly occur in growth position in the bioturbated sands.

These data suggest a shallow to very shallow marine zone of considerably fluctuating water energy, wherein infaunal elements largely homogenized the sediment in search of food. Periodic high energy waters disrupted bottom sediments and resedimented them as thick layers of skeletal rubble. The coquinas probably represent lag concentrates, moved onshore and buried by new sediment. The depositional environment, which may have been close to the intertidal zone, was protected from the open sea by offshore, subtidal sand bars, over which current velocities were high.

OVERLYING ROCKS. Rocks overlying those described here consist of thickly alternating (4 to 10 meters) cross-bedded, scoured and ripple-bedded, unfossiliferous drab sandstones and lenticular- and ripple-bedded, bioturbated red siltstones and mudstones. Cross-bedding in the sandstones is usually less than 1 meter in coset thickness, and sets and cosets are often thin and approximately parallel (Fig. 16–12). The sandy intervals grade upward into dominantly ripple-bedded to irregularly laminated siltstones and mudstones with sand flasers. These zones contain isolated vertical burrows (Fig. 16–13), and are mildly bioturbated throughout.

The fining-upward sequences are laterally discontinuous, and are highly variable between outcrops. These sequences may represent either point-bar and exposed floodplain deposition from meandering fluvial rivers on the delta top; or, equally

Figure 16-12 (left). Typical fining-upward sequence in overlying rocks. Boys stand at base; top is just below massive light sand in upper right. Darker colors of mudstones reflect red color.

Figure 16-13 (right). Vertical burrow structures in upper parts of mudstone interval in overlying rocks. Visible portion of pen is 9 cm long.

likely, intertidalite sequences (Klein, 1971) formed by deposition on intertidal flats fronting an open interdistributary bay. Such an environment contains lower sandy flats, medial mixed sandy and muddy flats, and upper mud flats; progradation of these environments would generate a progressively fining-upward sequence, and would account for the unusually parallel sets of cross-strata seen in the lower intervals.

Evidence for Tidal Origin

Rocks of this marginal marine zone cannot be directly interpreted as of tidal origin; incontrovertible evidence of diurnal rise and fall of sea level is lacking. However, sea level rise and fall is strongly suggested by associations of primary features, notably reversing current flow directions and mud cracks, and is compatible with the environmental interpretations presented above. Whether the rise and fall of sea level was in fact diurnal, or of longer and more irregular period, cannot be established on present information.

Summary

In summary, then, the series of traction-current-dominated, coastal environments interpreted here may be termed "tidal," if by "tidal" is understood the astronomically controlled periodic rise and fall of sea level. Such rise and fall need not necessarily be diurnal, but may be due to seasonal, annual, or longer term causes. The mean range of such tides, as inferred from the fining sequences of facies B (Klein, 1971), was 2 to 3 meters.

References

ALLEN, J. R. L. 1964. Fining-upwards cycles in alluvial successions. *Geol. J. 4*, 299–246.

EVANS, G. 1965. Intertidal flat sediments and their environments of deposition in the Wash. *Quart. J. Geol. Soc. London 121*, 209–245.

KLEIN, G. DEV. 1970. Depositional and dispersal dynamics of intertidal sand bars. *J. Sed. Petrol. 40*, 1095–1127.

——— 1971. A sedimentary model for determining paleotidal range. *Geol. Soc. Am. Bull. 82*, 2585–2592.

THOMPSON, A. M. 1968. Stratigraphy and sedimentology of Upper Ordovician clastic rocks in central Pennsylvania. Unpublished Ph.D. Dissertation, Brown Univ., 246 p.

STRAATEN, L. M. J. U. VAN. 1961. Sedimentation in tidal flat areas. *Alberta Soc. Petrol. Geol. 9*, 203–226.

WÜNDERLICH, F. 1970. Genesis and environment of the "nellenkopfschichten" (Lower Emsian) at locus typicus in comparison with modern environment of the German bay. *J. Sed. Petrol. 40*, 102–130.

17

Tidalites in the Eureka Quartzite (Ordovician), Eastern California and Nevada

George deVries Klein

Occurrence

The Eureka Quartzite (Ordovician) occurs as a well-exposed, cliff-forming sandstone in the Basin-and-Range Province of eastern California and Nevada over an area in excess of 200,000 sq km^2 and ranges in thickness from 50 to 150 meters (Ross, 1964; Webb, 1956). Selected outcrops (Fig. 17–1; Table 17–1) contain a variety of sedimentary and biogenic structures indicating the Eureka Quartzite to be a tidalite (cf. Klein, 1971; Swett et al., 1971). The Eureka is sandwiched between the intertidal carbonates of the Ely Springs Dolomite (Chamberlin, 1975) and the Pogonip Group (Stricker, 1973). The general stratigraphic relationships are summarized in Figure 17–2.

Research support for fieldwork in the Eureka Quartzite came from grants from the Department of Geology, University of Illinois and the Mobil Foundation. J. C. Harms is thanked for his constructive comments about the manuscript.

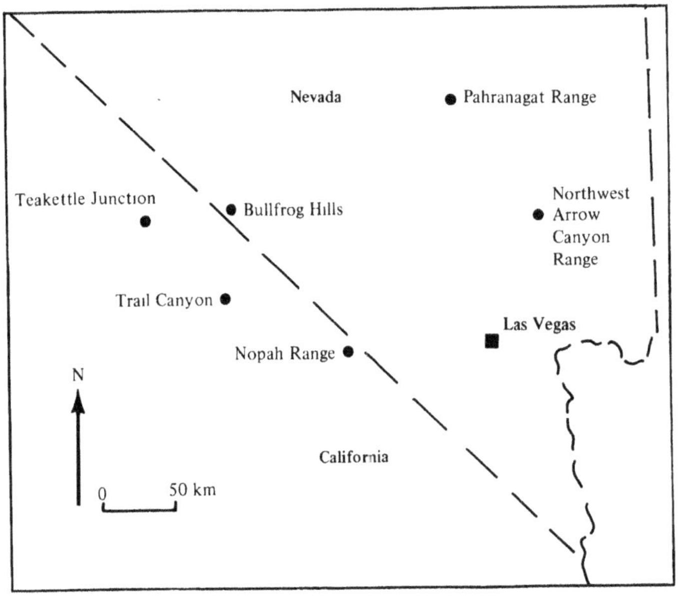

Figure 17-1. Location map showing outcrops mentioned in text (Table 17-1).

Sedimentary Features

The Eureka Quartzite consists dominantly of fine-grained, well-sorted, supermature, rounded quartz arenite and very fine-grained, silty, dolomitic quartz arenite. Very thin layers of dolomitic mudstone occur in accessory quantities. The quartz arenites are dominated by burrowing structures or are massive. Cross-stratified and parallel-laminated sandstones comprise about 20 percent of the Eureka. Fining-upward arenitic chan-

Table 17–1 *Outcrops of Eureka Quartzite (Ordovician)*

Teakettle Junction, Inyo County, California	SE¼, Sec 20 & NE¼, Sec 29, T13S, R41E (ext)
Trail Canyon, Inyo County, California	SW¼, Sec 2, T19S, R46E
Nopah Range, Inyo County, California	NE¼, Sec 12, T12S, R7E
Bullfrog Hills, Nye County, Nevada	Center Sec 2, T12S, R45E
Northwest Arrow Canyon Range, Clark County, Nevada	NE¼, Sec 14, T14S, R63E
Pahranagat Range, Lincoln County, Nevada	SE¼, Sec 31, T6S, R59E (ext)

nel fills occur at Teakettle Junction, California, and the Northwest Arrow Canyon and Pahranagat ranges in Nevada. The Eureka Quartzite is a tidalite deposited mostly under shallow, subtidal, tide-dominated conditions, although local zones of intertidal deposition occur in the middle and lower parts of the succession at the three localities mentioned.

Figure 17-2. Generalized stratigraphic column for Eureka Quartzite and sedimentary log, Teakettle Junction, California (*A*), and sedimentary log, Pahranagat Range, Nevada (*B*). *T*, tidal bedding; *B*, brachiopods.

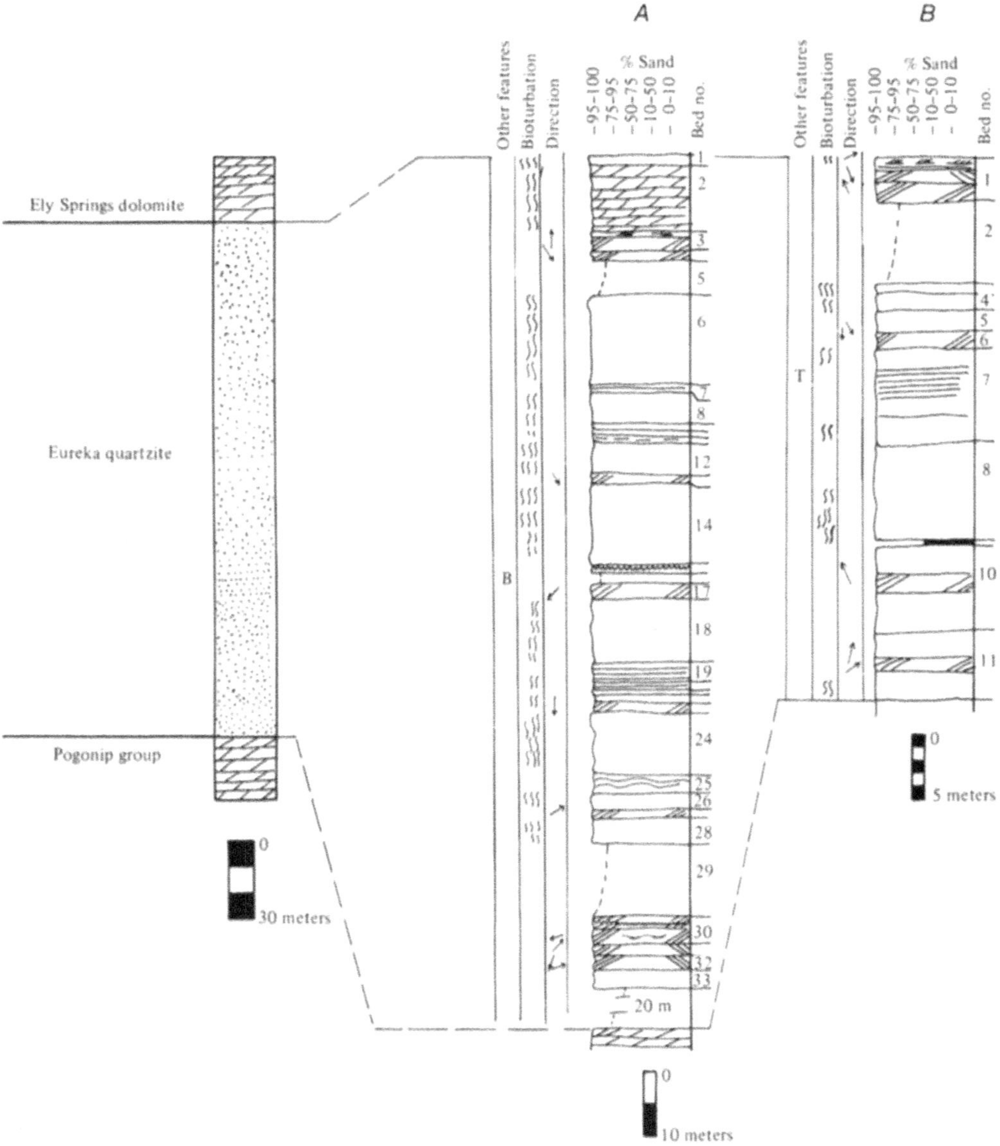

Evidence for Tidal Origin

Table 17–2 summarizes the sedimentary and biogeneic structures and textural attributes common to the Eureka Quartzite and the interpreted flow processes that produced them. Noteworthy as indicators of tidal depositional processes are herringbone cross-stratification and herringbone, micro cross-laminae (Reineck, 1963); multimodal frequency distributions of set thicknesses of cross-strata and bimodal frequency distributions of cross-strata dip angles (Klein, 1970a); B–C sequences consisting of cross-stratification (interval B) overlain by variably-oriented, micro cross-laminae (interval C) (Klein, 1970b); dolomitic mudstone drapes mantling current ripples (Fig. 17–3); flaser bedding (Fig. 17–4); and lenticular bedding and burrowing (Fig. 17–5).

The burrowed horizons in the Eureka are extremely well developed and occur in intervals 10 to 80 cm thick (40 cm

Table 17–2 *Association of sedimentary features in Eureka Quartzite, and interpreted flow models*

Features	Flow model
Herringbone cross-stratification and micro cross-laminae Parallel laminae (Fig. 17–8) Supermature grain roundness	Tidal current bedload transport with bipolar reversals of flow direction
Multimodel frequency distribution of set thicknesses of cross-strata Bimodal frequency distributions of dip angles of cross-strata	Time-velocity asymmetry of tidal current bedload transport
Current ripples (Fig. 17–3) B-C sequences (of cross-strata overlain by micro cross-laminae Current ripples oriented at 90° and 180° with respect to underlying cross-strata	Late-stage emergence outflow with sudden changes in flow directions at extremely shallow water depths (less than 2 meters)
Cross-stratification with flasers (Fig. 17–4) Lenticular bedding Tidal bedding (Fig. 17–9) Mudstone drapes over current ripples (Fig. 17–3)	Alternation of tidal current bedload transport with suspension deposition during slack water periods
Mudchip conglomerates at base of washouts and channels (Fig. 17–7)	Tidal scour
Mud cracks (Fig. 17–6)	Exposure
Deep burrows (Figs. 17–5, 17–8)	Burrowing

Figure 17-3 (left). Mudstone (dolomitic) drapes over current ripples in micro cross-laminated dolomitic quartzite, Bed 1, NW Arrow Canyon Range, Nevada. Scale in centimeters.

Figure 17-4 (right). Simple flaser bedding in cross-stratified quartzite, Bed 32, Teakettle Junction, California. Scale in centimeters.

average). This range of thickness is within the intertidal and extremely shallow subtidal burrowing depths reported by Rhoads (1967). X-ray radiography of samples that appear massive in outcrop are extensively burrowed (Fig. 17–5).

At Teakettle Junction, both the preservation of mud cracks (Fig. 17–6) and a pebble conglomerate (Fig. 17–7) at the base of a channel scour indicate local developments of tidal flats and tidal channels in the lower and middle parts of the sequence (Fig. 17–2*A*). Similar channel lag concentrates occur at the base of fining-upward channel sequences in the northwestern part of the Arrow Canyon Range and the West Pahranagat Range (Fig. 17–2*B*). These channel sequences again suggest local tidal-flat development characterized by a mature system of channels.

Discussion

The Ordovician Eureka Quartzite appears to be one of the very few documented Ordovician tidalites. Pryor and Amaral

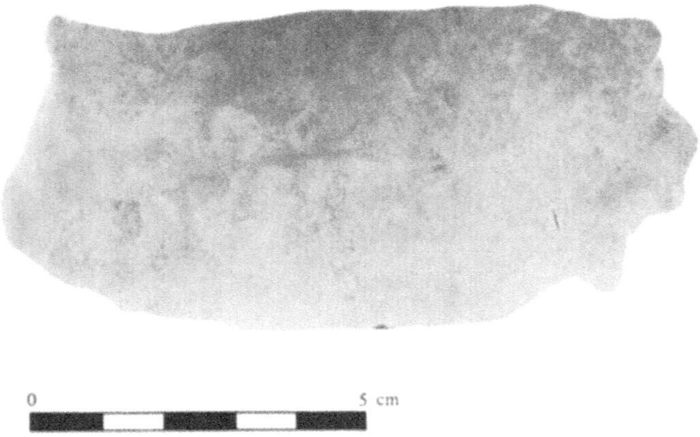

Figure 17-5. X-ray radiograph showing burrowed nature of massive slab of dolomitic quartzite, Bullfrog Hills, Nevada.

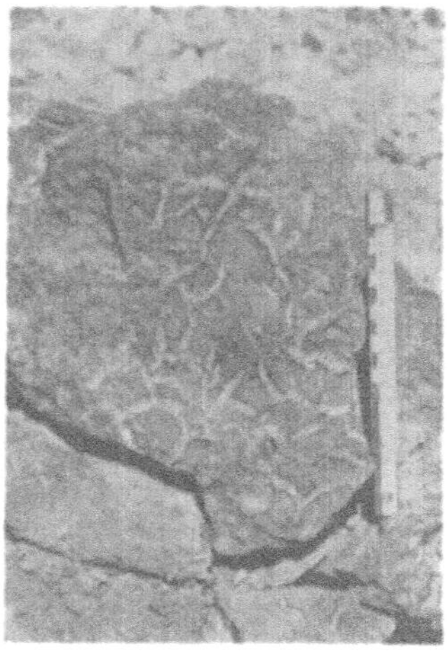

Figure 17-6. Mud cracks, Bed 30, Teakettle Junction, California. Scale in centimeters and decimeters.

Figure 17-7. Mudchip conglomerate at base of tidal channel, Bed 30, Teakettle Junction, California. Scale in centimeters and decimeters.

(1971) proposed a model for the St. Peter Sandstone (Ordovician) of Wisconsin that is analagous to the shallow, subtidal, tide-dominated sand bodies and associated sand waves described from the North Sea by Houbolt (1968). The Eureka is very similar to the Cambrian Eriboll Sandstone (Swett et al., 1971, Table 1). A model of a coalesced series of shallow, sub-

Figure 17-8. Parallel laminae, burrowing, and herringbone microcross laminae, Bed 1, NW Arrow Canyon Range, Nevada. Scale in centimeters.

Figure 17-9. Tidal bedding in dolomitic quartzite, Trail Canyon, California. Scale in centimeters and decimeters.

tidal, tide-dominated sand bodies and channelized intertidal sand flats appears to explain best the association of sedimentary features documented herein from the Eureka Quartzite.

References

CHAMBERLIN, T. L. 1975. Stratigraphy of Upper Ordovician Ely Springs Dolomite in the southern Great Basin of Utah and Nevada. Unpublished Ph.D. dissertation, Univ. Illinois, Urbana.

HOUBOLT, J. J. H. C. 1968. Recent sediments in the southern bight of the North Sea. *Geol. Mijnbouw 47*, 245–273.

KLEIN, G. DEV. 1970a. Depositional and dispersal dynamics of intertidal sand bars. *J. Sed. Petrol. 40*, 1095–1127.

——— 1970b. Tidal origin of a Precambian Quartzite—The Lower Fine-grained Quartzite (Dalradian) of Islay, Scotland. *J. Sed. Petrol. 40*, 973–985.

——— 1971. A sedimentary model for determining paleotidal range. *Geol. Soc. Am. Bull. 82*, 2585–2592.

PRYOR, W. A., and AMARAL, E. J. 1971. Large-scale cross-stratification in the St. Peter Sandstone. *Geol. Soc. Am. Bull. 82*, 239–244.

RHOADS, D. C. 1967. Biogenic reworking of intertidal and subtidal sediments in Barnstable Harbor and Buzzards Bay, Massachusetts. *J. Geol. 75*, 461–476.

REINECK, H.-E. 1963. Sedimentgefuge im Bereich der Sudliche Nordsee. *Abhandl. Sencken. Nat. Gesell. 505*, 1–138.

ROSS, R. J., JR. 1964. Relations of middle Ordovician time and rock units in Basin Ranges, western United States. *Am. Assoc. Petrol. Geol. Bull. 48*, 1526–1554.

STRICKER, G. D. 1973. Petrography and microfacies of the Pogonip Group, Arrow Canyon Range, Nevada. Unpublished Ph.D. dissertation, Univ. Illinois, Urbana.

SWETT, K., KLEIN, G. DEV., and SMIT, D. E. 1971. A Cambrian tidal sand body—Eriboll sandstone of northwest Scotland: An ancient-Recent analog. *J. Geol. 79*, 400–415.

WEBB, G. W. 1956. Middle Ordovician detailed stratigraphic sections for western Utah and eastern Nevada. *Utah Geol. Min. Sur. Bull. 57*.

18

Tidal Deposits in the Monkman Quartzite (Lower Ordovician) Northeastern British Columbia, Canada

Lubomir F. Jansa

Occurrence

The Monkman Quartzite (late Lower Ordovician; Slind and Perkins, 1966) crops out in thrust-faulted sheets of the Main Range of the Canadian Rocky Mountains (Fig. 18–1). The quartzite was deposited on the eastern flank of the eastern Cordilleran geosyncline. The quartz arenite unit reaches a maximum thickness of 532 meters at Mount Hunter and thins both southward and westward (Table 18–1). In discontinuous outcrops, the Monkman Quartzite can be followed for a distance of 250 km. Sedimentary and biogenic structures, texture, composition, and facies relationships indicate that the Monkman

The project was initiated during tenure of a National Research Council Fellowship (1969–1970), held at the Institute of Sedimentary and Petroleum Geology, Geological Survey of Canada in Calgary. G. S. Taylor introduced the author to the geology of the study area, and B. S. Norford kindly identified collected megafossils. Discussion with P. Schenk and J. Wade and critical comments by R. G. Walker led to the improvement of the manuscript.

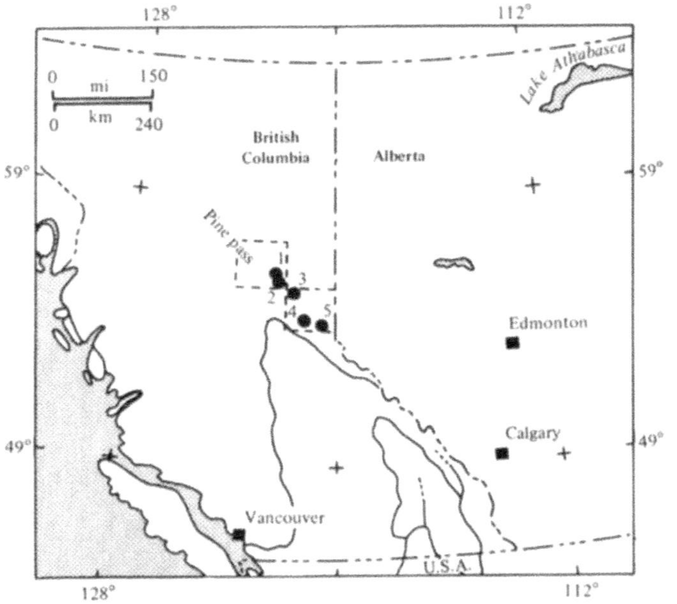

Figure 18-1. Map showing locations of sections Table 18–1.

Quartzite is a tidal deposit in the broadest sense, including shallow subtidal, intertidal, and questionable supratidal and eolian deposits. It represents a shoaling-upward sequence with records of a bidirectional water movement (herringbone cross-stratification) related to the tides.

Conformably below the Monkman Quartizite, the Lower Ordovician Chuchina Formation (Slind and Perkins, 1966) is composed of alternating argillaceous limestone-calcareous shale-dolomite sequences, partly of an intertidal origin. Transitionally above the Monkman Quartzite, the overlying Middle Ordovician Skoki Formation (Fig. 18–2*A*) is composed of dark gray fossiliferous dolomites deposited in a shelf-lagoon (Jansa, 1970).

Table 18–1 *Location of Ordovician sections*

Location	Topographic coordinates	Monkman Quartzite thicknesses, meters
1	55°17′ N, 122°25′ N (Mount Hunter)	532
2	55°03′ N, 122°05′ W	384
3	54°56′ N, 121°40′ W	128
4	54°17′ N, 121°18′ W	0
5	54°10′ N, 120°37′ W	68

Composition

The Monkman Quartzite consists dominantly of coarse to fine-grained quartz arenites with less than 1 percent of siltstone and shale interbeds (Fig. 18–2*B*).

FINE-GRAINED QUARTZ ARENITE. These arenites comprise 70 percent of the Monkman Quartzite. Occurrences of a large variety of depositional structures—including thin, horizontal bedding, parallel lamination, and small-scale cross-stratification

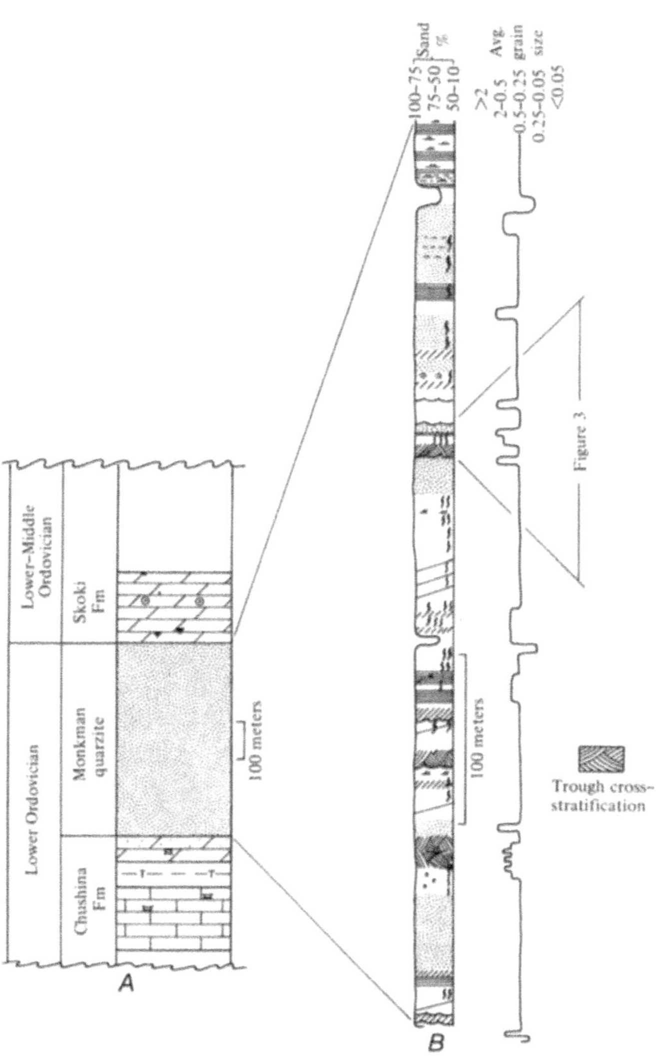

Figure 18-2. *A*. Generalized stratigraphic column of the Ordovician in Pine Pass area, British Columbia. *B*. Sedimentary log, Mt. Hunter.

Figure 18-3. *Scolithus* burrows zone overlain by herringbone small-scale cross-stratification, locally reworked by *Scolithus* (Table 18-1, loc. 2).

—are frequent. Twenty percent of the Monkman Quartzite shows bioturbation by *Scolithus* (Fig. 18–2*B*). Burrows (from 15 to 30 cm deep) form intensively reworked zones 30 to 60 cm thick (Fig. 18–3), which often coalesce into 3- to 10-meters thick units, occurring repeatedly in a vertical profile in approximately 70-meter intervals. Herringbone cross-stratification (Reineck, 1963) (average thickness of sets 70 cm) and small-scale trough cross-stratification (Fig. 18–4*C*) are in places obliterated by bioturbation (Fig. 18–3). Fine-grain quartz arenite, medium to thick bedded, without any internal structure is common; it is composed of well-sorted, well-rounded grains.

Silica cements the arenites; dolomite cements both the bioturbated and herringbone cross-stratified arenites.

Medium-grained quartz arenite. These arenites are trough cross-stratified with thickness of sets ranging between 60 and 100 cm (Figs. 18–5*A* and 18–4*A*). Reactivation sur-

Figure 18-4. Sequence of sedimentary features in the tidal cycle at locality #1; for environmental interpretation see Fig. 18–8. A. Outcrop view middle part of cycle. B. Zone of *Scolithus* burrows. C. Small-scale, trough cross-stratification overlain by zone of *Scolithus* burrows. D. Thin sets of trough cross-stratification with pebbles lining lower trough and overlain by a large-scale trough with angle of dip steepening upward; see Fig. 18–8 for location in sequence. E. Lag of pebbles and cobble in tidal channel, see Fig. 18–8 for location in sequence.

faces, (Klein, 1970) scours (Raff and Boersma, 1971), tabular cross-stratification with steep inclined beds (30° to 40°) are rare. *Scolithus* bioturbated zones are frequent. Large-scale, convex-up, cross-stratification is exceptionally rare; where present these units are up to 3 meters thick, have sharp boundaries, and show convex-up, large-scale foreset cross-stratification, whose dip-angle decreases upward (Fig. 18–6). Quartz grains in the medium-grain arenites are rounded and poor to moderately sorted. Skeletal fragments, mainly of echinoderms and trilobites, pellets, and glauconitic grains are rare.

COARSE-GRAINED PEBBLY QUARTZ ARENITES. Coarse-grained, pebbly quartz arenites are lensoid-shaped and cut the underlying beds. Flat pebbles averaging 3 cm long and cobbles averaging 15 cm long of quartz arenites were formed by subaqueous erosion. Clasts are concentrated at the bottom of the channels as a lag deposit (Fig. 18–4*E*) and/or are embedded into cross-stratified beds (Fig. 18–4*D*). Trough cross-stratification is frequent and sets are up to 1 meter thick (Fig. 18–7). Trough cross-stratified coarse-grained arenites are frequently bimodal, with a secondary mode of medium-grained sand. Grains are of high sphericity; cement is either dolomite or silicia. Echinoderm fragments rarely occur near the base of the scours.

Environmental Interpretation

Sedimentary structures—predominantly good sorting, high maturity, composition, and character of the surrounding sediments—indicate a sand barrier origin for the Monkman Quartzite, with periods of tidal-flat development. The fining-upward sedimentary sequence of a tidal cycle in the Monkman Quartzite is documented in Figure 18–4, and environmental interpretation is summarized in Figure 18–8.

The occurrence of herringbone cross-stratification (Reineck, 1963) and cross-stratification with reactivation surfaces (Klein, 1970) indicate bedload transport with bipolar reverse of flow directions. The well-developed *Scolithus*-burrowed zones (Fig. 18–4*B*) overlying the herringbone cross-stratified sediments are believed to have formed in the intertidal zone. The thickness of these burrowed horizons is within the range of intertidal burrowers (Rhoads, 1967). In the Monkman Quartzite, the intertidal flats must have been locally semiprotected for formation of the *Scolithus* burrows. Tidal flats exposed to the force of an open sea lack the *Scolithus* bioturbated zone. Thin-bedded, horizontally-stratified arenites that commonly overlie the *Scolithus* zones are of well-sorted, well-rounded quartz grains and are interpreted as foreshore beach deposits.

The genesis of the massive, medium- to thick-bedded sandstones forming most of the Monkman Quartzite is not well understood; they could be beach, washover fan, or lower intertidal to shallow subtidal sheet sand deposits. The large-scale, convex-upward, cross-stratified arenites (Fig. 18–6) may be eolian dunes of spillover lobes similar to eolian ridges in the Bahamas (Ball, 1967); however, a subaqueous spillover sandbar origin is possible. Local development of tidal channels and re-

Figure 18-5. Large-scale, steep-inclined cross-stratification with sharp set boundaries. Small-scale cross-stratification developed near the top of the picture. Note a shallow channel fill underneath the hammer (loc. 2).

Figure 18-6. Large-scale, convex-upward cross-stratification with decrease in the angle of dip upward, overlain by a medium-bedded dolomitic sandstone unit with small-scale current ripples and a thin bedded unit with small bipolar cross-stratification. Some of the beds are reworked by *Scolithus* burrows.

Figure 18-7. Trough cross-stratified quartzite beds with a set of a tabular cross-stratification in the lower half of the picture. Note obliquely downcutting surface in the middle of the picture (loc. 1).

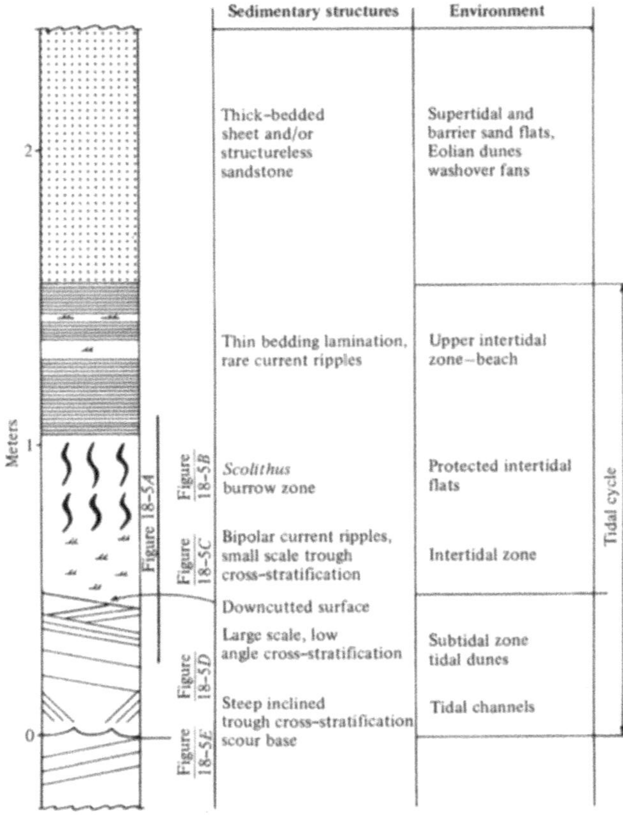

Figure 18-8. Sedimentary features and environmental interpretation of the tidal cycle in the Monkman Quartzite.

working of tidal surface are indicated by scour surfaces with pebbles at the base of the channel scours. The channels were 30 to 100 cm deep, with variable width up to 10 meters.

The trough, cross-stratified, pebbly sandstones represent subaqueous, gently sloping sandbar sequences (Klein, 1970). The bipartite sandstones with large-scale cross-stratification in the lower part, bounded by an inclined set boundary and grading upward into a small-scale, cross-stratified interval (Fig. 18–5, indicate shoaling tidal deposition. The lower interval was laid down by flood currents which prevailed over those of the ebb. The upper interval was deposited from alternating ebb and flood currents of about equal velocity (Raff and Boersma, 1971). The dominant transporting mechanisms for the Monkman Quartzite was bedload transport; the suspension component is lacking as a result of permanent water agitation and medium to high energy of the depositional system.

The sedimentary features indicate that the Monkman Quartzite represents a shallow, subtidal sand barrier belt, composed of coalesced sand bodies, which locally were built into the intertidal zone, followed by construction of subaerially exposed sand flats exposed to eolian reworking. Five to six periods of intra- to supratidal deposition can be recognized in the vertical profile of the Monkman Quartzite, which are characterized by sedimentary features summarized in Figure 18–2. Barrier sand of similar thickness exceeding 300 meters is known from the Recent barrier islands of southeastern Queenland (Hails and Hoyt, 1968).

References

BALL, M. M. 1967. Carbonate sand bodies of Florida and the Bahamas. *J. Sed. Petrol. 37*, 556–591.

HAILS, J .R., and HOYT, J. H. 1968. Barrier development on submerged coasts: Problems of sea-level changes from a study of the Atlantic Coastal Plain of Georgia, U. S. A., and parts of the East Australian Coast. *Z. Geomorphol. 7*, 24–55.

JANSA, L. F. 1970. Stratigraphy and sedimentology of the Ordovician and Lower Devonian strata in Pine Pass map-area, British Columbia. *Geol. Sur. Can. Paper 70-1 Pt. B*, 83–88.

KLEIN, G. DEV. 1970. Depositional and dispersal dynamics of intertidal sand bars. *J. Sed. Petrol. 40*, 1095–1127.

——— 1973. Determination of Paleotidal Range in Clastic Sedimentary Rocks. *XXIV Intern. Geol. Congr. Proc. Sec. 6*, 397–405.

RAFF, J. F. M., and BOERSMA, J. R. 1971. Tidal deposits and their sedimentary structures. *Geol. Mijnbouw 50*, 479–504.

REINECK, H.-E. 1963. Sedimentefüge im Bereick südliche Nordsee. *Senckenbergischen Naturf. Gesell. No. 505*, 138.

RHOADS, D. C. 1967. Biogenic reworking of intertidal and subtidal sediments in Barstable Harbour and Buzzards Bay, Massachusetts. *J. Geol. 75*, 461–476.

SLIND, O. L., and PERKINS, G. D. 1966. Lower Paleozoic and Proterozoic sediments of the Rocky Mountains between Jasper, Alberta and Pine River, British Columbia. *Bull. Can. Petrol. Geol. 14*, 442–468.

19

Tidal Deposits in the Zabriskie Quartzite (Cambrian), Eastern California and Western Nevada

John J. Barnes and
George deVries Klein

Occurrence

The Zabriskie Quartzite (Cambrian) occurs as a well-exposed, cliff-forming sandstone over an area of 36,000 sq km in the Basin-and-Range Province of eastern California and western Nevada, and ranges in thickness from 0 to 300 meters (Stewart, 1970). Selected outcrops (Fig. 19–1; Table 19–1) of this formation contain a variety of sedimentary and biogenic structures indicating that much of the Zabriskie is an intertidalite (Klein, 1971, 1972). The Zabriskie Quartzite is overlain by the mudstones and intertidal carbonates of the Carrara Formation (Reynolds, 1969; Stewart, 1970), and is underlain by the tidalites of the Wood Canyon Formation (Klein, Chapter 20). The stratigraphic relationships are summarized in Figure 19–2.

Research support for this program came from grants from the American Philosophical Society, Department of Geology, University of Illinois, the Mobil Foundation, and the Penrose Bequest of the Geological Society of America (to Barnes).

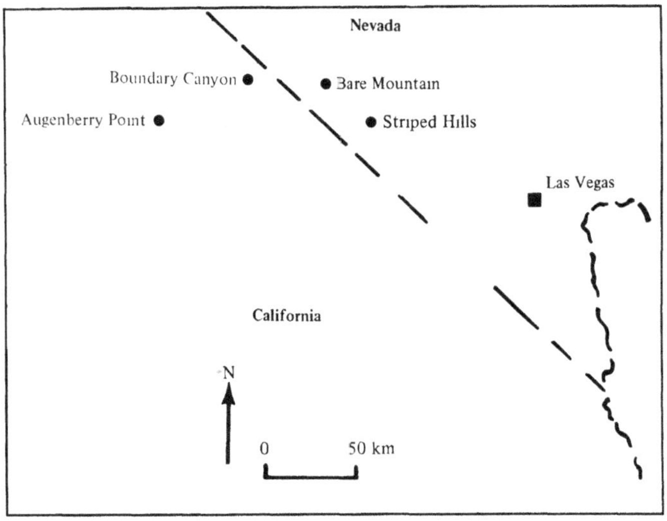

Figure 19-1. Location map showing outcrops mentioned in text (Table 19-1).

Sedimentary Features

The Zabriskie Quartzite consists dominantly of fine- to medium-grained, well-sorted, supermature, rounded quartz arenite with accessory interbedded siltstone and mudstone (less than 2 percent). The quartz arenites are either massive, massive and burrowed, or cross-stratified and rippled (Fig. 19–2). The uppermost Zabriskie is also characterized by many features indicating exposure and later emergence runoff common to intertidal environments (Table 19–2). The remainder of the Zabriskie Quartzite appears to represent a coalescing complex of subtidal, tide-dominated sand bodies such as those reported from the North Sea (cf. Houbolt, 1968).

Table 19–1 *Outcrops of Zabriskie Quartzite (Cambrian)*

Aguerberry Point, Inyo County, California	NE¼, NE¼, Sec. 24, T18S, R45E (ext)
Titus Canyon, Inyo County, California	SW¼, SW¼, Sec. 18, and NE¼, NE¼, Sec. 19, T13S, R46E
Daylight Pass, Inyo County, California	SW¼, SW¼, Sec. 25, T13S, R46E
Bare Mountain, Nevada	SE¼, NW¼, Sec. 12, T13S, R47E
Striped Hills, Nevada	SW¼, SE¼, Sec. 15, T15S, R50E
Nopah Range, California	Northern ½, Sec. 10, T20N, R8E
Winters Pass, California	Center, Sec. 33, T19N, R12E

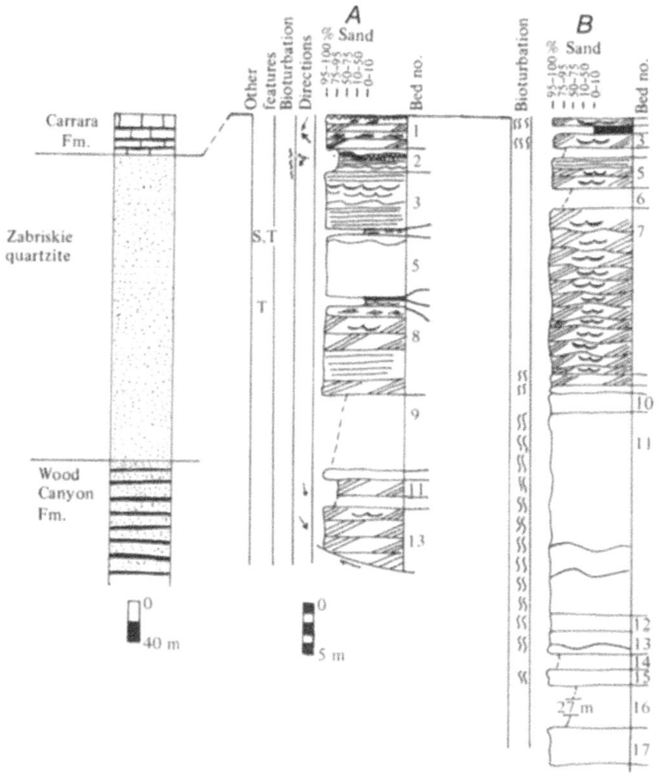

Figure 19-2. Generalized stratigraphic column for Zabriskie Quartzite. *A*. Sedimentary log, Titus Canyon, California. *B*. Sedimentary log, Striped Hills, Nevada. Scales for *A* and *B* are same. *S*, siltstone; *T*, tidal bedding.

Figure 19-3. Sawed slab showing cross-stratification (interval *B*), micro cross-laminae (interval *C*), and current-ripple surface—representing a B-C sequence, Bed 1, Titus Canyon, California.

Evidence for Tidal Origin

Table 19–2 summarizes the sedimentary, textural, and biogenic features occurring in the Zabriskie Quartzite and the interpreted tidal-flow processes that produced them. Noteworthy as indicators are supermature, rounded quartz grains (Balazs and Klein, 1972); herringbone cross-stratification (Reineck, 1963); reactivation surfaces and associated multimodal distributions of set thicknesses of cross-strata (Klein, 1970a); development of B-C sequences (Fig. 19–3) of cross-strata overlain by variably oriented micro cross-laminae (Klein, 1970b); dunes (Klein, 1970a) and preserved scour pits (Fig. 19–4); flaser bedding; and both *Monocraterion* escape structures (Fig. 19–5) and *Scolithus* (Fig. 19–6) with burrow zones up to 40 cm deep. Both types of burrows appear to be of marine aspect, and their association with beds containing mud cracks (Fig. 19–7) and preserved raindrop impressions (Fig. 19–8) pinpoints part of the upper 12 meters of the Zabriskie as an intertidalite. Deposition appears to have occurred both

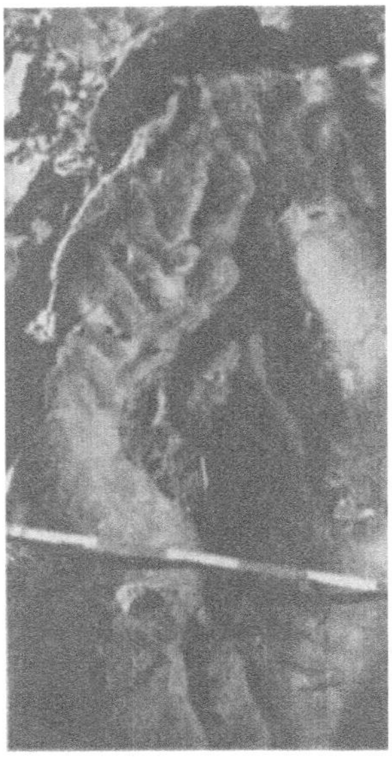

Figure 19-4. Dune, scour pit, and superimposed current ripples, top of Bed 1, Titus Canyon, California. Scale in centimeters.

Table 19–2 *Association of sedimentary features in Zabriskie Quartzite, and interpreted flow models*

Features	Flow model
1. Herringbone cross-stratification Parallel laminae Supermature grain roundness	Tidal-current bedload transport with bipolar reversals of flow directions
2. Reactivation surfaces Multimodal distribution of set thicknesses of cross-strata Supermature grain roundness	Time-velocity asymmetry of tidal current bedload transport
3. Current ripples (Figs. 19–8, 19–9) Current ripples superimposed at 90° on dunes (Figs. 19–4) Interference ripples (Fig. 19–10) Current ripples superimposed on scour pits associated with dunes (Fig. 19–4) B-C sequences of cross-strata overlain by micro cross-laminae (Fig. 19–3) Washout structures (Fig. 19–11)	Late-stage emergence outflow and emergence with sudden changes in flow directions in extremely shallow water (depths less than 2 meters)
4. Cross strata with flasers Simple flaser bedding Tidal bedding Clay drapes over ripples	Alternation of tidal-current bedload transport with suspension deposition during slack-water periods
5. Mud cracks Raindrop imprints (Fig. 19–8)	Exposure
6. Tracks and trails *Monocraterion* escape structures (Fig. 19–5) *Scolithus* (Fig. 19–6) Burrows (including those piercing ripples) (Fig. 19–8)	Burrowing
7. Pseudonodules (Fig. 19–12)	Differential compaction and loading due to rapid sediment deposition

on intertidal flats and under shallow, subtidal, tide-dominated conditions. The associations of structures texture and burrowing features (Table 19–2) fit very closely with a variety of paleotidal flow models.

Discussion

The Cambrian Zabriskie Quartzite appears to be one of many Cambrian tidal quartzites. Other Cambrian examples documented as tidal in origin include the Eriboll Sandstone of

Figure 19-5 (left). *Monocraterion*-type escape structure, Aguerberry Point, California. Scale in centimeters.

Figure 19-6 (right). Top of surface burrowed by *Scolithus*, Striped Hills, Nevada.

Scotland (Swett et al., 1971) and both the Bradore Formation of Newfoundland and the Kløftelv Formation of east Greenland (Swett and Smit, 1972). These other Cambrian quartzites are similar to the Zabriskie in containing structures common to Associations 1, 2, and 3 (Table 19–2) and contain both *Monocraterion* and *Scolithus*. A critical difference, however, is that the Zabriskie is characterized by more features indicating late emergence runoff (cf. Table 19–2 with Swett et al., 1971, Table 1), alternation of bedload and suspension deposition, exposure, and differential compaction and loading. Swett et al., (1971) interpreted the Cambrian Eriboll Sandstone to be predominantly a coalesced sheet of subtidal sand-bodies with some intertidal sand-body components where runoff structures occurred. The Zabriskie, in contrast, appears to contain a larger proportion of intertidal sedimentary rocks that are coalesced with fewer subtidal sand-body components.

Figure 19-7 (left). Mud cracks in clay mantling current ripples, Titus Canyon, California. Scale in centimeters.

Figure 19-8 (right). Current ripples, with fecal pellets and small raindrop imprints, Titus Canyon, California. Scale in centimeters.

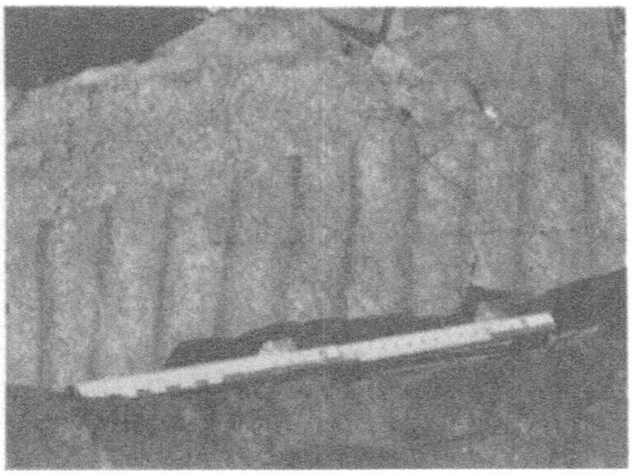

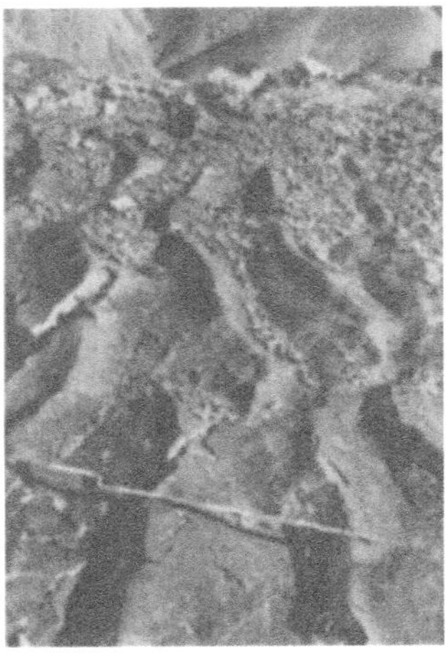

Figure 19-9 (above). Dunes with alternation in slip face orientation, Bed 1, Titus Canyon, California. Scale in decimeters.

Figure 19-10 (above right). Interference ripples, Bed 3, Titus Canyon, California. Scale in centimeters.

Figure 19-11 (right). Straight-crested current ripples and washout structure, top of Bed 1, Titus Canyon, California. Stadia scale is 2 meters long.

Figure19-12 (below). Pseudonodules in upper 15 meters of Zabriskie, Bare Mountain, Nevada. Scale in decimeters.

References

BALAZS, R. J., and KLEIN, G. DEV. 1972. Roundness-mineralogical relations of some intertidal sands. *J. Sed. Petrol. 42*, 425–433.

HOUBOLT, J. J. H. C. 1968. Recent sediments in Southern Bight of the North Sea. *Geol. Mijnbouw 47*, 245–273.

KLEIN, G. DEV. 1970a. Depositional and dispersal dynamics of intertidal sand bars. *J. Sed. Petrol. 40*, 1095–1127.

——— 1970b. Tidal origin of a Precambrian quartzite—The Lower Fine-grain Quartzite (Dalradian) of Islay, Scotland. *J. Sed. Petrol. 40*, 973–985.

——— 1971. A sedimentary model for determining paleotidal range. *Geol. Soc. Am. Bull. 82*, 2585–2592.

——— 1972. A sedimentary model for determining paleotidal range: Reply. *Geol. Soc. Am. Bull. 83*, 539–546.

REINECK, H.-E. 1963. Sedimentgefüge im Bereich der Südliche Nordsee. *Abhandl. Senckenberg. Gesell. 505*, 1–138.

REYNOLDS, M. W. 1969. Stratigraphy and structural geology of the Titus and Titanothere Canyons area, Death Valley, California. Unpublished Ph.D. dissertation, Univ. California, Berkeley.

STEWART, J. H. 1970. Upper Precambrian and Lower Cambrian strata in the southern Great Basin, California and Nevada. *U.S. Geol. Sur. Prof. Paper 620.*

SWETT, K., KLEIN, G. DEV., and SMIT, D. E. 1971. A Cambrian tidal sand-body—Eriboll Sandstone of northwest Scotland. *J. Geol. 79*, 400–415.

SWETT, K., and SMIT, D. E. 1972. Cambro-Ordovician shelf sedimentation of western Newfoundland, northwest Scotland and central east Greenland. *XXIV Intern. Geol. Congr. Proc. Sec. 6*, 33–41.

20

Paleotidal Range Sequences, Middle Member, Wood Canyon Formation (Late Precambrian), Eastern California and Western Nevada

George deVries Klein

Occurrence

The Middle Member of the Wood Canyon Formation (Precambrian) is well exposed in the Basin-and-Range Province of eastern California and western Nevada over an area of approximately 36,000 sq km (Stewart, 1970) and ranges in thickness from 150 to 550 meters. Selected outcrops (Fig. 20–1; Table 20–1) of this member show a variety of sedimentary structures and sedimentary sequences indicating that the Wood Canyon is an intertidalite (cf. Klein, 1971, 1972a, 1972b). The Upper Member (Cambrian) and the Lower Member (Precambrian) of the Wood Canyon also have features indicating a tidal depositional history. The Wood Canyon is overlain by the tidalites of the Cambrian Zabriskie Quartzite (Barnes and Klein, Chapter 19) and is underlain by the Precambrian Sterling Quartzite (Fig. 20–2).

Research support came from grants from the American Philosophical Society and the Department of Geology, University of Illinois.

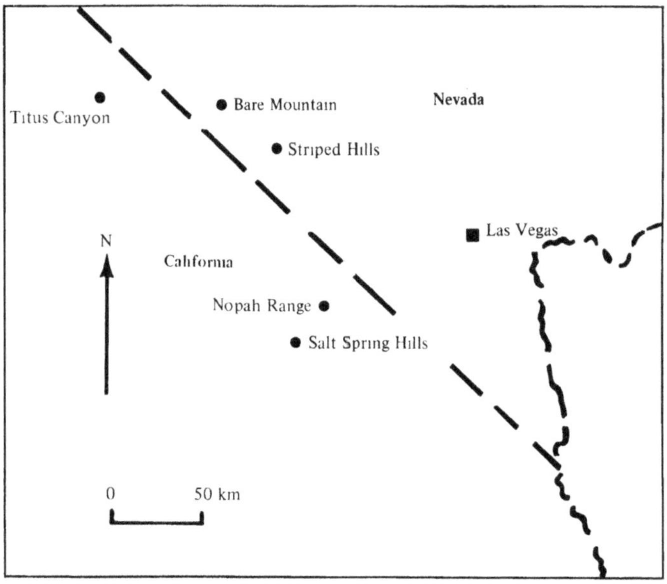

Figure 20-1. Location map showing outcrops mentioned in text (Table 20-1).

Sedimentary Features

The Middle Member of the Wood Canyon Formation consists of interbedded quartz arenites, siltstones, mudstones, and claystones. These lithologies are arranged in a preferred cyclic sequence or paleotidal range fining-upward sequence (Fig. 20–2) with a basal quartz arenite, an interbedded silty mudstone-sandstone interval, and a capping claystone. Complete paleotidal range sequences are rare, ranging from approximately 6 percent of the tidal thickness of the entire Middle Member at Nopah Range to 10 percent at Striped Hills. Partial sequences, lacking the uppermost claystone, are commoner (Fig. 20–2).

Table 20–1 *Outcrops of Middle Member, Wood Canyon Formation (Precambrian)*

Titus Canyon, Inyo County, California	E½, Sec. 16, T13S, R45E (ext)
Tungsten Canyon, Nye County, Nevada	SE¼, SW¼, Sec. 18, Y13S, R58E (ext)
Striped Hills, Nye County, Nevada	SW¼, SW¼, Sec. 15, T15S, R50E
South Nopah Range, Inyo County, California	N½, Sec. 10, T20N, R8E
Salt Spring Hills, San Bernardino County, California	N½, Secs. 17 and 18, T18N, R7E

Paleotidal range sequences are confined to the middle, lower-upper, and upper-lower portions of the Middle Member of the Wood Canyon Formation.

Evidence for Tidal Origin

Table 20–2 summarizes the associations of sedimentary structures in the Middle Member of the Wood Canyon Formation and the interpreted flow processes that produced them.

Noteworthy as particularly good indicators of tidal sedimentation processes are herringbone cross-stratification (Reineck,

Figure 20-2. *A.* Generalized stratigraphic column for Wood Canyon Formation. *B.* Sedimentary log, Salt Spring Hills, California. *C,D,E.* Detailed logs for selected intervals showing complete paleotidal range sequences. *R*, runzel marks.

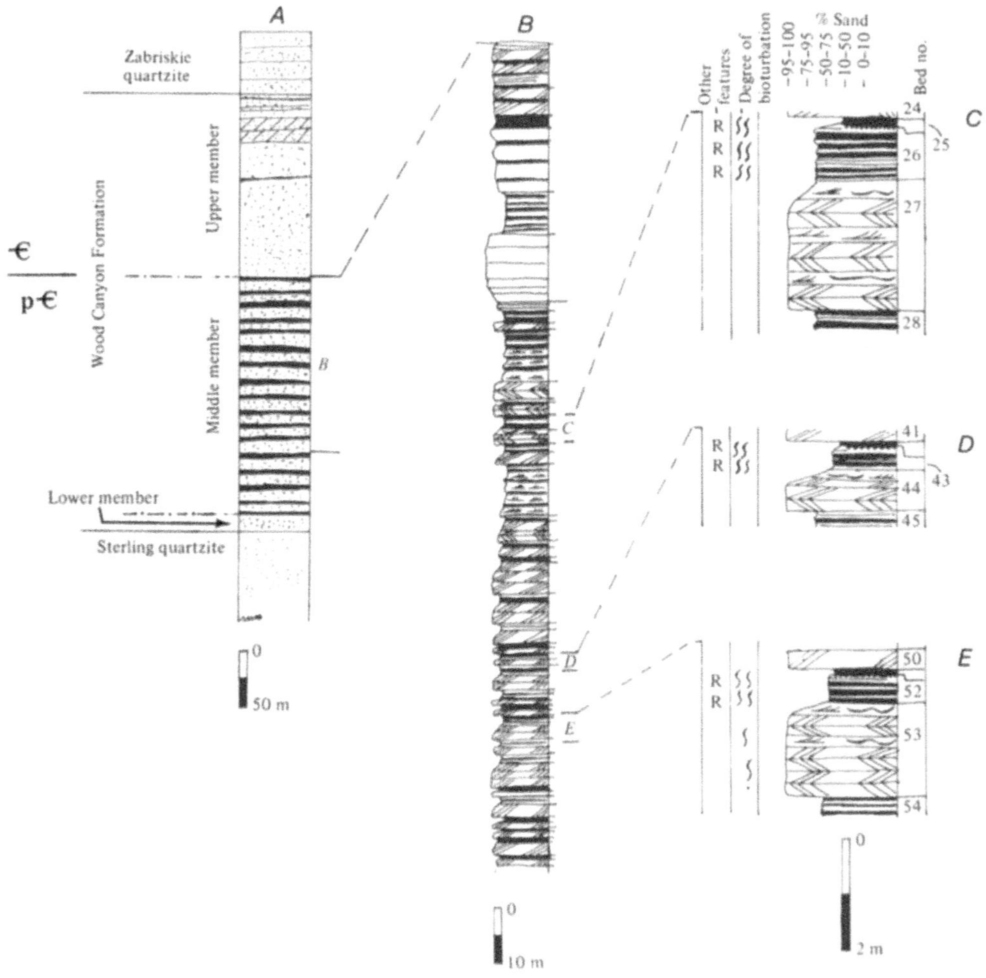

1963); the development of reactivation surfaces (Fig. 20–3) and associated multimodal distribution of set thicknesses of cross-strata (Klein, 1970a); the relationship of current ripples to underlying cross-strata into B-C sequences in which cross-strata (interval B) are overlain by micro cross-laminae (interval C) as outlined by Klein (1970b); superimposed current ripples

Table 20–2 *Associations of sedimentary structures, Middle Member, Wood Canyon Formation (Precambrian), and interpreted flow models*

Structures	Flow model
1. Herringbone cross-stratification Parallel laminae	Tidal current bedload transport with bipolar reversals of flow directions
2. Reactivation surfaces (Fig. 20–53) Multimodal distribution of set thicknesses of cross-strata	Time-velocity asymmetry of tidal current bedload transport
3. Current ripples (Fig. 20–11) Current ripples superimposed at 90° on current ripples Current ripples superimposed at 90° on cross-strata Interference ripples B-C sequences of micro cross-laminae overlying cross-strata	Late-stage emergence outflow and emergence with sudden changes in flow directions at extremely shallow water depths (less than 2.0 meters)
4. Cross-strata with flasers (Fig. 20–4) Simple flaser bedding Wavy bedding Isolated, thin lenticular bedding Tidal bedding (Figs. 20–9, 20–10)	Alternation of tidal current bedload transport with suspension deposition during slack-water periods
5. Flaser and lenticular bedding (Fig. 20–4)	Tidal slack-water mud deposition
6. Mudchip conglomerates at base of washouts	Tidal scour
7. Mud cracks Runzel marks (Fig. 20–7)	Exposure
8. Tracks and trails *Monocraterion* escape burrows (Figs. 20–5, 20–6) Burrows (Fig. 20–8)	Burrowing
9. Load casts Pseudonodules Convolute bedding	Differential compaction and loading due to rapid deposition
10. Paleotidal range sequences	High rate of prograding tidal flat sedimentation

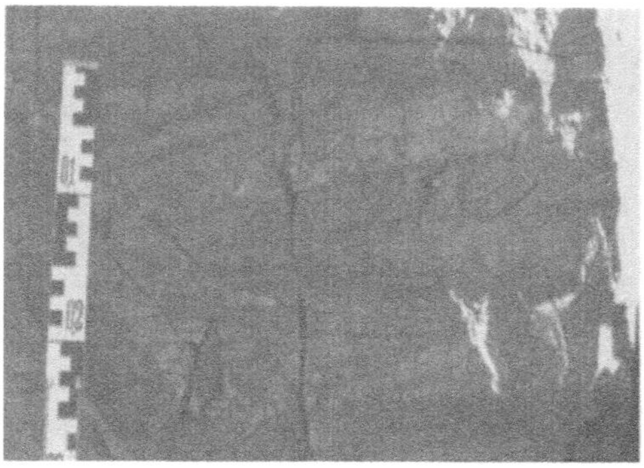

Figure 20-3. Cross-stratification with simple flaser bedding and reactivation surfaces, Bed 74, Salt Spring Hills, California. Scale in centimeters.

Figure 20-4. Cross-stratification with simple flaser bedding, Bed 72, Salt Spring Hills, California. Scale in centimeters.

Figure 20-5. Monocraterion escape structures, intersecting cross-strata with simple flaser bedding, Bed 86, South Nopah Range, California. Scale in centimeters.

Figure 20-6 (right). *Monocraterion* escape structures, Bed 194, Striped Hills, Nevada. Scale in centimeters.

on larger current ripples (Klein, 1970a, 1970b), flaser (Fig. 20–4) and lenticular bedding (Reineck and Wünderlich, 1968); *Monocraterion* escape structures (Figs. 20–5, 20–6), which occur in the lowermost part of the Middle Member; and the preservation of paleotidal range sequences (Klein, 1971a). The incorporation of mud cracks and runzel marks (Fig. 20–7) (Reineck, 1969) in claystones and interbedded thin sandstones, mudstones, and siltstones confirms exposure of these rocks during deposition. Associated burrows and tracks and

Figure 20-7 (about left). Runzel marks, Bed 17, Middle Member, Salt Spring Hills, California. Scale in centimeters.

Figure 20-8 (above right). Burrowed surface, Bed 16, Salt Spring Hills, California. Scale in centimeters.

trails indicate this exposure to occur in a tidal flat for most of the Middle Member.

Discussion

The Middle Member of the Wood Canyon Formation appears to be one of many Late Precambrian tidalites occurring within stable platform successions. Other Late Precambrian examples documented by several workers include the Lower Fine-grained Quartzite of Scotland (Klein, 1970b), the Late Precambrian Nyborg and Stappogiedde Formations of Norway (Reading and Walker, 1966; Banks, 1970), and the quartzites of Telemark, Norway (Singh, 1969). The Lower Fine-grained Quartzite shows many features in common with the Middle Member of the Wood Canyon Formation, whereas the quartzites at Telemark show more in common with the sandstone beds of the Wood Canyon only. It would appear that the stable platform and mioclinal sea environment of the Precambrian favors the development and preservation of intertidalites and tidalites.

Figure 20-9 (below left). Tidal bedding, Bed 27, Middle Member, Salt Spring Hills, California. Scale in centimeters.

Figure 20-10 (below right). Tidal bedding and flat, isolated lenticular bedding, Bed 18, Middle Member, Salt Spring Hills, California. Scale in centimeters.

Figure 20-11. Current ripples with *Diplocraterion*-type burrows, Bed 16, Middle Member, Salt Spring Hills, California. Scale in centimeters.

References

BANKS, N. L. 1970. Trace fossils from the late Precambrian and lower Cambrian of Finnmark, Norway. *In* Crimes, T. P., and Harper, J. C., eds., *Trace Fossils*. Liverpool, Seel House Press, pp. 19–34.

KLEIN, G. DEV. 1970a. Depositional and dispersal dynamics of intertidal sand bars. *J. Sed. Petrol. 40*, 1095–1127.

——— 1970b. Tidal origin of a Precambrian quartzite—the Lower Fine-grained Quartzite (Dalradian) of Islay, Scotland. *J. Sed. Petrol. 40*, 1095–1127.

——— 1971. A sedimentary model for determining paleotidal range. *Geol. Soc. Am. Bull. 82*, 2585–2592.

——— 1972a. A sedimentary model for determining paleotidal range: Reply. *Geol. Soc. Am. Bull. 83*, 539–546.

——— 1972b. Determination of paleotidal range in clastic sedimentary rocks. *XXIV Intern. Geol. Congr. Proc. Sec. 6*, 397–405.

READING, H. G., and WALKER, R. G. 1966. Sedimentation of Eocambrian tillites and associated sediments in Finnmark, North Norway. *Paleogeog. Paleoclimatol. Paleoecol. 2*, 177–212.

REINECK, H.-E. 1963. Sedimentgefüge im Bereich der Südliche Nordsee. *Abhandh. Senckenb. Naturf. Gesell. 505*, 1–138.

——— 1969. Die enstehung von Runzelmarken. *Natur. Mus. 97*, 193–197.

REINECK, H.-E., and WÜNDERLICH, F. 1968. Classification and origin of flaser and lenticular bedding. *Sedimentology 11*, 99–104.

SINGH, I. B. 1969. Primary sedimentary structures in Precambrian quartzites of Telemark, southern Norway and their environmental significance. *Norsk. Geol. Tids 49*, 1–31.

STEWART, J. H. 1970. Upper Precambrian and lower Cambrian strata in the southern Great Basin, California and Nevada. *U.S. Geol. Sur. Prof. Paper 620*, 206 pp.

21

A Pennsylvanian Interdistributary Tidal-Flat Deposit

Glenn S. Visher

Setting

The outcrop is located in SW-1/4 NW-1/4 NW-1/4 Sec. 20, T. 7 N., R. 19 E., Haskell County, Oklahoma. It occurs in a spillway, and approximately 16 meters of section is exposed (Visher, 1968). Extensive bedding surfaces are exposed in the middle and upper portions of the section. The Bluejacket-Bartlesville section is underlain by shales and thin sandstones of the Savannah Formation, and is overlain by the shales, coal, and sandstones of the Boggy Formation.

Occurrence

The Middle Pennsylvania depositional framework is one of deltaic progradation from the north (Visher, 1971). The Bluejacket-Bartlesville sandstone of Desmoinesian age is one of these deltaic complexes. This sandstone covers some 10,000 sq mi, and is composed of deposits that represent a wide variety

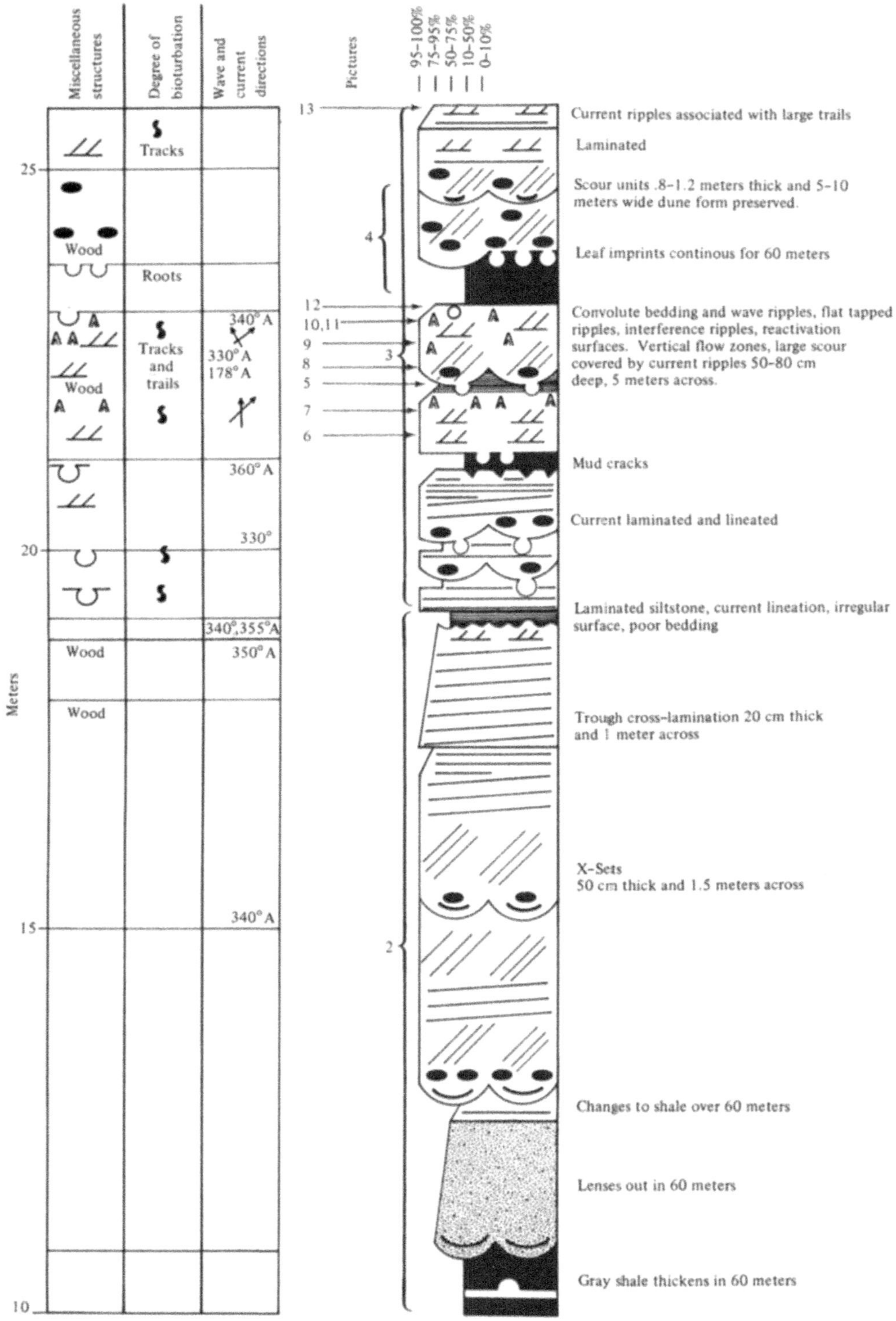

Miscellaneous structures
Degree of bioturbation
Wave and current directions
Pictures
95–100%
75–95%
50–75%
10–50%
0–10%
Meters
25
20
15
10
Tracks
Wood
Roots
Tracks and trails
Wood
340° A
330° A
178° A
360° A
330°
340°,355°A
Wood
350° A
Wood
340° A
13
12
10,11
9
8
5
7
6
4
3
2
Current ripples associated with large trails
Laminated
Scour units .8–1.2 meters thick and 5–10 meters wide dune form preserved.
Leaf imprints continous for 60 meters
Convolute bedding and wave ripples, flat tapped ripples, interference ripples, reactivation surfaces. Vertical flow zones, large scour covered by current ripples 50–80 cm deep, 5 meters across.
Mud cracks
Current laminated and lineated
Laminated siltstone, current lineation, irregular surface, poor bedding
Trough cross-lamination 20 cm thick and 1 meter across
X-Sets
50 cm thick and 1.5 meters across
Changes to shale over 60 meters
Lenses out in 60 meters
Gray shale thickens in 60 meters

Figure 21-1 (opposite). Vertical section of the outcrop showing the lithologies and sedimentary structures. The percent sand in each bed is indicated by the scale at the top.

of sedimentary environments. Within a well-defined time-stratigraphic interval, successive periods of progradation produced separate, elongated deltaic lobes that together form a complex of sandstones. Within each interval, the initial sand deposits were modified as the deltaic complex prograded; pro-delta sediments were overlain by deltaic, marginal marine, and shoreline deposits. Locally, the complex was reworked by distributary or tidal channels.

On the southern margin of the subaerial part of the deltaic complex, the environments are dominantly marine and associated with distal pro-delta environments of the outer continental shelf. The outcrop described in this paper is a marginal marine unit separating the dominantly distributary and interdistributary units from the dominantly marine pro-delta units.

Vertical Sequence

The sequence consists of two genetic units: a basal unit 13 meters thick separated by 10 cm of light-colored shale from an upper unit 2 meters thick (Fig. 21–1). For the following description, the location of the photographs is indicated on the section of Figure 21–1.

The basal part of the outcrop has large-scale, cross-bed sets, with scour surfaces separating depositional units (Fig. 21–2). The associated shales are gray, with thin, fine-grained sand, and silt layers; current ripples are rare and wave ripples absent. The scale of sedimentary structures decreases upward and the associated structures include current lineation and lamination and

Figure 21-2 (left). View of parallel-bedded and festoon cross-bedded zone.

Figure 21-3 (right). Large scour surfaces covered by current ripples.

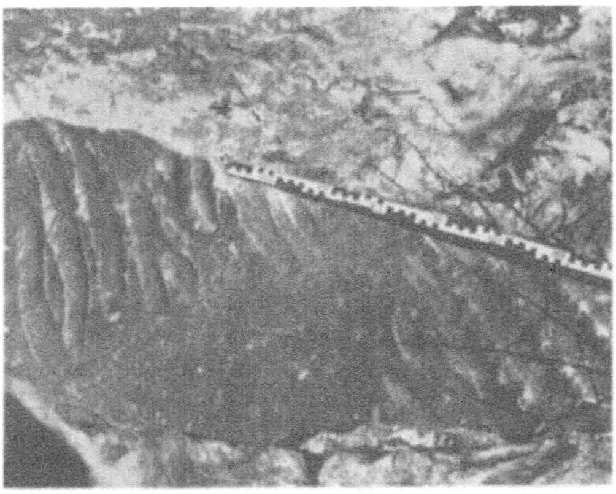

Figure 21-4. Rippled surface truncated by scour.

Figure 21-5. Interference ripples (sometimes called rhomboid ripples).

Figure 21-6. Burrowed surface with a large lebenspuren.

Figure 21-7. View of rippled and laminated upper zone. Center of photograph shows a scour overlain by light-colored shale.

low-angle, trough cross lamination. Overlying this basal part is a zone in which the predominant structures are current and wave ripples (Figs. 21–3 and 21–4), interference ripples (Fig. 21–5), reactivation surfaces, flat-topped ripples, and tracks and trails on bedding surfaces (Fig. 21–6). There is some tabular cross-bedding, a single large scour whose surface is covered by wave and current ripples (Fig. 21–3), and, locally, mud cracks. Overlying the rippled scour surface is a light-colored silty shale with plant impressions and possible root casts.

The upper unit of the outcrop has large-scale scour surfaces covered by current ripples (Figs. 21–7 and 21–8) like those of the basal unit; this zone of scour surfaces is laterally equivalent to beds with current lineation and parallel lamination.

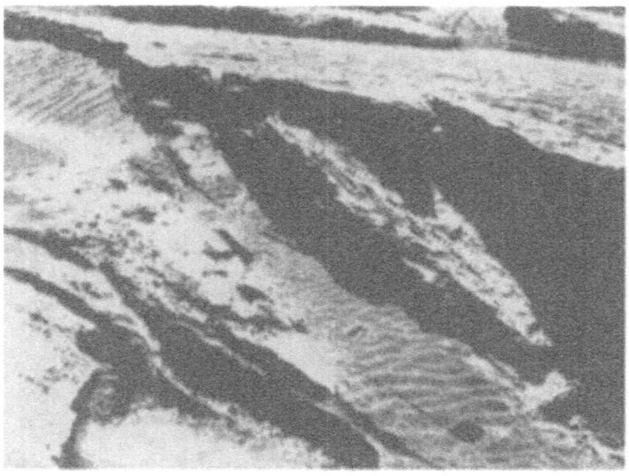

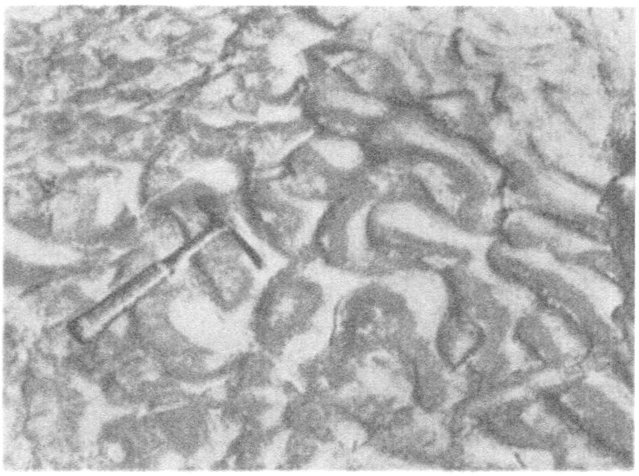

Figure 21-8. Intersecting scour surfaces covered by wave and current ripples.

Figure 21-9. Current rippled surface with troughs locally filled with detrital fecal pellets.

Figure 21-10 (left). Ladder-backed ripples.

Figure 21-11 (right). Oblique view of ladder-backed ripples showing truncated ripple crests.

Evidence for Tidal Origin

Tidal origin of the sequence is indicated by

1. The association of wave and current ripples, abundant tracks and trails, interference ripples, ladder ripples, flat-topped ripples (scuffed), and reactivation surfaces
2. Changing direction of wave ripples, and the general northerly trend of current structures and wave ripples, in opposition to the dominant northern source and transport direction of the Bluejacket-Bartlesville sandstone
3. The generally shallowing upward sequence with indications of subaerial exposure, that is, mud cracks, fecal pellet accumulations in ripple troughs, combination of current and wave characteristics, a possible marsh de-

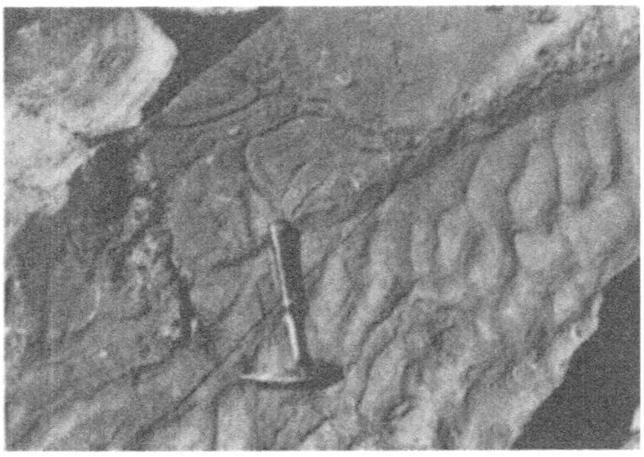

Figure 21-12 (left). Symmetrical wave ripples.

Figure 21-13 (right). Current ripples with large trails on a common surface. Some bioturbation.

posit with roots and carbonaceous plant impressions, and the association of current structures and scour units with indicators of marine environment (for example, wave ripples, and tracks and trails on bedding planes)

4. The heavy concentration of wood, plant debris, and shale clasts suggesting close proximity to a terrestrial source.

These combined factors indicate strandline deposition. Features of the 4-meter thick unit below the light-colored shale suggest very shallow water deposition, and the 2-meter thick unit between the mud-crack horizon and the shale is interpreted as intertidal. Figures 21-4, 21-6, and 21-9 to 21-13 illustrate structures common to the intertidal environment (Reineck, 1972).

Discussion

The vertical sequence suggests a tidal channel, probably an abandoned distributary, filled by tidal-current deposits. Lower units were deposited in a channel 10 meters deep, and successive units were deposited as channel fill. The top portion is interpreted as intertidal channels associated with the marine fringe of the deltaic complex. The 70-cm thick unit at the top of this portion of the section may represent upper tidal flat or coastal marsh deposition. The uppermost 2 meters could have been deposited in shallow water by tidal currents in channels developed on the marsh following compaction and subsidence of the lower units. This unit is possibly another genetic unit, and does not represent the supratidal, intertidal, and channel-fill sequence of environments developed in the lower portion of the outcrop .

The paleogeography at the time of the channel filling was possibly an inlet into a marginal bay or lagoon, or an abandoned distributary on the southern margin of the Bluejacket-Bartlesville deltaic complex, a situation similar to a modern delta with a moderate tidal range.

References

REINECK, H.-E. 1972. Tidal flats. *In* Rigby, J. K., and W. K. Hamblin, eds., *Recognition of Ancient Sedimentary Environments*. Tulsa, Okla. Soc. Econ. Paleontol. Mineral. *Spec. Publ. 16*, 146–159.

VISHER, G. S. 1968. Geology of the Bluejacket-Bartlesville Sandstone, Oklahoma, Oklahoma City, Oklahoma. Oklahoma City Geol. Soc., 72 pp.

———, SAITTA B., and PHARES, R. S. 1971. Pennsylvanian delta patterns and petroleum occurrences in eastern Oklahoma. *Am. Assoc. Petrol. Geol. Bull. 55*, 1206–1230.

22

Sedimentary Sequences of Lower Devonian Sediments (Uan Caza Formation), South Tunisia

Antonio Rizzini

Occurrence

In wells of the Tunisian Djeffara (south Tunisia) a Paleozoic sequence ranging from Silurian (Gothlandian) to Permian has been recognized (Furon, 1964). This sequence can be correlated with that outcropping in the Libyan Fezzan, the type localities of the formations (Desio, 1936; Sacal, 1963; Burollet, 1967; Klitzsch, 1969) (Fig. 22–1).

This paper deals with cores of the Ech Chouech No. 2 well (EC–2; lat. 31°02′28″ N; long. 9°28′09″ E) from 3,460 to 3,485 meters, which is part of the Uan Caza Formation of Lower Devonian (Emsian) age (Fig. 22–2). To frame the environment of deposition of the Uan Caza Formation, one has to start with the Siegenian, during which time a new depositional cycle was initiated. The new cycle followed the regressive marine sandstones of the Acacus Formation with a

Thanks are due to the managements of AGIP AMI, and SITEP for permission to publish the data of well EC-2 and to Walter O. Gigon of AGIP-Shell for critically reading the manuscript.

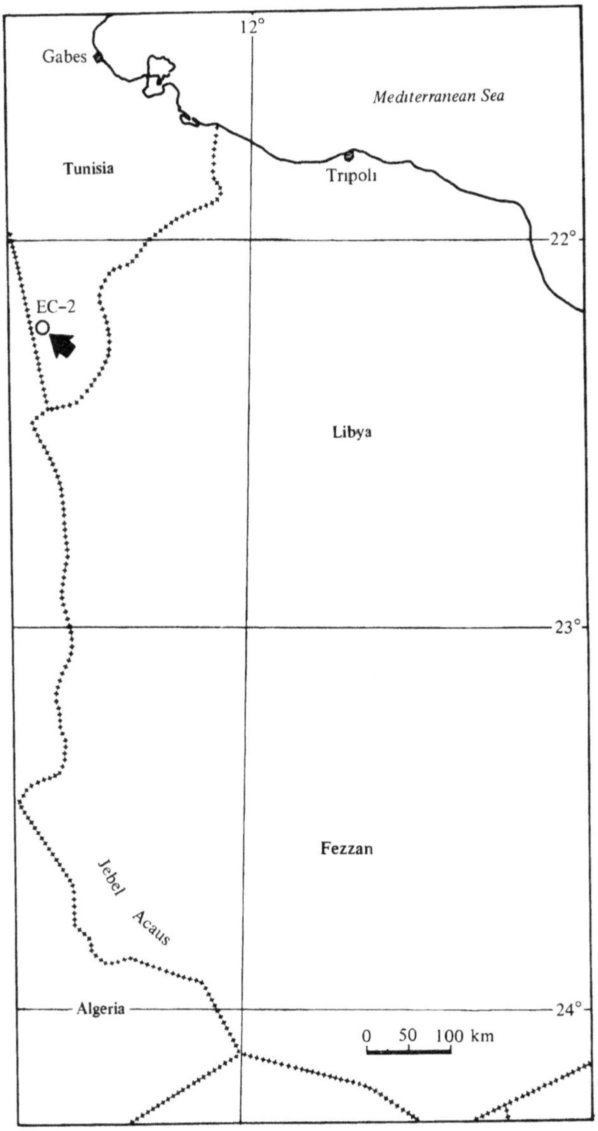

Figure 22-1. Location map.

thick cover of continental sandstones (Tadrart Formation) that reflects the Ardennian phase of the Caledonian orogeny. The Tadrart sandstones form a transition in the Emsian to shallow marine facies consisting of gray shales with siltstones and sandstones containing brachiopods, tentaculites, and styliolines (Uan Caza Formation). Sedimentation then continued without great variation for the rest of the Devonian, a period of tectonic quiescence. The marine sandstones and shales of the Aouinet Ouenine Formation were laid down at this time.

Facies

The Uan Caza Formation in well EC–2 consists of irregular alternations of whitish sandstones and silts with blackish shales (Fig. 22–3). The sandstones are quartzose, fine grained with siliceous cement, and in the finer fractions they have dolomite and argillaceous cements. There are also thin beds (1 to 3 cm thick) or nodules of microcrystalline siderite. This siderite and the associated ferruginous oolites weather rapidly. The argillaceous intercalations are rich in silt and are composed of mixtures in equal parts of chlorite, kaolinite, and illite with about 10 percent mixed layers illite-montmorillonite. The sandy and argillaceous alternations appear in various rhythmic sequences. The complete ideal sequence is composed, from bottom to top, of the following intervals (Fig. 22–4):

1. Thin, uniform beds, 2 to 5 cm thick, of fine shell conglomerate consisting of rounded brachiopod fragments and shale flakes with arenaceous matrix (Fig. 22–5 and 22–6). The base of the bed is erosional.

2. The transition from the conglomerates to the overlying sandstone beds is often gradual (Fig. 22–5), but it can also be erosional. Single sandstone beds, up to 30 cm thick, occur separately or in groups up to 2 meters thick. Within these groups, but always at the base of a single bed, are sometimes shell conglomerates and layers with shale flakes (Figs. 22–9 and 22–10). The contacts between the sandstone beds are usually erosional, but they can also be gradual with an intercalation of some centimeters of shale (Fig. 22–10). In general, the sandstone beds show a massive structure or planar cross bedding; there are also parallel laminations and trough cross-bedding, especially in the thinner beds. There is a tendency to have graded bedding within each bed (Figs. 22–7, 22–9, and

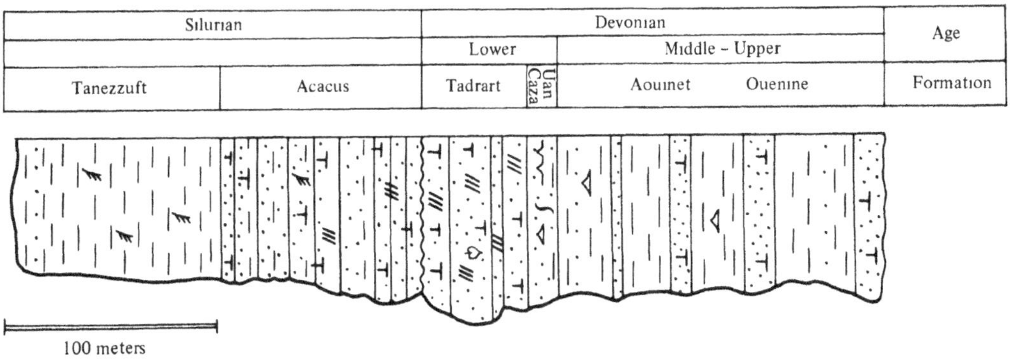

Figure 22-2. Composite Paleozoic section in Fezzan (Libya).

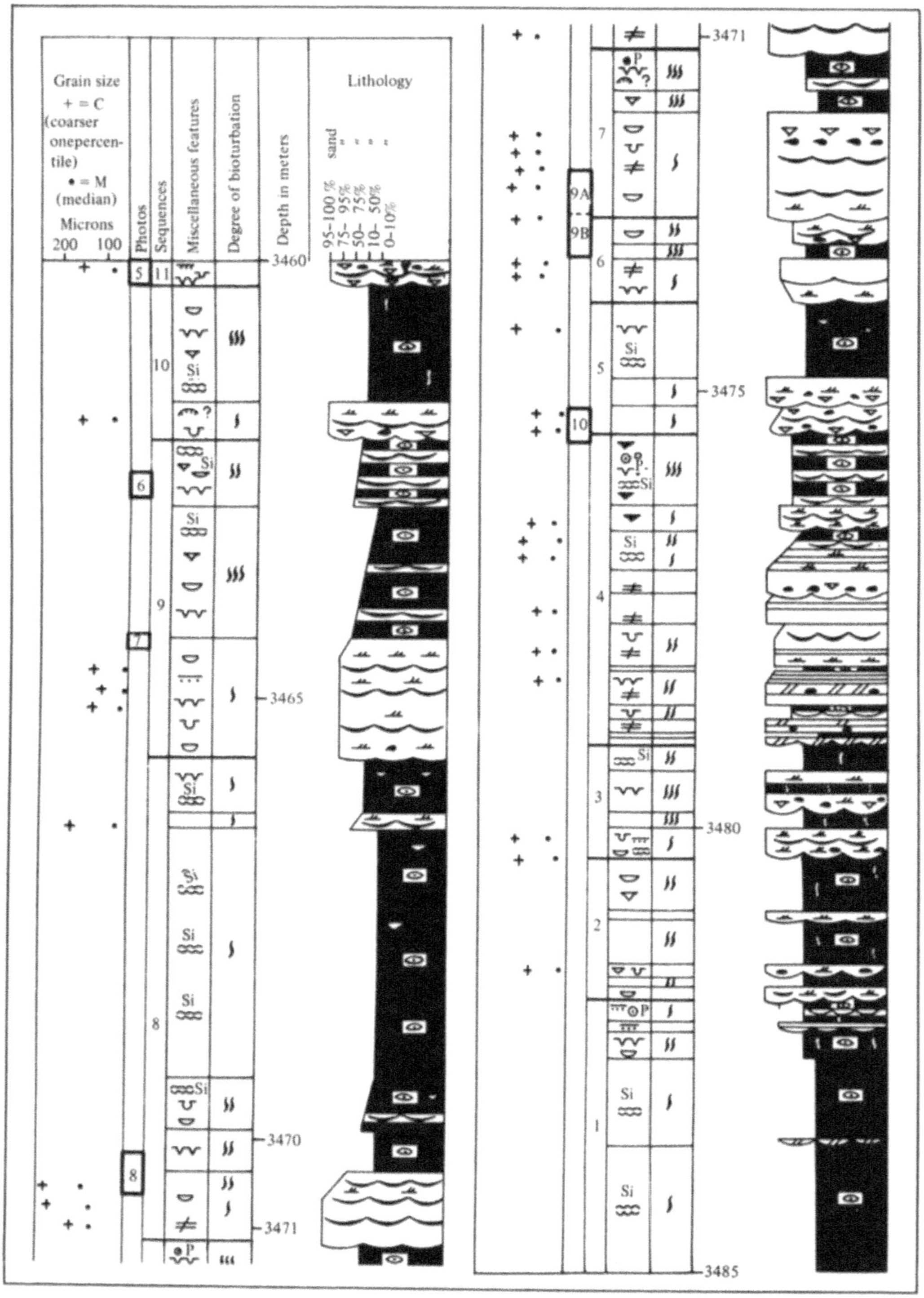
Grain size
+ = C
(coarser
onepercen-
tile)
• = M
(median)
Microns
200 100
Photos
Sequences
Miscellaneous features
Degree of bioturbation
Depth in meters
Lithology
95–100 % sand
75– 95%
50– 75%
10– 50%
0–10%
3460
3465
3470
3471
3475
3480
3485
9A
9B
10
Si
P

22–10). The erosion phenomena at the base of the beds are often emphasized by load casts. Bioturbation is conspicuous, especially in the more argillaceous parts.

3. Thin sandstone beds alternate with shaly beds a few centimeters thick, and the unit tends to become more shaly toward the upper part. The sandstones are characterized by planar and trough cross bedding and some flaser bedding. The sandstones, commonly graded, are much more churned by organisms than the ones in the previous intervals. In some layers the activity of the mud-dwellers has destroyed most of the preexisting sedimentary structures (lower part of Fig. 22–8, Fig. 22–9*B*). Mud cracks (Fig. 22–9*A*), nodules, and thin layers of siderite are common. Shales occur as drapes on some sandstone ripples. The fossils, particularly within the shales, are less broken but they may be present also in the shell conglomerate beds (Fig. 22–6).

4. Shales with flaser-bedded silt are completely churned by mud-dwellers, and it becomes difficult to recognize original sedimentary structures (upper part of Fig. 22–8). There are still mud cracks, nodules, and thin layers of siderite, flaser, and lenticular bedding in the silts; fossils are mostly whole within the shales. There are some structures in the silty fractions that could be interpreted as stromatolites.

5. Nearly pure shales with little churning, rare silt laminae rich in bedded and nodular siderite, and with well-preserved fossils. The silts show lenticular bedding.

Figure 22-3 (opposite). Section, well EC-2, depth 3,460 to 3,485 meters. *P*, pyrite; *Si*, siderite.

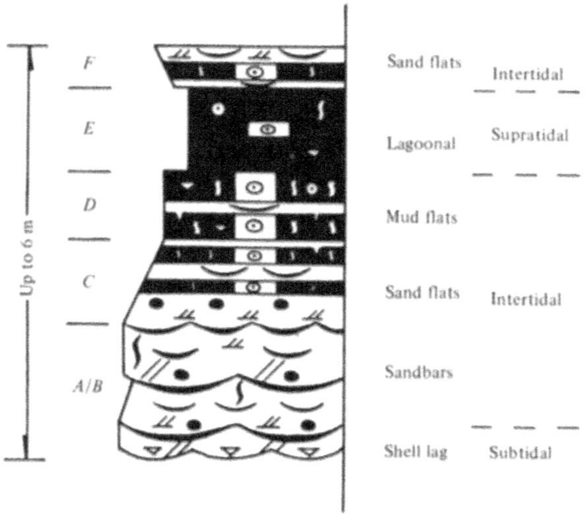

Figure 22-4. Idealized tidal cycle of Uan Caza Formation.

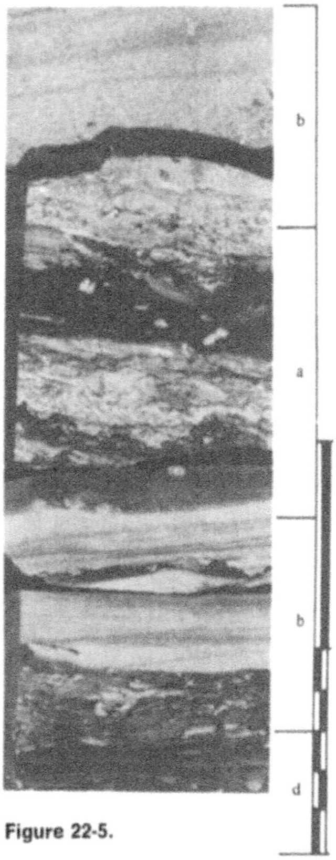

Figure 22-5.

Figures 22-5 to 10. Slabbed cores of the well EC-2, the position of the photos in the cored section is shown in Fig. 22-3. Letters on the right side of photos refer to the intervals of the tidal sequence. Photos are described in the text. Scale in centimeters.

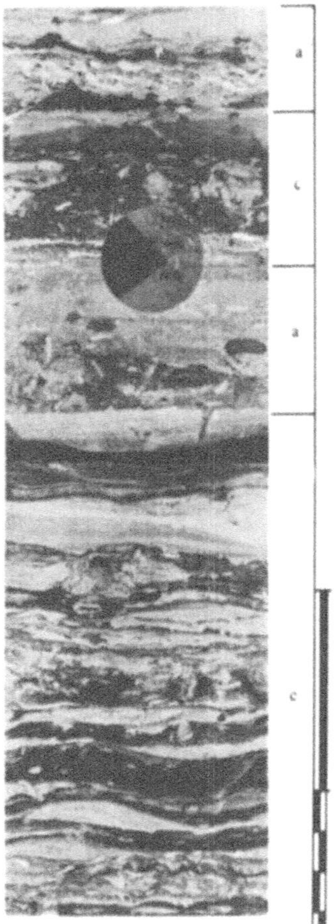

Figure 22-6.

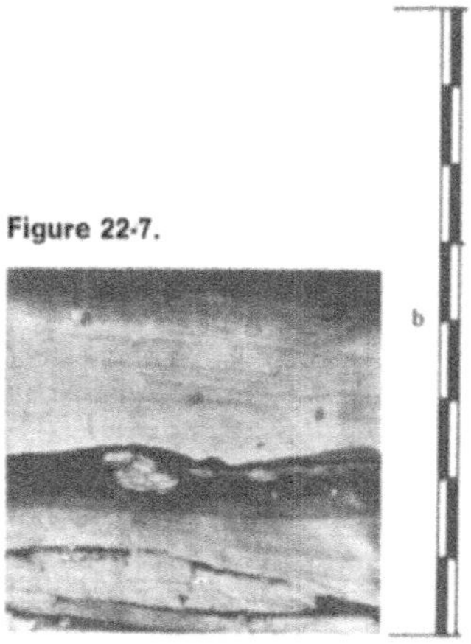

Figure 22-7.

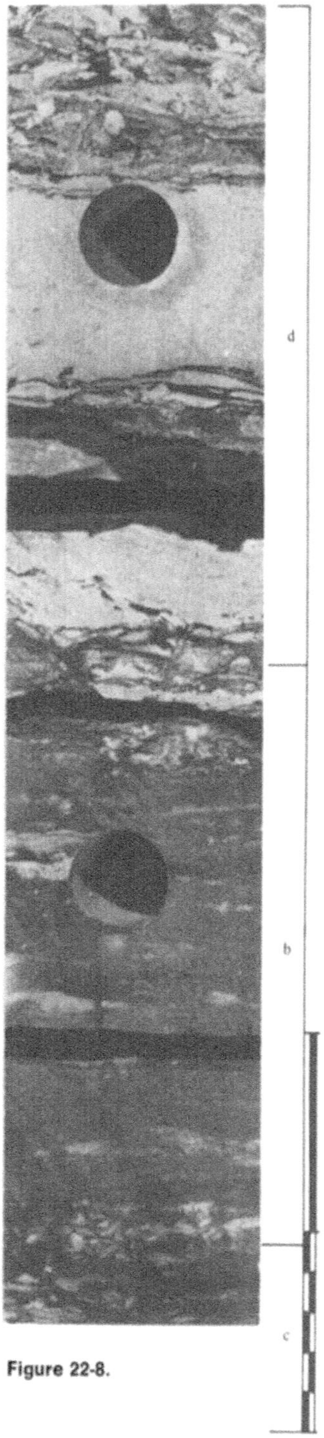

Figure 22-8.

Figure 22-9A.

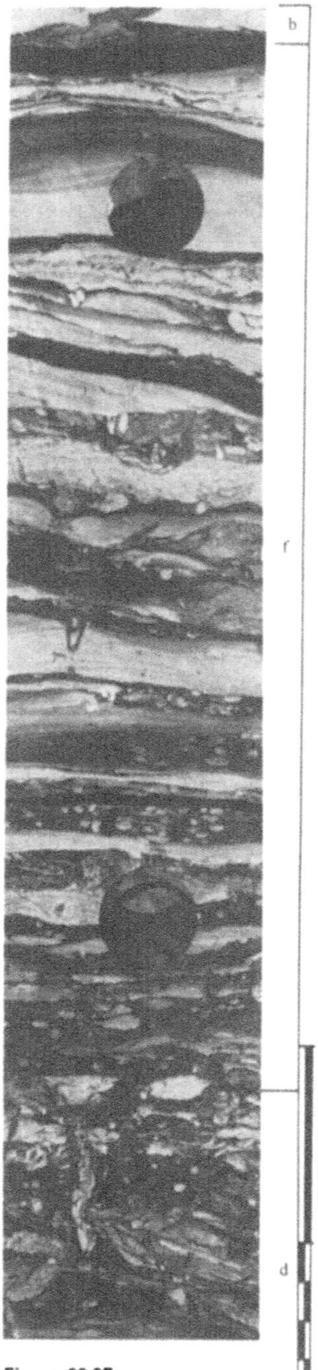

Figure 22-9B.

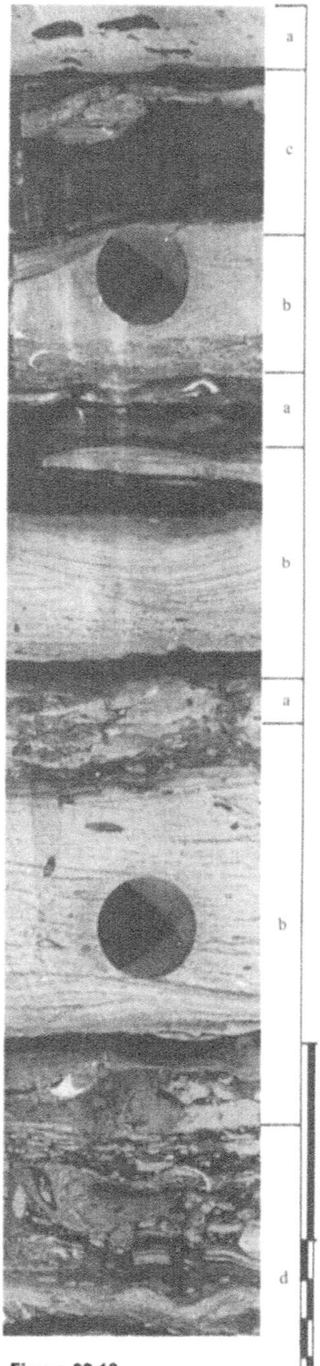

Figure 22-10.

6. A facies analogous to (3) in lithology and sedimentary structures but is coarsening upward instead of fining (Fig. 22–9*B*).

The type of section described above is not always complete: only two sequences (1 and 8 of Fig. 22–3) contain facies (5); five sequences of the ten examined have at the top the interval coarsening upward (6); only one sequence, 8, has no shell conglomerate but it has the thickest interval (5), that is, 2.8 meters. The minimum thickness of a sequence is 0.95 meters (sequence 6); the maximum is 5.5 meters (sequence 8); and the average thickness is 2.5 meters. Considering the entire 25 meters examined, one recognizes lithologic variations of a higher order than the ones described above. The lower part has predominantly argillaceous sediments (sequences 1 and 2) and in the central part the sandstones prevail (sequences 3 to 7, lower part 8); then there is an increase of shales (middle-upper part sequence 8) and at the top the mixed facies prevail (sequences 9 to 11).

Evidence for Tidal Origin

This study has been carried out on cores, and therefore the orientation of the sedimentary structures, the lateral variations, and the typical surface structures of the beds could not be observed. The presence of rhythmic sequences fining upward with erosional contacts, shell lag, and mud-pebble conglomerates at the base followed by uniform or planar cross-bedded sandstones indicates clearly that these were deposited by tractional currents in a marine environment. The second part of the sequence is characterized by alternations of sandstone beds with small trough cross-bedding, shales with flaser and lenticular bedding (Reineck and Wünderlich, 1968), burrowing structures, mud cracks, and well-preserved fossils; these structures indicate alternations of subaerial exposure with deposition from tractional currents and slack water. In addition there were periods of precipitation of sideritic muds and ferruginous oolites that occur normally in environments near the coast. The purely shaly intervals (Interval 5), are little churned, rich in siderite nodules, and without tractional current deposits or mud cracks. They can be interpreted as deposits of small saline lagoons that formed on tidal flats at the end of the regressive cycles.

Therefore it appears that these cycles could be attributed to tidal fluctuations.

Discussion

The sequences are similar to many siliclastic tidal sequences already described in the literature (Straaten, 1954; Reineck, 1963; Klein, 1963; Evans, 1965), but are particularly interesting because of two additional characteristics. The more important one is the presence in half the sequences of the coarsening-upward interval (6) within the tidal fining-upward sequence. The fact that interval 6 is at the end of the sequence, and therefore directly underlying the thicker sandstone beds of the following sequence may be the result of sand deposited by traction currents that increase in intensity. This increase could be related to the phenomenon that tidal flats receive less sediments the higher they are and then tend to subside, as a result of the general subsidence of the area, until they are again invaded by the sea and a new cycle starts gradually.

Also, the presence of interval 5, that is, the existence of small iron-concentrating lagoons in the higher tidal flats, is noteworthy.

References

BUROLLET, P. F. 1967. Sédimentologie du Dévonien inférieur en Libye. *Mem. Bureau Recher. Geol. Min. 33*, 205–213.

DESIO, A. 1936. Riassunto sulla costituzione geologica del Fezzan. *Boll. Soc. Geol. Ital. 55*, 319–356.

EVANS, G. 1965. Intertidal flat sediments and their environments of deposition in The Wash. *Geol. Soc. London Quart. J. 121*, 209–245.

FURON, R. 1964. *Le Sahara, géologie, ressources minerales.* Paris, Payot, 313 pp.

KLEIN, G. DEV. 1963. Bay of Fundy intertidal zone sediments. *J. Sed. Petrol. 33*, 844–854.

KLITZSCH, E. 1969. Stratigraphic section from the type areas of Silurian and Devonian strata at western Murzuk Basin (Libya). *In* Kanes, R., ed., *Geology, Archeology and Prehistory of the Southwestern Fezzan, Libya.* Tripoli, Petrol. Expl. Soc. Libya, pp. 83–90.

REINECK, H.-E. 1963. Sedimentgefüge im Bereich der Südliche Nordsee. *Abhandl. Sencken. Nat. Gesell. 505*, 1–138.

REINECK, H.-E., and WÜNDERLICH, F. 1968. Classification and origin of flaser and lenticular bedding. *Sedimentology 11*, 99–104.

SACAL, V. 1963. Microfaciès du Paléozoique saharien. *Compagnie Francais Petroles Notes Memoires 6*, 1–30.

STRAATEN, L. M. J. U. VAN. 1954. Composition and structure of recent marine sediments in the Netherlands. *Leids. Geol. Mededel. 19*, 1–110.

Section III

Recent Carbonate Examples

By the early 1900s, if not before, it was generally accepted that many fossil carbonate sediments accumulated in water so shallow that they were often exposed.* This simplistic view prevailed until the 1950s, when the change began toward the present conception of tidal-flat deposition. The singular characteristic of the rapid developments over the past twenty years is the continuing interaction between research on Recent and on ancient examples.

An early advance in specifying the environment of deposition was the realization that many algal stromatoids† are guide structures for the intertidal zone. This conclusion came initially from the observed range of modern algal-laminated sediments in the Bahamas (Black, 1933) and in Florida (Ginsburg et al., 1954), but it was supported by the association of fossil stromatoids with evidence of intermittent exposure, desiccation cracks, and intraformational conglomerates. The subsequent discoveries of subtidal stromatoids (Playford and Cockbain, 1969; Hoffman, 1974) indicate that not all fossil algal stromatoids are restricted to the intertidal zone; there are, however, many growth forms that remain reliable guides to the intertidal zone.

In the 1960s a series of major advances in understanding tidal-flat deposition in carbonates was characterized by the mutual stimulation of research on Recent and on ancient examples: Fine-grained dolomite so common in ancient shoaling carbonates was discovered on the surfaces of three widely separated modern tidal flats (see papers in Pray and Murray, 1965); the intertidal origin of birdseye voids (loferites, fenestral pores) proposed by Fischer (1964) from his study of the alpine Triassic was confirmed in Recent sediments by Shinn (1968); a simple geometric classification of algal stromatoids related variations in growth form to subenvironments of the tidal zone (Logan et al., 1964;) modern nodular evaporites, both gypsum and anhydrite, which closely resemble common fossil occurrences, were discovered on the supratidal flats, sabkha, of the Persian Gulf (Curtis et al , 1963; Kinsman, 1966); synsedimentary cementation of intertidal and subtidal sediments in the Persian Gulf (Taylor and Illing, 1969; Shinn, 1969) and in the Bahamas (Shinn et al., 1965) limits erosion and leads to the formation of intraclasts like those that are so common in ancient intraformational conglomerates (flat-pebble and edgewise conglomerates). It is interesting to note in retrospect that the interpretation of the environments of these structures derived from fossil examples could have been used as a kind of treasure map to find their recent analogs.

Each of the above discoveries of the recent analog of a common fossil sedimentary feature added another potential criterion for identifying ancient tidal-flat deposits. There is, however, a natural tendency to exaggerate the reliability of an individual feature, particularly in the early stages of application. All laminated carbonates are not necessarily stromatolitic and thus indicative of the intertidal zone; laminated evaporites may accumulate in environments other than sabkhas (Davies and Ludlam, 1973); there is reason to suspect that fenestral voids (birdseyes, loferites) may develop in subtidal sediments. The way to avoid the expectable uncertainties of interpretations based on individual sedimentary features is to use an association of features or, better still, a vertical sequence of features like that in siliciclastic examples.

The first contribution in Section III, by Ginsburg and Hardie, gives the vertical sequence of sedimentary features from a Bahaman tidal flat; the fluctuations in sea level are expressed by the percent time exposed for all levels (see also Larsonneur, Chapter 3). The second contribution on a tidal flat in the Persian Gulf by Schneider provides another example of the tidal zonation of sediments, structures, organisms, and diagenetic minerals; here the intertidal and subtidal structures and sediments resemble those in the Bahaman example, but the supratidal deposits rich in evaporites are entirely different, the result of an arid climate. The third contribution, by Hagan and Logan, summarizes the lateral and vertical sequences of a stunning array of sediments and sedimentary structures from a hypersaline embayment of Shark Bay, Western Australia; the coquinas and muddy skeletal sands of the lower intertidal zone, the cross-bedded ooid sands and molluskan coquinas of the supratidal zone, and the rich variety of algal-laminated and cryptalgal structures of the intertidal zone are all common features of ancient carbonates. The final contribution, by Woods and Brown, describes the tidal-flat deposits in another hypersaline embayment of Shark Bay. In this example protection from the prevailing wind favors the formation of

* A statement by Rice (1915) is typical of this attitude: "The study of the sedimentary rocks which cover our existing continents shows that almost all of them were deposited in shallow water; many of the strata, indeed, in waters so shallow that the layers of mud and sand were from time to time exposed by the receding tide of the subsiding freshet, to dry and crack in the sun or to be pitted by raindrops."

† As pointed out by Hoffman (1973), stromatoid is the correct term for individual structures, stromatolite for a rock in which they are abundant.

well-cemented crusts near high water, crusts which are broken by wave action to produce extensive intraclasts in the intertidal and lower supratidal zones.

REFERENCES

Black, M. 1933. The algal sediments of Andros Island, Bahamas. *Phil. Trans. Roy. Soc. London Ser. B 222*, 165–192.

Curtis, R., Evans, G., Kinsman, D. J. J., and Shearman, D. J. 1963. Association of dolomite and anhydrite in the recent sediments of the Persian Gulf. *Nature 197*, 679–680.

Davies, G. R., and Ludlam, S. D. 1973. Origin of laminated and graded sediments, Middle Devonian of western Canada. *Geol. Soc. Am. Bull. 84*, 3527–3546.

Fischer, A. G. 1964. The Lofer cyclothems of the Alpine Triassic. *In* Merriam, D. F., ed., Symposium on Cyclic Sedimentation. *Kansas Geol. Surv. Bull. 169*, 107–151.

Ginsburg, R. N., Isham, L. V., Bein, S. J., and Kuperberg, J. 1954. Laminated algal sediments of South Florida and their recognition in the fossil record. *Marine Lab. Univ. Miami Rept. 54-21*, 33 pp.; available from author.

Hoffman, H. J. 1973 Stromatolites: Characteristics and utility. *Earth-Sci. Rev. 9*, 339–373.

Hoffman, P. 1974. Shallow and deepwater stromatolites in a Lower Proterozoic platform-to-basin facies changes, Great Slave Lake, Canada. *Am. Assoc. Petrol. Geol. Bull. 58*, 856–867.

Kinsman, D. J. J. 1966. Gypsum and anhydrite of Recent age Trucial Coast, Persian Gulf. *Second Symposium on Salt.* Cleveland, Northern Ohio Geol. Soc., Vol. 1, pp. 302–326.

Logan, B. W., Rezak, R., and Ginsburg, R. N. 1964. Classification and environmental significance of algal stromatolites. *J. Geol. 72*, 68–83.

Playford, P. E., and Cockbain, A. E. 1969. Algal stromatolites: Deepwater forms in the Devonian of Western Australia. *Science 165*, 1008–1010.

Pray, L. C., and Murray, R. C. eds. 1965. Dolomitization and limestone diagenesis. *Soc. Econ Paleontol. Mineral. Spec. Publ. 13*, 180 pp.

Rice, W. N. 1915. *In* Schuchert, C., ed., *Problems of North American Geology.* New Haven, Yale Univ. Press, p. 13.

Shinn, E. A. 1968. Practical significance of birdseye structures in carbonate rocks. *J. Sed. Petrol. 38*, 215–223.

———. 1969. Submarine lithification of Holocene carbonate sediments in the Persian Gulf. *Sedimentology 12*, 109–145.

Shinn, E. A., Ginsburg, R. N., and Lloyd, R. M. 1965. Recent supratidal dolomites from Andros Island, Bahamas. *In* Pray, L. C., and Murray, R. C., eds., Dolomitization and limestone diagenesis. *Soc. Econ. Paleontol. Mineral. Spec. Publ. 13*, 112–124.

Taylor, J. C. M., and Illing, L. V. 1969. Holocene intertidal calcium carbonate cementation, Qatar, Persian Gulf. *Sedimentology 12*, 69–109.

23

Tidal and Storm Deposits, Northwestern Andros Island, Bahamas

Robert N. Ginsburg and
Lawrence A. Hardie

Occurrence

Holocene tidal and storm deposits extend for 160 km along the western shores of Andros Island. The wedge of soft sediments, up to a maximum of 4 meters thick near the shoreline, accumulated in the last 5,000 years, either during or following the rise of sea level to its present position. The area studied is located near Point Simon, lat. 25° N, 78° W (see also Shinn et al., 1969).

Hydrography and Climate

Tides are semidiurnal but strongly influenced by wind direction and velocity. Mean spring range is 41 cm; mean neap range, 17 cm; and the extreme range (1969–1971), 95 cm. Sustained onshore winds of 15 knots or more increase the level of high waters and reduce the range; winds of similar velocity

This research was supported by National Science Foundation Grant GA 1345. We thank Owen Bricker, Peter Garrett, and Harold Wanless for their active collaboration in the field observations and for access to their unpublished results. Additional reports on specific aspects of the study are in preparation.

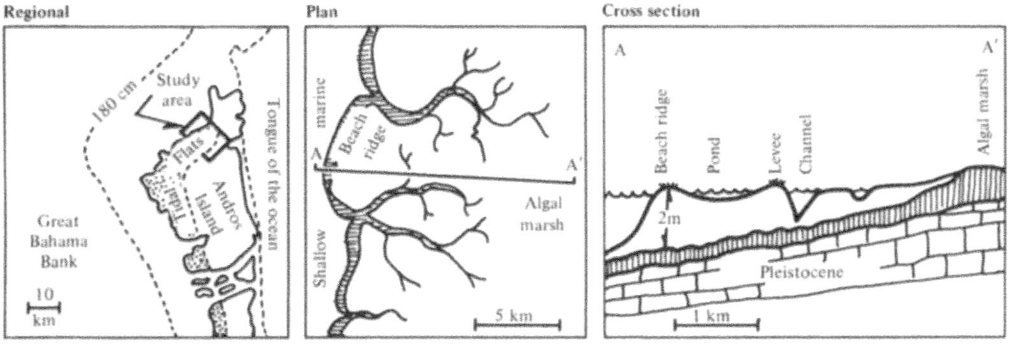

Figure 23-1. Setting, generalized plan, and cross section of study area.

blowing offshore produce extremely low waters and reduced range. The climate is subtropical, humid, with average annual rainfall of 114 cm, largely in thunderstorms from June to October. The prevailing easterly trade winds are gentle to moderate in spring and summer, but frequently fresh (>15 knots) in fall and winter. An average of 40 high pressure fronts per year from the continent brings strong winds from the north and northwest quadrants.

Figure 23-2. Physiographic-hydrographic subdivisions. Shinn et al. (1969) recognized three major subdivisions: (1) shallow marine—the inner part of Great Bahama Bank; (2) channeled belt—inward branching tidal channels, levees, and intervening ponds; (3) algal marsh exposed most of the year and flooded by fresh water during the rainy season.

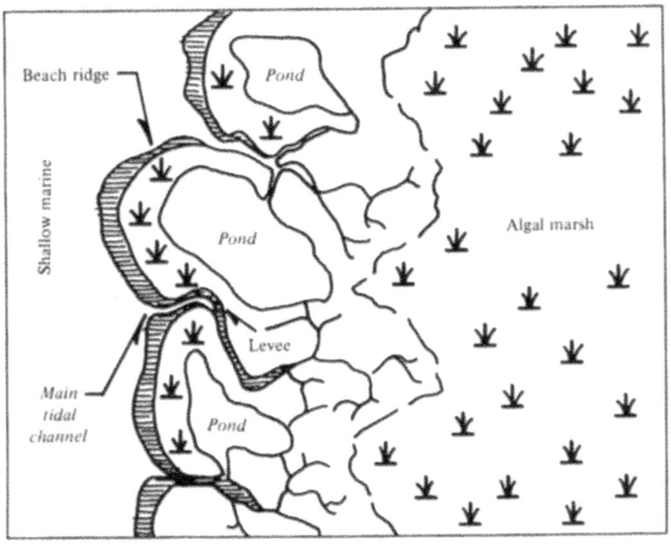

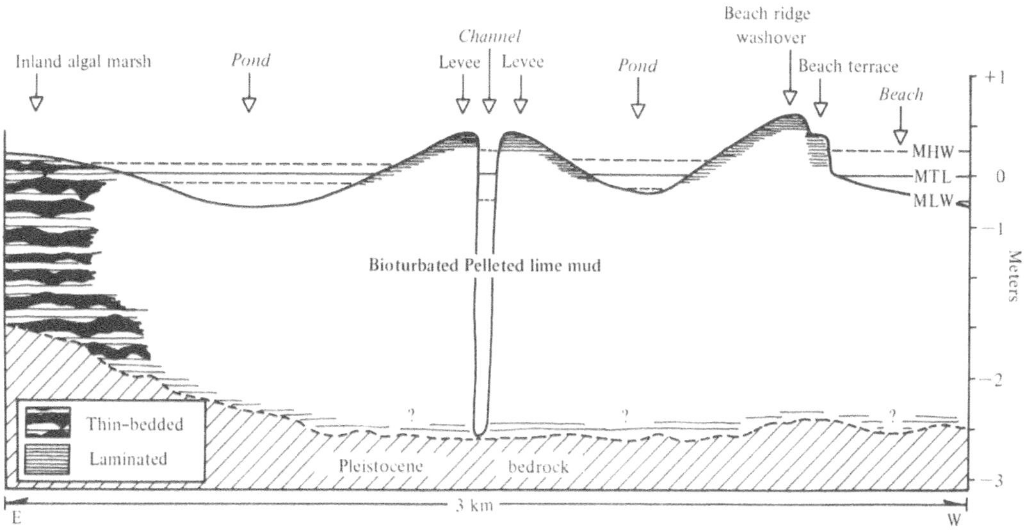

Figure 23-3. Generalized distribution of sedimentary structures, Three Creeks Area. *Right*, Great Bahama Bank; *left*, Andros Island.

Figure 23-4. At the left the curve shows the exposure index of levels within the channeled belt; at the right the columns show zonal distribution of the major sedimentary features.

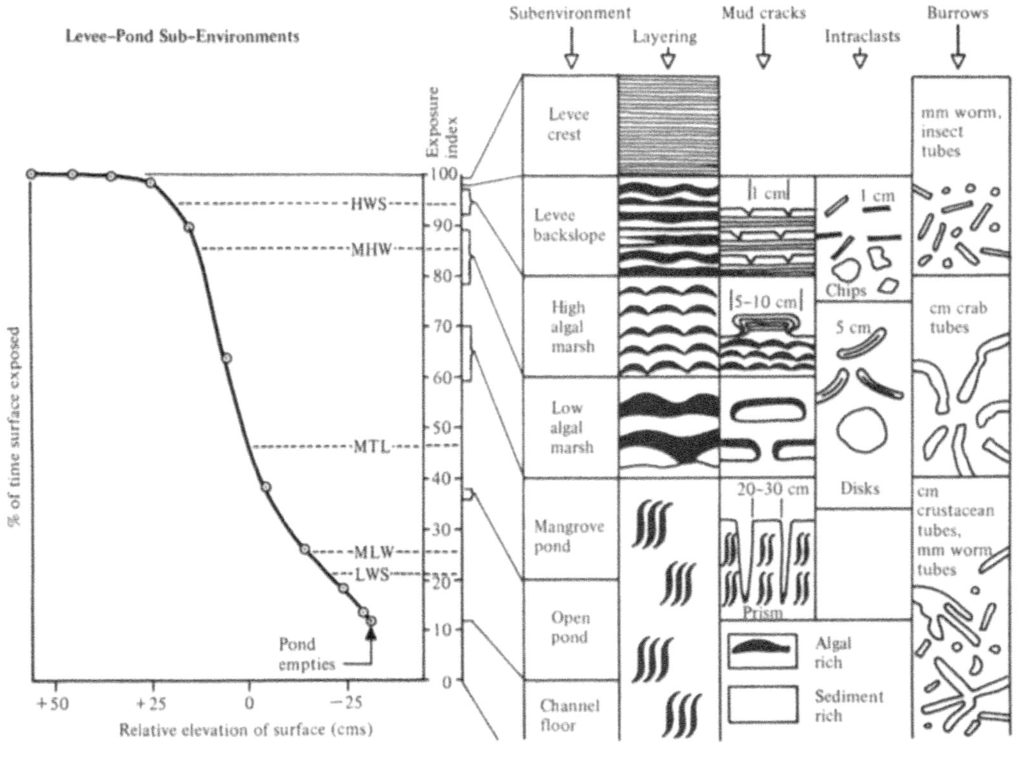

Figure 23-5 (top left). View looking across the crest of a natural levee along one of the main tidal channels, Three Creeks area. This zone is exposed more than 95 percent of the year (Fig. 23-4); it is flooded only as a result of persistent and strong onshore winds. The laminated structure of the levee shown here in the core held by Benjamin Lewis is seen closeup in Fig. 23-6.

Figure 23-6 (left). Section of a resin-impregnated core of the levee. The laminations range in thickness from 0.1 to 1.0 mm; there are two kinds: (1) laterally discontinuous laminae of peloids 50 to 150 μm; and (2) thinner, more continuous laminae of silt and clay-sized carbonate and filaments of blue-green algae (see also Shinn et al., 1969). The small cavities cutting the lamination in the lower part of this core are insect and earthworm (?) burrows.

Figure 23-7 (bottom left). View of the levee backslope here flooded, but normally exposed more than 90 percent of the year (Fig. 23-4). The dark surface is a smooth mat of the blue-green alga *Scytonema* sp. The handle of the knife in the foreground is 10 cm. The stratification of this zone is shown in Fig. 23-8.

Figure 23-8 (bottom right). Section of a resin-impregnated core of the levee backslope. The darker laminations are buried mats of *Scytonema* sp.; the lighter laminations are alternations of lime mud and peloids like those of Fig. 23-6. The low amplitude undulations and crinkling opposite the 1-centimeter scale division and opposite 2 to 2.5 cm are characteristic of this zone where deposition from flooding of the levee alternates with the growth of the *Scytonema* mat.

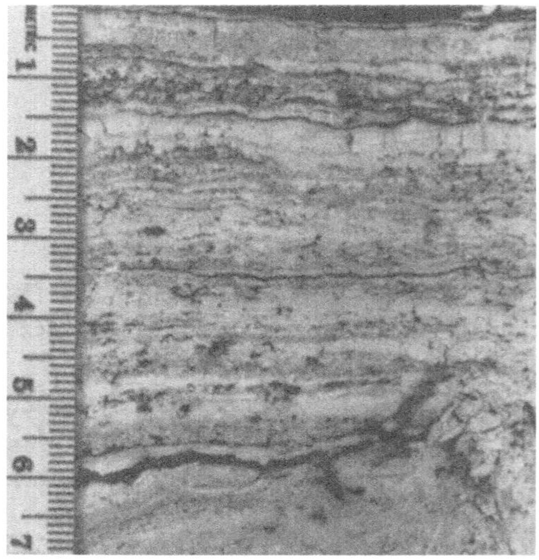

Figure 23-9 (top right). View of the inland algal marsh at the landward side of the study area (see Fig. 23-2) that is exposed much of the year and only flooded normally with fresh water during the rainy season. The surface is covered with a centimeter-thick mat of *Scytonema* sp., locally cemented to a friable crust. The stratification of this zone is shown in Fig. 23-10.

Figure 23-10 (right). Section of a resin-impregnated core from the inland algal marsh (Fig. 23-9). The dark layers are degraded mats of *Scytonema* sp.; the light layers are predominantly fine-sand-sized peloids deposited during the most severe storms or hurricanes.

Figure 23-11 (bottom left). Small-scale mud cracks on the surface of the levee backslope (see Fig. 23-4). The cracked layer of peloids and lime mud is 2 to 3 cm thick and overlies a *Scytonema* mat. Erosion of this cracked surface produces granule and sand-sized interclasts (chips) shown in Fig. 23-12. Scale in millimeters.

Figure 23-12 (bottom right). Surface view of the levee backslope showing interclasts (chips) formed by erosion of small-scale, mud-cracked polygons similar to those in Fig. 23-11. These interclasts are not cemented, but air-drying makes them coherent; the thicker clasts show lamination like that in Fig. 23-6. Metric scale.

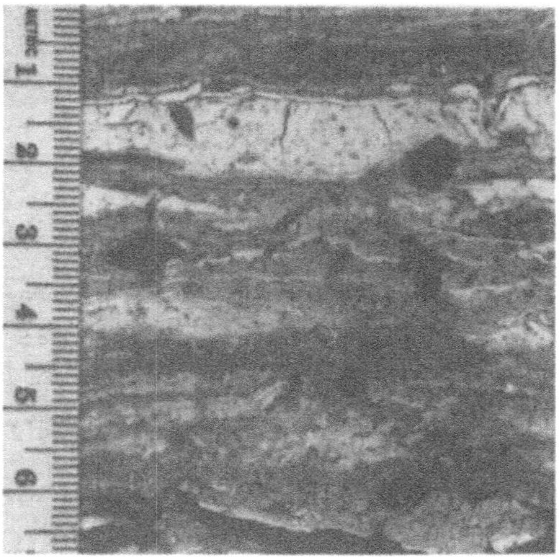

Figure 23-13. Surface of a pond showing prism cracks; the long dimension of the can is 16 cm. Cracks of this size appear on the floors of ponds and channels as soon as they are drained, indicating that the shrinkage is a result of dewatering rather than desiccation. The conical mounds are fecal pellets of the deep-burrowing polychaete worm *Marphysa* sp. (Garrett, 1972). The larger cracks are at least 30 cm deep; they heal when the surface is reflooded.

Figure 23-14. Surface of the levee backslope showing the cemented crust that is locally dolomitic (see Shinn et al., 1965). This kind of crust occurs low on the backslope and behind beach ridges near an exposure index of 95 (see Fig. 23-4). The crust has stratification similar to that shown in Fig. 23-8; here it is sufficiently cemented to break up into large, thin slabs. The pencil in the right foreground is 15 cm long.

Distribution of Sedimentary Structures

The first-order distinction in sedimentary structures is between (1) bioturbated pelleted lime mud with minor admixtures of skeletal debris, tests of high-spired gastropods (Cerrithidea) and Foraminifera; and (2) laminated and thin-bedded, pelleted lime mud and algal-rich layers (Fig. 23–3). The bioturbated muds occur below tide levels; the stratified deposits are intermittently flooded and exposed.

The sedimentary structures, both physical and organic, are restricted to surprisingly narrow zones determined by the duration of exposure or flooding. We determined the annual exposure index, percent of year exposed, from analysis of almost two years of continuous records of two tide gauges (Fig. 23–4). We established the zonal range of sedimentary structures by leveling several occurrences of each structure to the tidal datums (Fig. 23–4). Small variations in stratification preserve a surprisingly sensitive and faithful record of exposure index (Figs. 23–4, 23–5 to 23–10). To a large degree, variations in

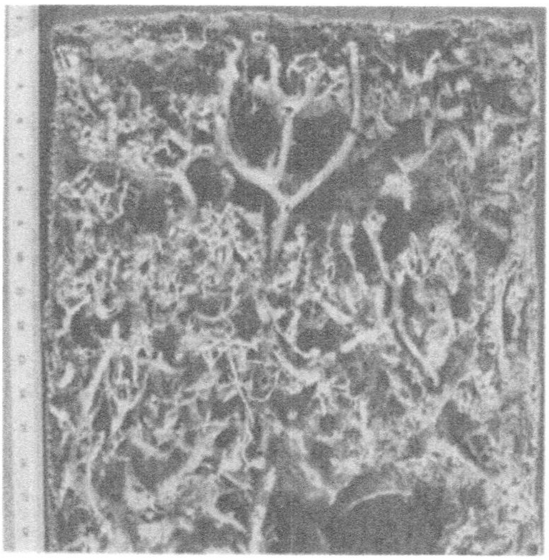

Figure 23-15. Resin cast of the burrows in a box core taken in the open pond subenvironment (see Fig. 23-4). This box core was impregnated wet so that the resin penetrated only the open burrows. After the resin hardened the intervening sediment was jetted out with water The larger branched burrows are made by the polychaete *Dasybranchus* sp.; the small burrows by other worms. *Dasybranchus* burrows in section are shown in Fig. 23-16. Metric scale.

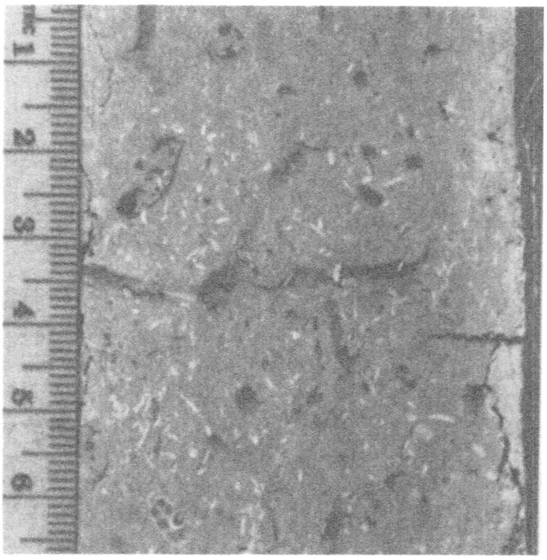

Figure 23-16. Section of a resin-impregnated core taken in a secondary tidal channel. The sediment is pelleted lime mud with numerous tests of the foram *Peneroplis* sp. The open burrows are those of *Dasybranchus* sp., the commonest polychaete in the study area (see Fig. 23-15; and Garrett, 1972).

the omnipresent mats of blue-green algae are responsible for this sensitivity, but parallel variations in physical sedimentation and desiccation are also evident (Figs. 23–4, 23–5 to 23–14). The burrows of animals and insects are also restricted to zones of exposure index (Figs. 23–4, 23–15 and 23–16; Garrett, 1972).

Significance

The distribution of structures and organisms in the Andros area shows that in an area of low tidal range, climate, wind, and rainfall play a much larger role than in areas of larger tidal range. The remarkably sensitive zonation of sedimentary and organic structures (Fig. 23–4) is related to the regimen of local sea-level fluctuations, of which only a part is the semi-diurnal tides. Almost all deposition is the result of storm-generated waves and flooding. The luxuriant and widespread *Scytonema* mat (Figs. 23–7 to 23–10) is the result of the high rainfall.

Because of the complex topography there is no unique sequence of sedimentary structures for the entire area (see Fig. 23–3; Shinn et al., 1969). However, the distribution of structures has been used to guide the interpretations of ancient shoaling carbonates (Roehl, 1967; Lucia, 1972).

References

GARRETT, PETER. 1972. Distribution and sedimentary record of organisms on a Bahamian tidal flat. Ph.D. thesis, Johns Hopkins Univ., Baltimore, p. 192.

LUCIA, F. J. 1972. Recognition of evaporite-carbonate shoreline sedimentation. *In* Rigby, J. Keith, and Hamblin, William Kenneth, eds., Recognition of Ancient Sedimentary Environments. *Soc. Econ. Paleontol. Mineral. Spec. Publ. 16*, 160–191.

ROEHL, P. O. 1967. Stony Mountain (Ordovician) and Interlake (Silurian) facies analogs of Recent low-energy marine and subaerial carbonates, Bahamas. *Am. Assoc. Petrol. Geol. Bull. 51*, 1979–2032.

SHINN, E. A., GINSBURG, R. N., and LLOYD, R. M. 1965. Recent supratidal dolomite from Andros Island, Bahamas. *In* Pray, L. C., and Murray, R. C., eds., *Dolomitization and Limestone Diagenesis, A Symposium. Soc. Econ. Paleontol. Mineral. Spec. Publ. No. 13*, 112–123.

SHINN, E. A., LLOYD, R. M., and GINSBURG, R. N. 1969. Anatomy of a modern carbonate tidal flat, Andros Island, Bahamas. *J. Sed. Petrol. 39*, 1202–1228.

24

Recent Tidal Deposits, Abu Dhabi, UAE, Arabian Gulf

Jean F. Schneider

Occurrence

Holocene tidal flats, sabkhas,* and lagoon complexes south of the city of Abu Dhabi in the Arabian Gulf near 24°30′ N, 54°20′E. Similar areas of tidal flats, sabkhas, and lagoons occur intermittently along the western coast of the Arabian Gulf (Persian Gulf) from Dubai on the east to Kuwait on the north.

According to Evans et al. (1969) the postglacial history of the Abu Dhabi Sabkha can be divided into two stages: (1) flooding and erosion of Pleistocene dune sands during the Holocene rise of sea level from about 7000 to 4000 years BP; and (2) accumulation of subtidal, intertidal, and supratidal deposits during an apparent fall in sea level of about 1 meter from 4000 years BP to the present. The apparent fall in sea level may be either an actual eustatic change, or, more likely, the result of reduced tidal range and wave heights at the shore produced by the growth of offshore topography, islands, and reefs (Evans et al., 1969).

* Sabkha (sabka, sabaka, sabakha, sebka) is the Arabic term for salt marsh, salt flat, or salt swamp. It is used in the geological literature for a supratidal flat in an arid climate.

Hydrography

The tides are diurnal with a spring range of 2.10 meters and a neap range of about 0.75 meter; the maximum range in the inner lagoon is 1.20 meters. The Shamal, storm winds from the north-northwest, can drive a sheet of water at times more than a meter thick over the sabkha, and flooding may reach several meters above mean sea level. The salinity of nearby open water is 38‰, but it is over 60‰ in the inner lagoon. The water temperature ranges from 23° to 32° C in the inner lagoon. The salinity of groundwater in the sabkha, as a result of the extensive evaporation and input of continental groundwater, can exceed the saturation value for halite, 830% of seawater.

Climate

The average maximum air temperature is 41° C and the average minimum is 13° C; the highest recorded temperature is 53° C and the lowest is below 0° C. The average rainfall is

Figure 24-1. Map of the Abu Dhabi area showing the physiographic-hydrographic subdivisions.

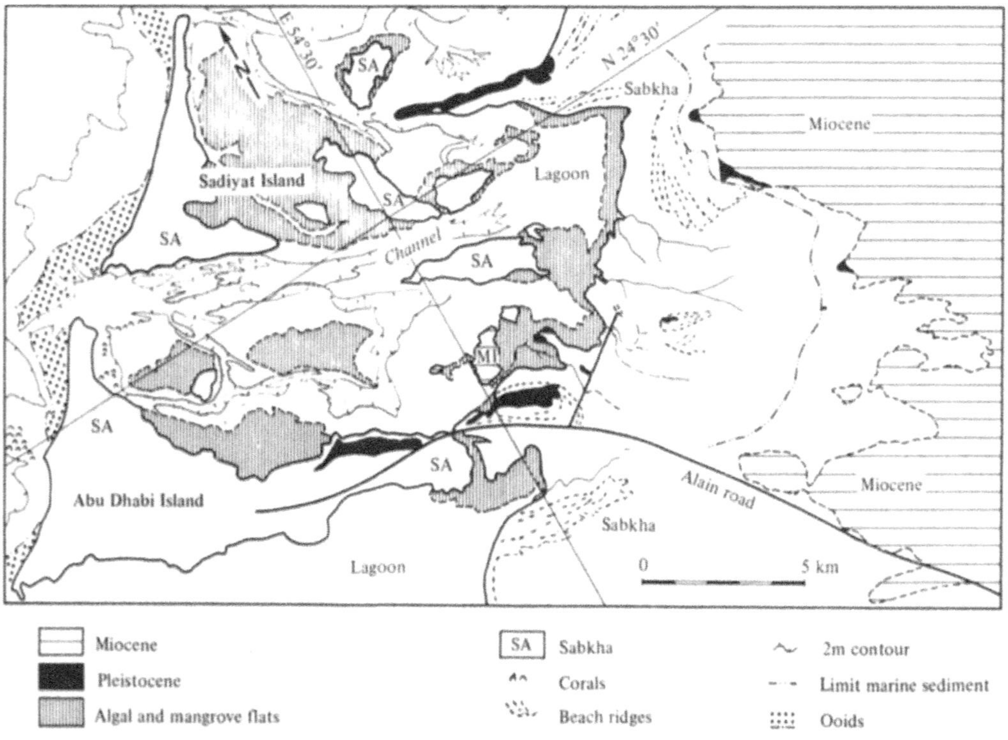

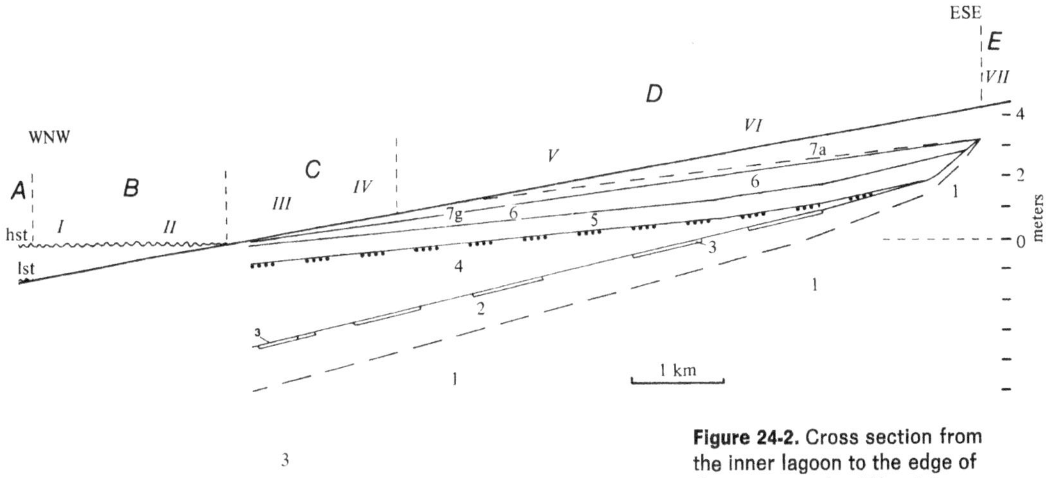

Figure 24-2. Cross section from the inner lagoon to the edge of the continental sabkha. *I*. Lower intertidal zone—lagoon terraces and mangrove flats. *II*. Upper intertidal zone—algal-laminated sediments and mangrove flats. *III*. Beach ridges. *IV*. Lower sabkha, frequency of flooding with seawater 10 to 20 times a year. *V*. Intermediate sabkha, frequency of flooding one to three times a year. *VI*. Upper sabkha, frequency of flooding less than once a year.

A. Subtidal facies; *B*. intertidal facies; *C*. coastal sabkha with groundwater derived from the sea; *D*. coastal sabkha with groundwater from the land; *E*. continental sabkha. *a*, anhydrite; *g*, gypsum; *four dots*, cemented horizon or crust. For descriptions of numbered units see p. 213.

less than 4 cm/year. The relative humidity averages 40‰ during the day and 90‰ during the night. The average daily evaporation is 25 mm and the maximum 55 mm (potential evaporation from a Piche evaporimeter). The winds from November to April are from the north-northwest (Shamal).

Physiographic-Hydrographic Subdivisions

The barrier islands, Abu Dhabi and Sadiyat of Figure 24–1, are triangular with beach ridges and dunes on their windward, north-northwest sides, and leeward sediment tails. On the shallow shelf seaward of the islands are coral reefs with *Acropora* sp., *Siderastrea* sp., and *Porites* sp. The islands, which have small sabkhas in their central parts, are separated by tidal channels that terminate in deltas of ooid sand (Figs. 24–1 and 24–3). In the transition zone between channels and the islands there are crab flats (*Scopimera* sp. and *Ocipoda* sp.), carbonate crusts, and flats with the black mangrove *Avicennia nitida* (Fig. 24–4). The inner lagoons grade into algal flats that are either sharply bounded by beach ridges stabilized by the halophyte *Arthroconeum glaucum* or slope imperceptibly into the lower sabkha (Figs. 24–1, 24–2).

Stratigraphic Sequence of the Sabkha, Southeast Abu Dhabi

The following summary is based on the report of Evans et al. (1969) and the author's observations. The basement is Miocene and Pleistocene carbonates. The overlying Holocene deposits can be divided into units numbered from the base upward;

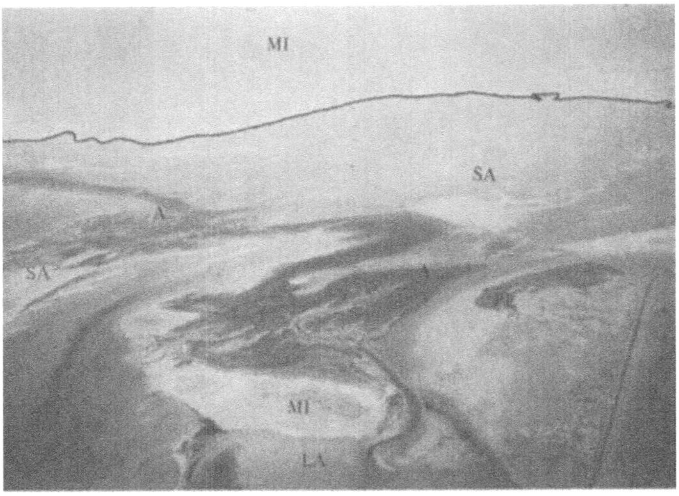

Figure 24-3. Aerial photograph, inner lagoon and sabkha of Abu Dhabi. View to southeast. Lower-right corner: road from Abu Dhabi to Al Ain oasis; compare with Fig. 24-1. *MI*, Miocene; *PL*, Pleistocene with waste deposit; *SA*, sabkha; *A*, algal flats; *LA*, lagoon.

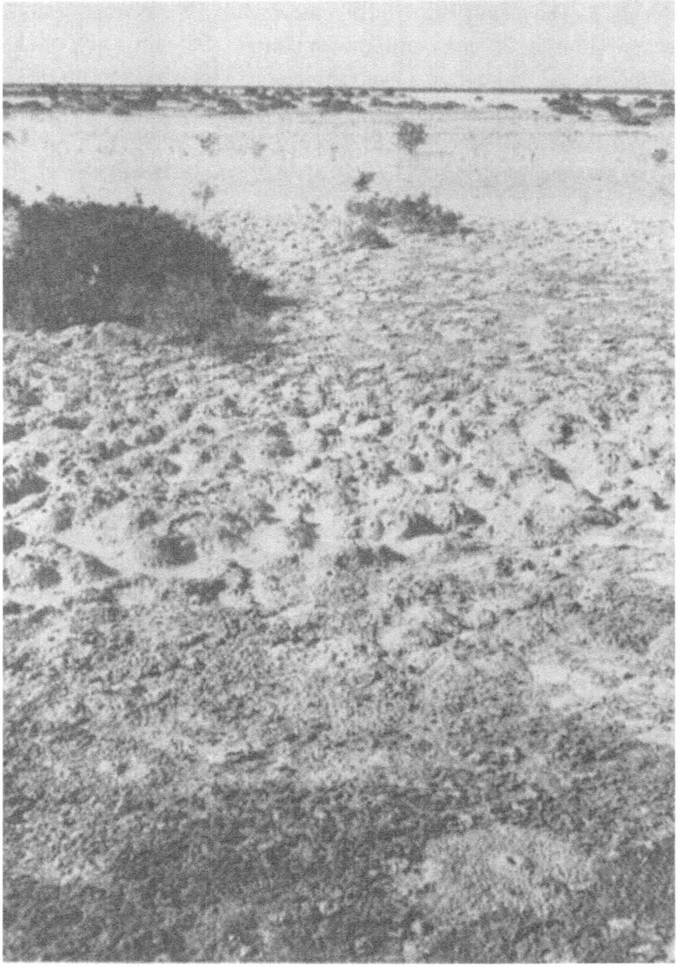

Figure 24-4. View of the margin of the central lagoon southeast of Abu Dhabi city. In the foreground is a surface cemented with aragonite; in the center crab burrows, pellets, and the halophyte *Arthroconeum glaucum*; in the backgorund a tidal channel; and on the horizon a swamp with black mangroves. The halophyte in the upper-left center is about 30 cm high.

Figure 24–2 shows their distribution and Figure 24–5 the structures of the upper units.

Unit 1. Up to 10 meters thick. Brown, aeolean, cross-bedded quartzose carbonate sand.

Unit 2. 50 to 160 cm thick. Gray, aeolean, cross-bedded quartzose carbonate sand, with bivalve and gastropod shells abundant in the upper part. Unit 2 has more fine carbonate material than Unit 1 and the percentage increases upward. Large gypsum disks are common.

Unit 3. 0 to 10 cm. Algal laminations and fine-grained, creamy carbonate mud, rich in cerithid shells and fecal pellets.

Units 2 and 3 represent the transgressive facies in the stratigraphic profile. The overlying units represent, first, the intermediate stage with subtidal facies and then the fully regressive development of lower and upper intertidal and supratidal facies.

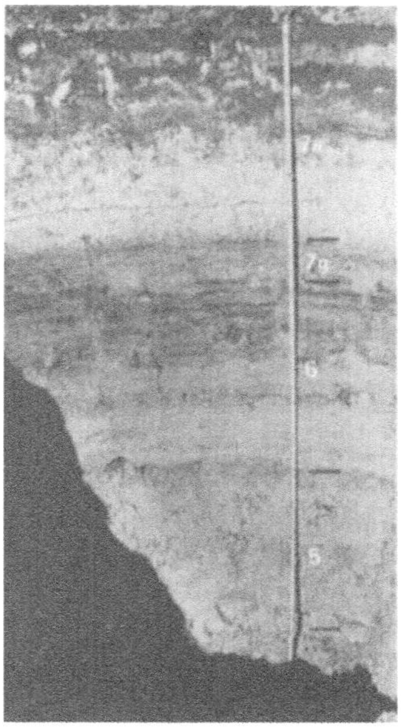

Figure 24-5. View of the section in a pit dug near the center of the Abu Dhabi sabkha, zone VI of Fig. 24-2. The total length of the scale is 1.8 meters, and the top of the section is indicated by the dashed line. Only the upper part of the lithified crust of unit 4 is visible, indicated by four dots. In the lower part of unit 7a, the anhydrite is layered (chicken-wire structure); in the upper part the anhydrite occurs as diapirs (nodular).

Unit 4. 0 to 300 cm. Gray-brown, muddy carbonate sand with more carbonate material than unit 2, but decreasing upward. Shells (gastropod and bivalve), shell fragments, pellets, and fragments of lithified crust increase upward. At the top there is a shelly coquina marked by a lithified horizon or crust cemented with gypsum or carbonate. Large gypsum disks in the nonlithified section are common.

Unit 5. Up to 60 cm. Gray, muddy bioclastic to pelletal carbonate sand and mud with abundant cerithids and small, thin-shelled bivalves. Gypsum disks are common. This unit is locally well dolomitized.

Unit 6. Up to 80 cm. Algal mats interlaminated with gray muddy bioclastic to pelletal carbonate sand and mud. Small gypsum crystals are abundant, and small gypsum disks are common. Coarse-grained skeletal sands are the high energy equivalent of the algal mat; locally cerithid shells are abundant. If the sequence is dolomitized, unit 6 is the most intensively dolomitized part of it (see Kendall and Skipwith, 1968).

Unit 7. Up to 80 cm. Brown, quartzose carbonate sand quite similar to that of unit 1, with the exception of the occurrence of both gypsum and anhydrite in the seaward areas and the predominance of anhydrite in the landward areas of the sabkha. In some areas magnesite forms up to half the carbonate fraction at the base of this layer.

Sedimentary Facies

Subtidal	Coral reefs, skeletal sands, deltas of ooid sand stabilized by the grass *Halodule* sp.
Lower intertidal	Lime muds burrowed by crabs and cerithids, black mangrove swamps, lithified crust, fecal pellets.
Upper intertidal	Algal mats (Schizophyta and Clorophycophyta).
Lower supratidal	Beach ridges formed principally of cerithid shells.
Upper supratidal	Aeolean sands.

Diagenetic Facies of the Abu Dhabi Sabkha

Upper intertidal	Gypsum mush, celestite, dolomite, calcite.
Lower supratidal	Lower sabkha: gypsum mush, bassanite, dolomite, calcite, nesquehonite ($MgCO_3 \cdot 3H_2O$)
Upper supratidal	Intermediate sabkha: anhydrite nodules, large gypsum disks, dolomite, magnesite, huntite ($Mg_3Ca(CO_3)_4$), halite, some polyhalite.
Continental sabkha	Anhydrite layer, secondary gypsum. Dolomite and anhydrite are secondary; only gypsum is a primary precipitate. Anhydrite can rehydrate to gypsum (see Bathurst, 1971 for review and references).

References

BATHURST, R. G. C. 1971. Carbonate sediments and their diagenesis. The Trucial Coast Embayment, Persian Gulf. *Develop. Sed. 12*, 178–212.

EVANS, G., SCHMIDT, V., BUSH, P., and NELSON, H. 1969. Stratigraphy and geologic history of the Sabkha, Abu Dhabi, Persian Gulf. *Sedimentology 12*, 145–159.

KENDALL, C. G., and SKIPWITH, P. A. 1968. Recent algal mats of a Persian Gulf lagoon. *J. Sed. Petrol. 38*, 830–840.

25

Prograding Tidal-Flat Sequences: Hutchison Embayment, Shark Bay, Western Australia

Gregory M. Hagan and
Brian W. Logan

Occurrence

The Hutchison Embayment is a Holocene-Recent tidal flat located on the eastern margin of Hamelin Pool, Shark Bay, Western Australia; Lat. 26°10′; Long. 114°20′ (Logan and Cebulski, 1970). See Figure 25–1, the locality map, for topographic and bathymetric features.

Climate

The average annual rainfall is 23 cm; the evaporation 220 cm. The average maximum temperature is 29° C, the average minimum 15° C. Southerly winds prevail with an average maximum velocity of 45 km/hr, and last 4 to 6 days; cyclonic winds up to 180 km/hr occur once every six years.

Hydrography

Tidal pond waters: 60 to 65‰; tidal groundwater: 80 to 90‰. Gradient 60 to 90‰ in lower intertidal zone and up to 300‰ in supratidal zone; brackish influx in supratidal zone. The surface sediments are kept moist by capilliary rise.

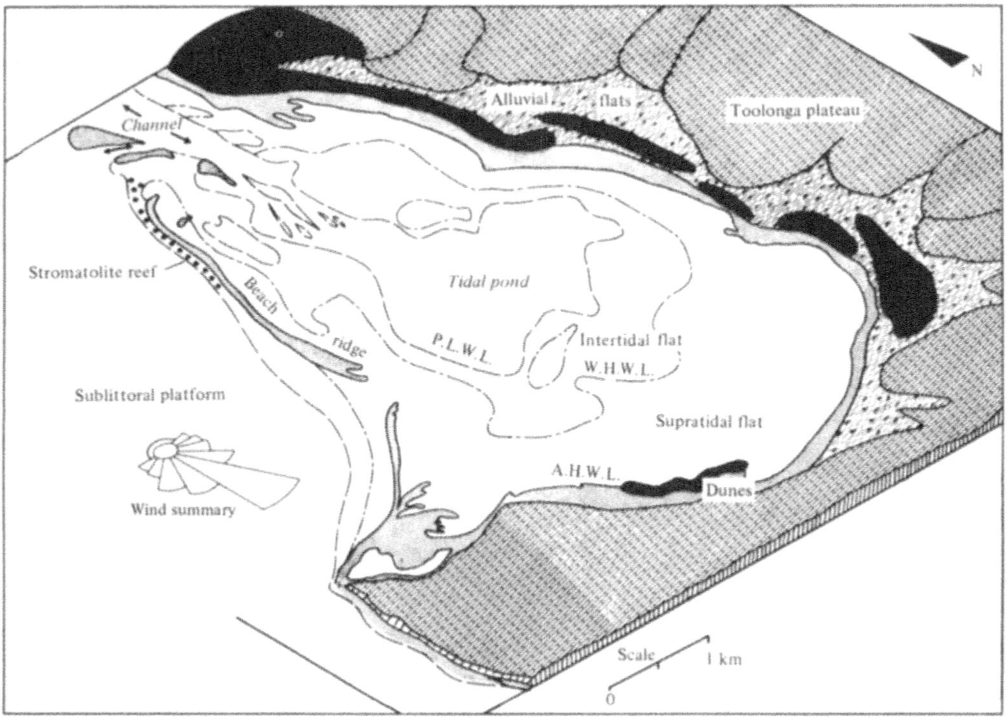

Figure 25-1. Topographic and bathymetric features of Hutchison Embayment.

The spring range is about 60 cm; the maximum range of approximately 3 meters is influenced by wind and other factors. The prevailing southerly waves on the sublittoral platform are refracted into easterly propagation on crossing the offshore platform; maximum wave parameters are H–1 m, L–12 to 15 m, P–4 to 6 seconds. Wave action is negligible in the embayment under prevailing conditions; cyclone-generated waves from north sweep the tidal-supratidal flat.

Physiographic-Hydrographic Subdivisions

The depth range of the *sublittoral platform* is from prevailing low water to 5 meters; the slope is 1 meter/km. *Tidal channel* depth is 0.3 to 3 meters. The *tidal pond*, 3.5 sq km, ranges from 0.3 to 2 meters deep; its slope is 0. The *intertidal zone*, an area of approximately 9 sq km, is up to 2 meters deep; the slope is 0.6 meters/km. With a depth range of up to 0.3 meters and a slope of 0.6 meter/km, the *supratidal zone* has an area of approximately 15 sq km. *Beach and storm ridges* are characterized by multiple ridges up to 4 meters above tidal-flat level. The *Toolonga Plateau* is up to 50 meters in elevation in the area of Hutchison Embayment; it is underlain by limestone and dolomite of Cretaceous age. The *Hutchison Embayment* is underlain by Pleistocene marine units, Dampier and Bibra Formations and Carbla Oolite (Logan et al., 1970).

Biota

The sublittoral platform is characterized by *Fragum* (pelecypod-foram) community; the tidal pond, by *Acetabularia sp.* and *colloform* algal mat. *Smooth* algal mat occurs in the lower intertidal zone. Middle and upper intertidal zones have *pustular, tufted,* and *gelatinous* algal mats. *Blister* (damp) and *film* (dry) mats characterize the supratidal zone.

Figure 25-2. Map of Hutchison Embayment showing distribution of surface sediments and lines of cross section.

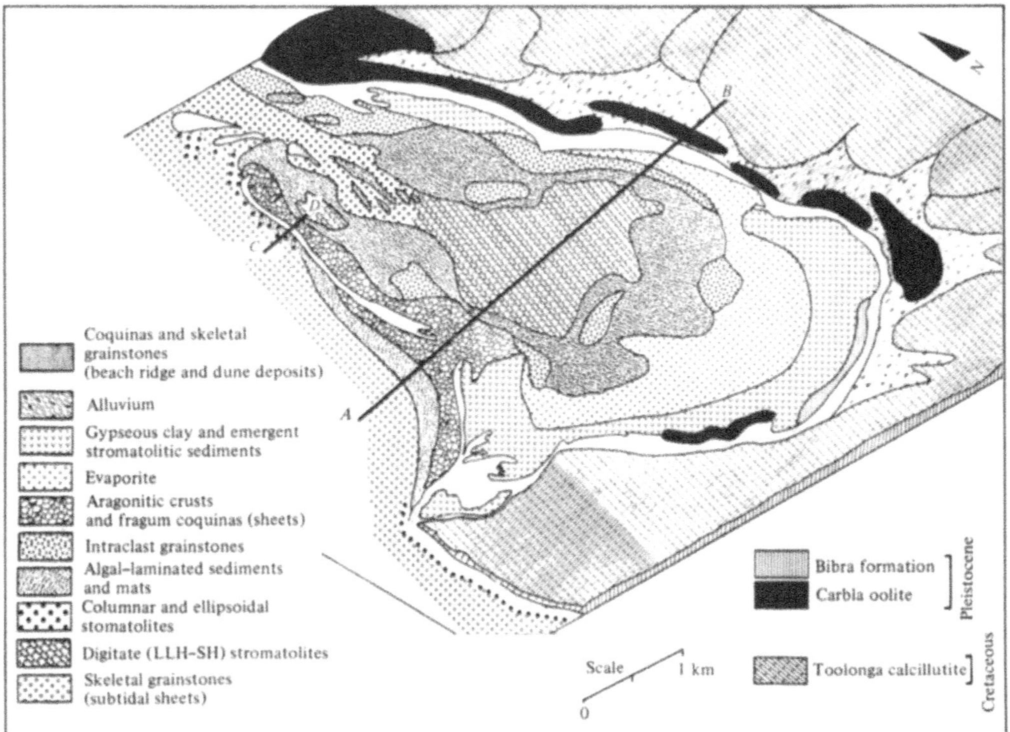

Contemporary Sediments and Sedimentary Structures

See Figure 25–2 for distribution map, Figure 25–3 for cross section.

SUBLITTORAL PLATFORM AND TIDAL CHANNEL. *Fragum* coquinas, foraminiferal microcoquinas, ooid grainstones; cemented with acicular aragonite; planar-bedded to cross-bedded; intraclast breccias of above lithologic types form superficial veneers.

BEACH AND STORM RIDGES. The outer barrier ridge is characterized by unconsolidated *Fragum* coquinas with large-scale inclined bedding dipping seaward at 10 to 15°.

INNER BEACH AND STORM RIDGES. Fragum coquinas, skeletal fragment grainstones; intraclasts and remanié Pleistocene fossils are abundant. They are inclined-bedded and cross-bedded with dips 5 to 10° toward the tidal flat.

TIDAL POND. A stratiform sheet composed of pellet wackestone, homogeneous to faintly laminated with interbedded compound cryptalgal structures (*Colloform* mat) consisting of an irregular basal column that gives rise to branched digitate columns (SH/LLH→SH); the columns have coarse laminoid-fenestral fabric, a diameter of 40 cm, relief of 20 cm and synoptic relief of 3 to 5 cm.

LOWER INTERTIDAL ZONE. Outer margin of barrier ridge: clusters of cryptalgal columns, ridge-rill structure, and sheets

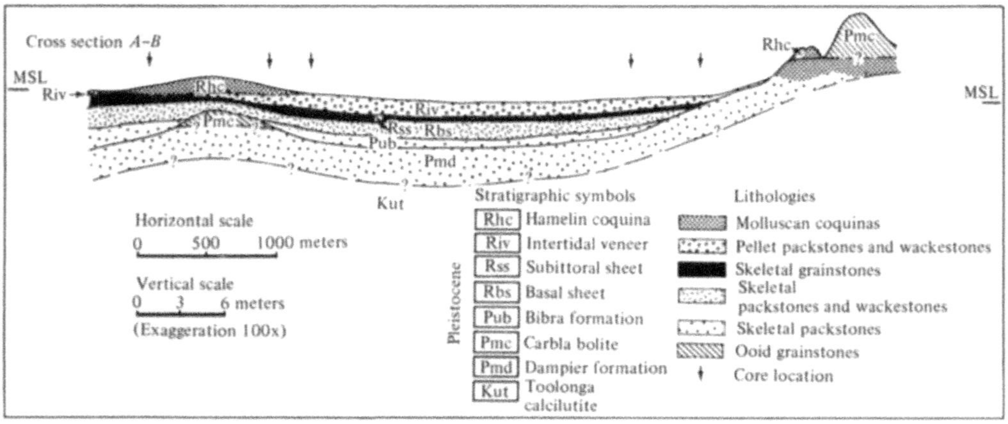

Figure 25-3. Cross section (*A-B*) showing Pleistocene and Holocene-Recent stratigraphic sequences.

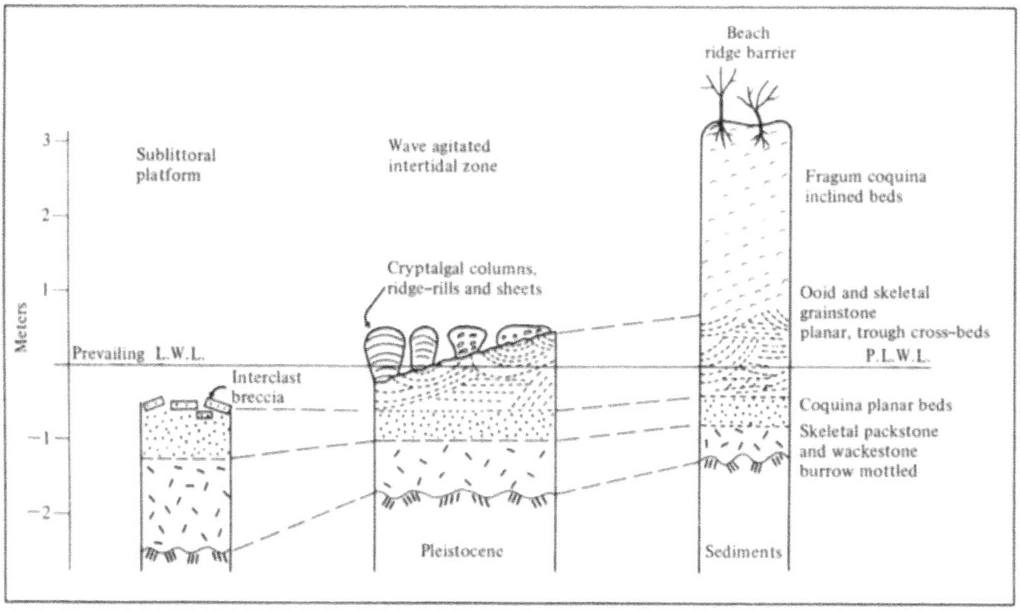

Figure 25-4. Cross section (*C-D*) showing the upper part of the Holocene-Recent sequence on the outer margin of Hutchison Embayment.

that form a seaward-thickening wedge; pellet grainstones and packstones consolidated with cryptocrystalline aragonite cement; fine laminoid-fenestral fabrics. Elongate cryptalgal structures are aligned normal to the shore (Fig. 25–4) and decline in height toward the beach ridge.

LOWER INTERTIDAL ZONE, MARGIN OF TIDAL POND. Stratiform cryptalgal laminite sheet composed of unconsolidated pellet packstone with fine laminoid-fenestral fabric.

MIDDLE TO UPPER INTERTIDAL ZONE, OUTER MARGIN OF BEACH RIDGE. Clusters of cryptalgal columns, ridge-rill structures, and sheets; consolidated, pellet packstone with irregular fenestral fabric. Elongate cryptalgal structures are aligned normal to the shore (Fig. 25–4) and decline in height toward the beach ridge.

MIDDLE TO UPPER INTERTIDAL ZONE, EMBAYMENT. Stratiform sheet of unconsolidated pellet packstone with medium, irregular fenestral fabric interstratified with scallop fabric and ribbon fabric having discoid to irregular cryptalgal structures and irregular fenestral fabrics; desiccation polygons, graded storm layers (intraclast grainstone), and overfolds (in ribbon-fabric sheets).

UPPER INTERTIDAL TO SUPRATIDAI ZONES. High groundwater table: Gypsum crystals in host sediments. Fluctuating high and low groundwater table: Intraclast breccia and grainstone sheets, 1 to 10 cm thick, overlying sediments in which gypsum crystals are growing. Low groundwater table: Intraclast breccia (well-drained locations) and grainstone sheets, or a deflation surface.

SUPRATIDAL ZONE. Has same characteristics as the upper intertidal to supratidal zones, as well as deflation surface with relict ring and crescent cryptalgal structures aligned normal to the shore (Fig. 25–5).

Holocene-Recent Sequence

1. Transgression with rising sea level until about 4000 to 5000 years BP; sea level maximum approximately 2 meters above PLWL marked by emergent ring and crescent cryptalgal structures.

2. Regression to present sea level.

3. Salinities of tidal waters increase from metahaline (40 to 53‰) to hypersaline (53 to 72‰) about 4000 years BP as a

Figure 25-5. Expanded Holocene-Recent stratigraphic columns from the eastern margin of the tidal pond along line *A-B* in Figure 25-3; inner Hutchison Embayment.

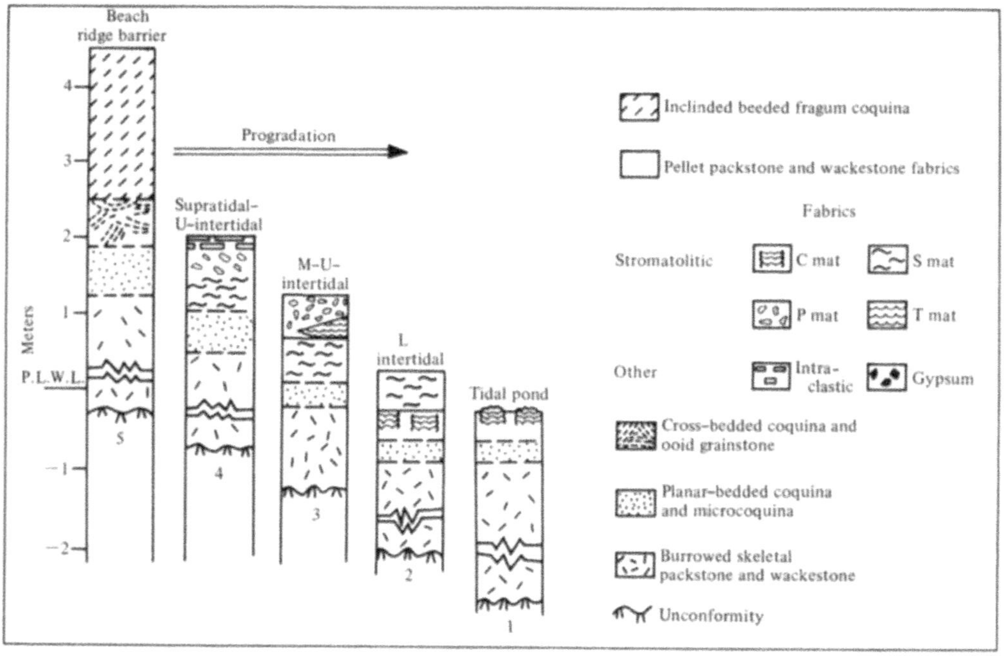

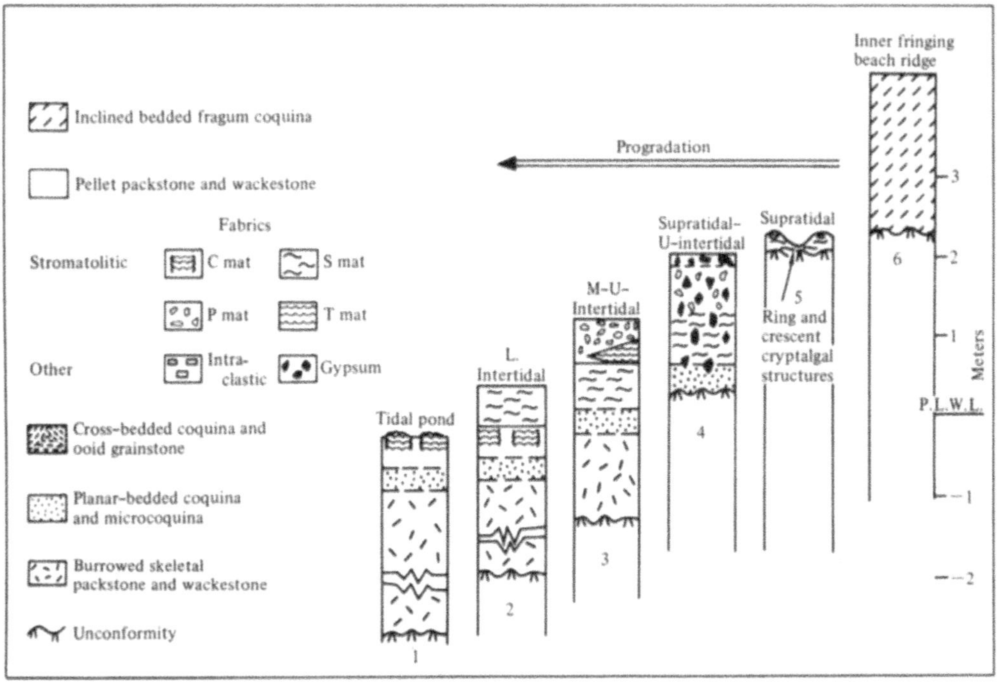

Figure 25-6. Expanded Holocene-Recent stratigraphic columns from the western margin of the tidal pond along line *A-B* in Figure 25-3; inner Hutchison Embayment.

result of shoaling and growth of a sea grass barrier bank (Faure Sill) to the north (Logan et al., 1970).

4. Salinities of tidal pond waters rise to aragonite-precipitation field concentrations (72‰ to 90‰) with restriction consequent on accretion of the beach ridge barrier and shoaling (about 2000 years BP).

5. Groundwaters reach aragonite and gypsum-precipitation field concentrations; concentration gradients increase with growth of tidal-supratidal surface and geochemical zones expand.

Stratigraphic Sequence

For the stratigraphic sequence see Figures 25–3 to 25–6.

References

LOGAN, B. W., and CEBULSKI, D. E. 1970. Sedimentary environments of Shark Bay, Western Australia. *In* Logan, B. W., et al., Carbonate sedimentation and environments, Shark Bay, Western Australia. *Am. Assoc. Petrol. Geol. Memoir 13*, 1–37.

LOGAN, B. W., READ, J. F., and DAVIES, G. R. 1970. History of carbonate sedimentation, Quaternary Epoch, Shark Bay, Western Australia. *In* Logan, B. W., et al., Carbonate sedimentation and environments, Shark Bay, Western Australia. *Am. Assoc. Petrol. Geol. Memoir 13*, 38–84.

26

Carbonate Sedimentation in an Arid Zone Tidal Flat, Nilemah Embayment, Shark Bay, Western Australia

Peter J. Woods and
Raymond G. Brown

Occurrence

A Holocene-Recent carbonate tidal flat at the southern end of Hamelin Pool, Shark Bay, Western Australia. Lat. 26°27′ S, Long. 114°5′ E.

The flat is situated at the southern, closed end of Hamelin Pool, a hypersaline basin 50 km long, 30 km wide, and averaging 7 meters in depth.

Climate and Hydrography

Rainfall is 23 cm/year, and evaporation is 220 cm/year. Maximum and minimum average temperatures are 29° and 15° C. respectively.

WINDS. Prevailing southerly winds, average velocity commonly 18 to 27 km/hour for 3 to 5 day periods in summer. Cyclonic winds average about once every 6 years, from northerly directions, with velocities 70 to 110 km/hour over 12-hour periods and gusts to 180 km/hour.

SALINITY. Salinity of basin waters ranges from 55 to 65‰; salinity of groundwaters increases from 200‰ in intertidal zone to 240‰ in supratidal zone and then decreases to 100‰ at beach ridges in response to rainwater seepage.

TIDES. Astronomic range is 0.6 to 0.9 meters; winds and storms may increase the range to 3 meters.

WAVES. Under prevailing winds, wave action is negligible except along the western margin when an easterly component is developed. Under cyclonic conditions, the flat is exposed to the effects of northerly winds operating over the full fetch of Hamelin Pool.

Sedimentation in Physiographic Subdivisions (see Figs. 26–1 and 26–2)

Sublittoral platform

MORPHOLOGY. The outer limit is a break in slope at 1.3 meters below MSL, the upper limit, extreme low water level; 0.5 km wide.

BIOTA. *Fragum* community (pelecypod-foram assemblage). *Smooth* algal mat, mainly at the inner and outer limits of the platform. (The algal-mat nomenclature used in this paper follows that of Logan et al., 1974).

SEDIMENTARY AGENTS. Wave action along outer edge on west side forms columnar stromatolites and storms sweep detritus shoreward. Tidal flow causes elongation of stromatolites normal to the shoreline and introduces cool basin waters to warm, shallow conditions, precipitating aragonite at outer edge as ooids and cement at the bases of algal columns.

Fragum community produces skeletal debris. Algal mats locally generate a laminated fabric and reducing conditions in the sediments (carbonate grains blackened by dark pigments) and induce alteration of skeletal grains and ooids to microcrystalline aragonite pellets.

SEDIMENTS. Gray to white, poorly bedded to laminated, skeletal pellet grainstones (unconsolidated sands, with auto-

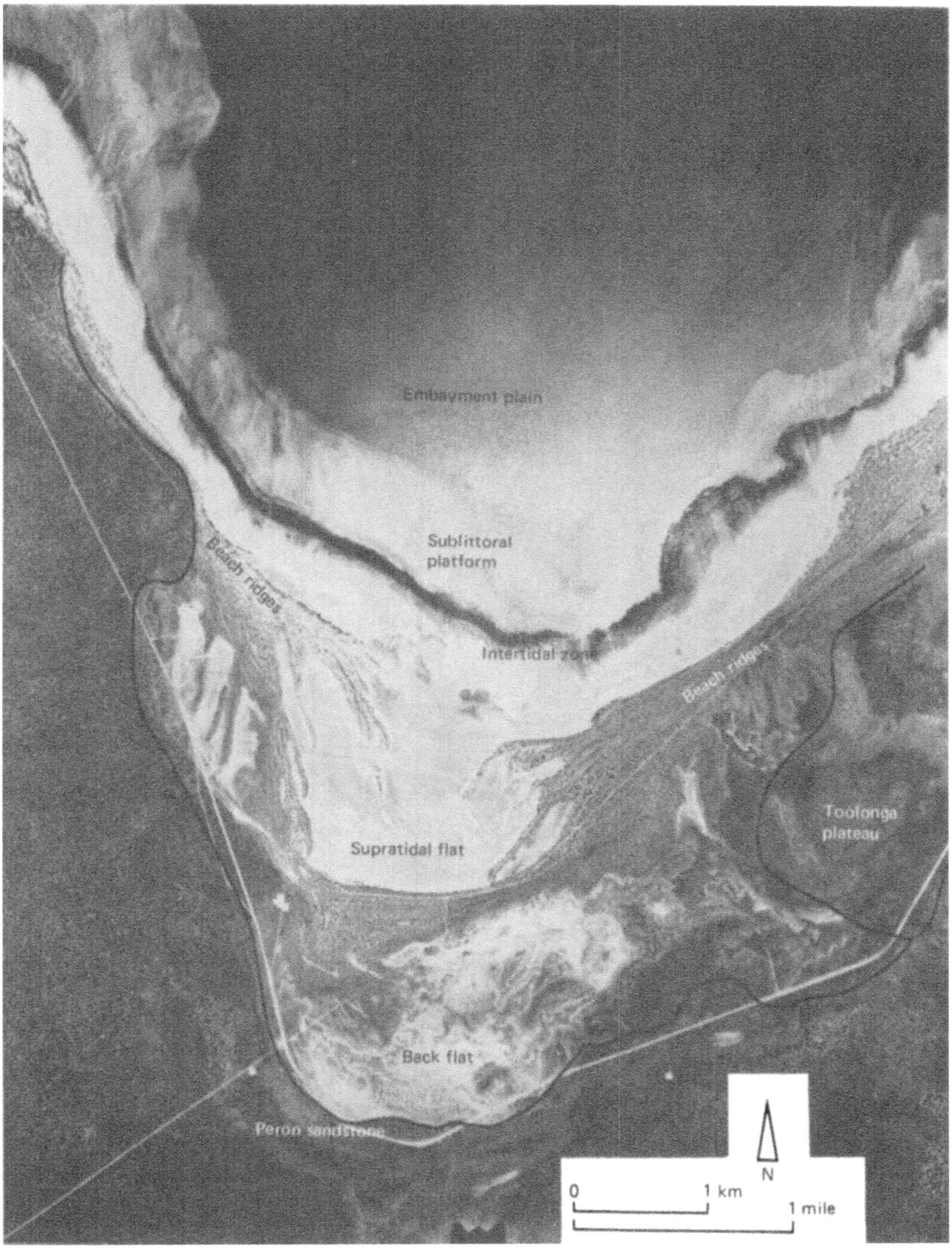

Figure 26-1. Vertical air photograph of Nilemah Embayment, showing major geomorphologic subdivisions. A dark band of algal mat approximately corresponds to the middle intertidal zone.

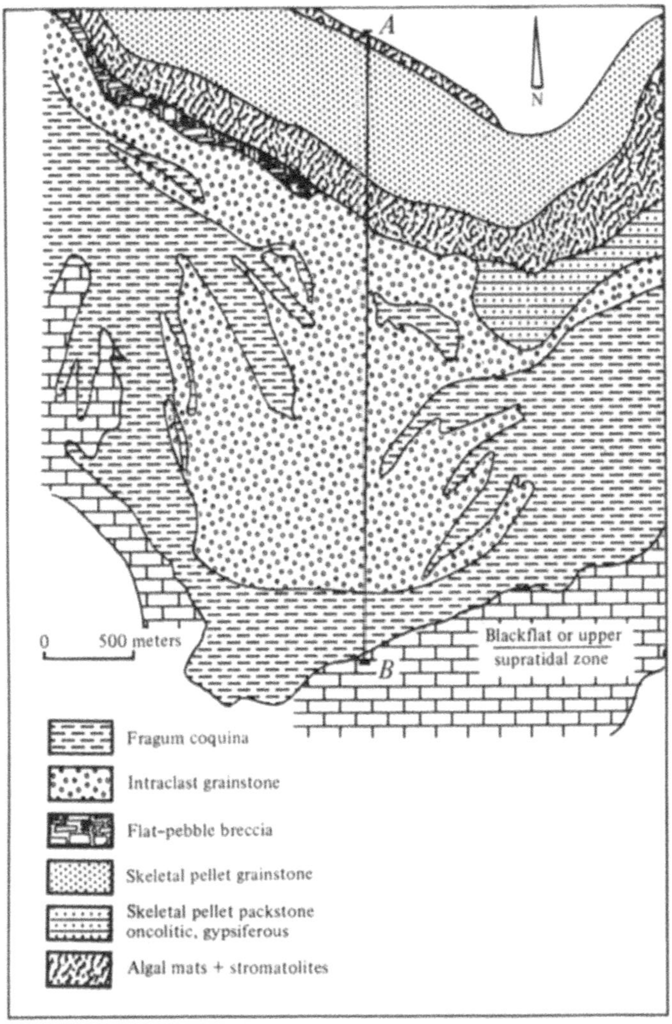

Figure 26-2. Distribution of surface sediment types, Nilemah Embayment.

chthonous forams, mollusk fragments, and pellets); minor proportions of intraclasts (intertidal origin), quartz, and coarse shell (*Fragum* valves).

Lower intertidal zone

MORPHOLOGY. Extreme low water to MLW; width 50 to 100 meters; rarely exposed.

BIOTA. Similar to that of the sublittoral platform, but *smooth* mats are more extensive and cover most of the zone.

SEDIMENTARY AGENTS. *Smooth* algal mats trap and bind sediment in a well-laminated sequence; they cause recrystallization of skeletal grains and ooids to form cryptocrystalline aragonite pellets, induce black pigmentation under reducing conditions, and create laminoid fenestrae in a weakly cemented fabric.

SEDIMENTS. Initially unconsolidated skeletal pellet grainstones, with increasing intraclasts shoreward; may grade to pellet packstone with recrystallization and cementation by cryptocrystalline aragonite; well laminated, with fine (<1 mm high) laminoid fenestrae.

Middle intertidal zone

MORPHOLOGY. MLW to MHW; width 100 meters. Short-term exposure is frequent in the lower parts, with progressively longer exposure upward.

BIOTA. Continuous sheets of algal mat (see Fig. 26–1), with some sinuous stromatolites (up to 45 cm relief and 7 meters long) along the west side. *Smooth* mat in lower parts, but predominately *pustular* mat.

SEDIMENTARY AGENTS. Wave and current action along the west side forms sinuous stromatolitic ridges elongated normal to the shoreline; elsewhere the algal mat is continuous. *Pustular* mat disrupts lamination in trapped sediments and produces irregularly dispersed or crudely laminar, millimeter-scale fenestrae. Frequent exposure and a porous fabric maintain oxidizing conditions.

SEDIMENTS. Cream-colored, intraclastic grainstones with pellets and skeletal grains, lightly to moderately indurated in a 10-cm crust below the algal mat; poorly laminated with a medium irregular fenestral fabric. Grainstones grade to packstone with development of cryptocrystalline aragonite cement and recrystallization of skeletal grains.

Upper intertidal zone

MORPHOLOGY. MHW to HWS; width 200 to 300 meters. Exposed for 10-day periods, diurnally submerged over 5-day periods.

BIOTA. *Pustular* mat developed in lower half, mainly continuous sheets. Tufted mat is a minor element in small depressions. *Film* mat is on stromatolitic heads and crusts in upper parts of the zone.

SEDIMENTARY AGENTS. Wave and current action as in middle intertidal zone; sinuous stromatolites continue upslope. Tidal scouring of partly indurated stromatolites produces "ring and crescent structures." Storms break up surface sediments and produce intraclasts.

Film mat is associated with pronounced cementation and recrystallization of carbonate grains in producing a strongly indurated crust of aragonitic packstone over upper parts of the zone. *Pustular* mat induces development of irregular fenestrae. Refluxing of groundwaters, evaporation during prolonged exposure, and the microenvironment beneath the algal mat are likely factors inducing precipitation of aragonite in an indurated layer below *pustular* mat; stromatolites become partly lithified.

SEDIMENTS. Yellow, intraclastic grainstones, with irregular fenestral fabric. In lower half of zone, a 10-cm layer of indurated grainstone overlies 20 cm of unlithified grainstone. In upper half, the indurated layer is exposed at the surface, colonized by *film* mat, and is altered to a 5-cm crust of strongly indurated pellet packstone, with pellets and altered skeletal grains densely cemented by microcrystalline aragonite.

Lower supratidal zone

MORPHOLOGY. HWS to extreme high water 2.5 meters above MSL; width 1.5 km. Submerged by storms, surface kept damp by capillary action.

BIOTA. *Film* mat in lower parts adjacent to the intertidal zone. *Blister* mat forms a discontinuous thin veneer in the upper half.

SEDIMENTARY AGENTS. Inundation accompanied by high energy wave and current action in storms produces scouring and rippling in loose sediment, and disrupts crusts (producing intraclasts) and algal mats (locally generating oncolites).

Film mat is associated with the generation of aragonitic packstone crusts in the lower parts of zone. Beneath *blister* mat, groundwaters precipitate gypsum as small crystals, which disrupt depositional fabrics.

SEDIMENTS. Yellow and tan, intraclastic grainstones, which are unconsolidated to semiconsolidated, poorly bedded, moderately well sorted; mainly clasts of aragonitic packstone crust, grading up-slope from flat-pebble breccias (clasts 10 to 30 cm) to progressively finer sands.

Beach ridges

MORPHOLOGY. Multiple ridges up to 5 to 6 meters above MSL in zones up to 1 km wide. Basal elevation of oldest ridges is 2 meters above MSL.

BIOTA. Land plants cover older ridges, but are sparse on younger ridges.

SEDIMENTARY AGENTS. Storm waves construct the ridges. On the oldest ridges incipient soils are developed and basal sediments are cemented.

SEDIMENTS. Thinly bedded *Fragum* coquina, with minor intraclasts.

Upper supratidal zone

MORPHOLOGY. An emergent flat, kept moist by capillary action, up to 1.5 km wide and 3 meters above MSL (see Fig. 26–1); bordered by the Toolonga Plateau (Cretaceous sediments) and the Plio-Pleistocene Peron Sandstone, and underlain by Pleistocene limestones.

BIOTA. Salt-tolerant land plants on better drained areas only.

SEDIMENTARY AGENTS. Wind forms small sand dunes and leaves a thin lag deposit of shells and lithoclasts reworked from underlying limestones. Evaporating terrestrial groundwaters precipitate gypsum in subsurface sediments.

SEDIMENTS. Dune sands in mounds up to 5 meters thick and thin (5 cm) sheets of shelly, lithoclast gravel overlying Pleistocene sediments. At a depth of 0.5 meters, gypsum is precipitating to form red gypsiferous limestones containing 30 to 50 percent gypsum. Small outcrops of indurated, well-laminated packstone indicate the presence of *smooth* algal mat and mark an early stage of the Holocene tidal-flat development.

Holocene-Recent Sequence

Stratigraphic Sequence (See Fig. 26–3.)

MAJOR EVENTS

1. Transgression across an irregular surface of Pleistocene sediments; sea level rose to a maximum 2 meters above present

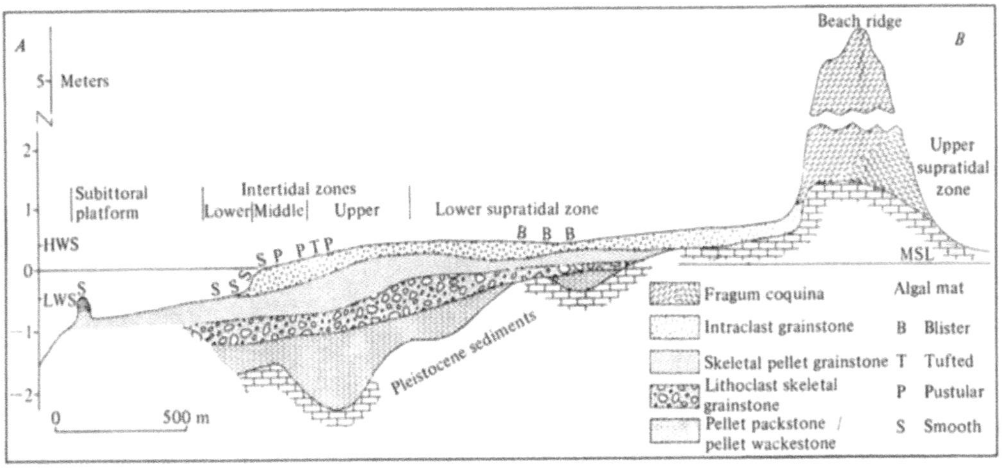

Figure 26-3. North-south stratigraphic cross section, Nilemah Embayment.

MSL. Irregular pods of pellet packstone and wackestone accumulated beneath a dense seagrass cover (salinities 40 to 53‰).

2. Salinity of basin waters increased to hypersaline range (beyond 53‰) and seagrasses disappeared. Calcrete clasts were spread basinward from an eroded shore as a prograding sheet of lithoclastic skeletal grainstone.

3. Lowering of sea level and progradation widened the intertidal flat, decreased the availability of lithoclasts, and progressively shifted beach-ridge deposition. Algal mats developed, and sublittoral to intertidal sand sheets prograded over lithoclastic grainstones.

4. With continued lowering of sea level, algal heads at the base of the latest beach ridge were exposed and modified to form ring and crescent structures; sediments beneath algal mats became cemented and desiccated, forming aragonitic crusts, which were disrupted by desiccation and storms to form intraclasts.

5. At present, as progradation proceeds, intraclasts are spreading across the flat as a sheet of landward-fining intraclast grainstone. The sublittoral platform is prograding slowly and is now subjected to partial exposure along its outer edge.

6. The development of a wide supratidal zone is creating conditions favoring the appearance of evaporitic minerals. Gypsum is a major component in the upper intertidal zone and a minor component in the lower supratidal zone; conditions more favorable for gypsum precipitation are developed locally (see Fig. 26–2*A*). During summer, halite forms on the lower supratidal zone, but is removed by rainwater and storms.

Distinctive Tidal-Flat Characteristics

1. Nilemah Embayment is sheltered from the prevailing regional wave action and exposed to little tidal variation. Although there is sufficient wave action in some localities to form stromatolitic heads (elongated by tidal action), continuous mats are a dominant feature of the intertidal zone. The low energy setting also favors development of continuous crusts of well-indurated sediments near the high water level.

2. Infrequent high energy episodes leave a strong imprint on the supratidal zone—indurated crusts are broken, and resulting intraclasts spread across the flat; ripples and other structures disappear between storms. Skeletal debris swept from the sublittoral zone is concentrated as beach-ridge coquinas at high water level.

3. Dominant grain types in the tidal-flat sediments are aragonitic pellets, partly altered skeletal grains, and intraclasts. Aragonitic pellets and altered skeletal grains are likely components of limestones forming in any environment where boring algae, and so forth can alter skeletal materials, and after diagenesis may be difficult to distinguish from calcrete grains.

Intraclasts are a major feature of the intertidal and lower supratidal zones. Their lithologies are distinctively intertidal, being either fragments of aragonitic crust or finely cemented skeletal pellet grainstone.

4. Algal mats provide distinctive criteria for the recognition of intertidal sediments. Well-laminated sediments with fine laminoid fenestrae characterize areas of *smooth* mat in the lower intertidal zone. Poorly laminated sediments with an irregular fenestral fabric produced beneath *pustular* mats characterize the middle to upper intertidal zone.

5. Induration is a distinctive feature, arising from prolonged subaerial exposure of the higher zones, which (1) results in the formation of a distinctive aragonitic pellet packstone beneath areas of *film* mat; (2) enhances the preservation of open fenestrae in algal-bound sediment; (3) is reflected by abundant intraclasts over parts of the flat; and (4) is one factor in the development of ring and crescent structures.

6. The creation of wide intertidal and supratidal flats provides a mechanism for the development of gypsiferous sediments without the presence of extreme salinities in adjacent basin waters. Gypsum may precipitate within existing sediments beneath the supratidal zone while lithologically comparable, but nongypsiferious sediments are accumulating in the intertidal and sublittoral zones in less saline waters.

Reference

LOGAN, B. W., HOFFMAN, P., and GEBELEIN, C. D. 1974. Algal mats, cryptalgal fabrics and structures, Hamelin Pool, Western Australia. Amer. Assoc. Petrol. Geol. Mem. 22, 140–194.

Section IV

Ancient Carbonate Examples

Vertical Sequence of Sedimentary Structures

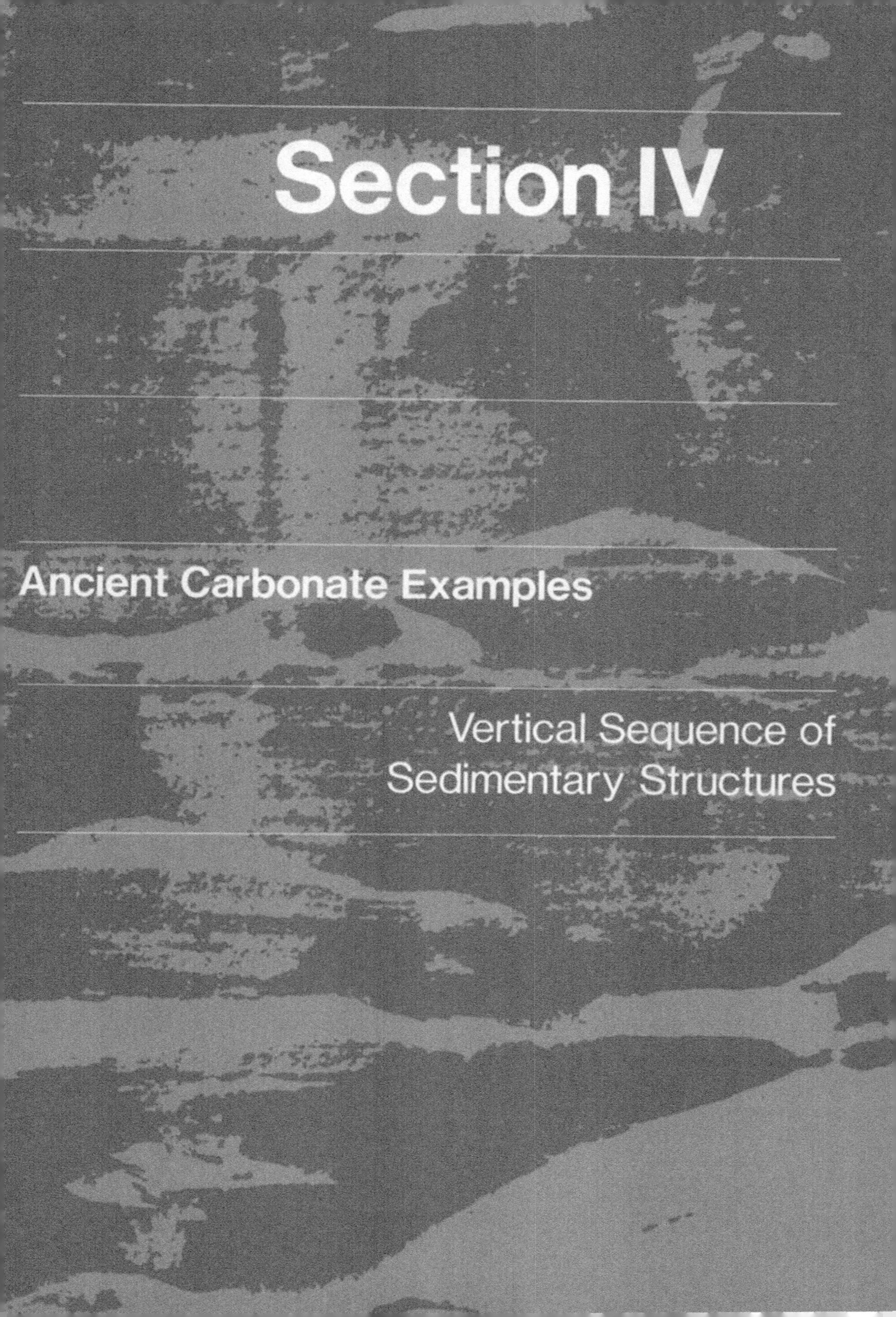

The summaries of the zonal distribution of sedimentary features in Section III are an outline of the state of the art. The inventory of features and vertical sequences, although incomplete, is widely used to interpret ancient carbonates. From this inventory the three major zones of accumulation can be characterized:

SUBTIDAL: *Permanent submerged.* The diagnostic features in this zone are the evidences of marine invertebrates: skeletons and debris, trace fossils, especially those made by burrowing forms. These indications of relatively abundant animal life may, in sandy sediments, be accompanied or replaced by bedforms and stratification made by waves or currents—ripples, planar and cross-stratification, and winnowed sand lenses.

INTERTIDAL: *Alternately flooded and exposed.* The variety of diagnostic features can be grouped into four process categories:

1. Intermittent exposure: desiccation cracks, fenestral pores (birdseyes, loferites), rain prints, animal tracks and trails, intraclasts of muddy sediments
2. Algal-laminated structures: stromatoids, crinkled laminations, cryptalgal structures
3. Alternating erosion and deposition, rapid changes in current or wave velocity, channels, scour-and-fill, accumulations of intraformational conglomerate (especially flat-pebble and edgewise complomerates), substantial changes in grain size from bed to bed or lamina to lamina.
4. Reversals of depositing currents: herringbone cross-stratification.

SUPRATIDAL: *Infrequent flooding.* The diagnostic features are those produced by prolonged exposure—nodular evaporites, cemented crusts (caliche), soil horizons, plant roots, and karsted surfaces.

Because the intertidal zone has so many diagnostic structures it is the zone most often identified; a similar bias of criteria for the intertidal zone in siliciclastic examples has also been noted. Deposits considered intertidal, are often used by themselves as evidence of tidal-flat deposition, but such an interpretation is substantially strengthened when the underlying beds are clearly subtidal and the overlying ones are supratidal. The use of vertical sequence, sub-, intra-, and supratidal features, is illustrated in the four examples of Section IV. The first two, by Fisher and Laporte, are summaries of pioneer works that have strongly influenced subsequent research. In the third contribution, Read describes a Devonian succession in which cryptalgal-fenestral carbonates succeed stromatoporoid-rich beds; in the fourth, Hoffman analyzes a Precambrian section in which the useful evidences of animal life are necessarily absent.

27

Tidal Deposits, Dachstein Limestone of the North-Alpine Triassic

Alfred G. Fischer

Occurrence

Late Triassic (Norian-Rhaetian) limestones, 1,000 to 1,500 meters thick, consist of interbedded lagoonal limestones (90 percent) and intertidal-supratidal dolomitic limestones (10 percent). The belt was originally some 20 km wide, extending some 250 km from the Loferer and Leoganger Steinberge in the west, through the region of Berchtesgaden and Salzburg, toward Vienna. It is bounded on the south by a belt of reefs defining the southern edge of the Dachstein bank, and on the north by the Hauptdolomit ultrabackreef facies. The deposit is well exposed in the many mountain faces and high plateaus of the region; lower and readily accessible exposures south of Salzburg include cuts and a quarry near Golling (Pass Lueg), and at Berchtesgaden, the road from Obersalzberg to Hitler's "Eagle's Nest" on Mt. Kehlstein. Rocks are moderately deformed, with primary porosity lost by cementation.

Facies

An ideal representation of the Lofer cycle (Sander, as modified by Fischer, 1964) is shown in Figure 27–1. A weathered and solution-riddled surface of limestone (member C) is overlain (and penetrated) by red or green argillaceous material (member A), which may include limestone cobbles, and is interpreted as a modified terrestrial soil. Member B (average 50 cm) consists of the intertidal beds to which this paper is devoted. Member C (average 5 meters) is a generally unbedded,

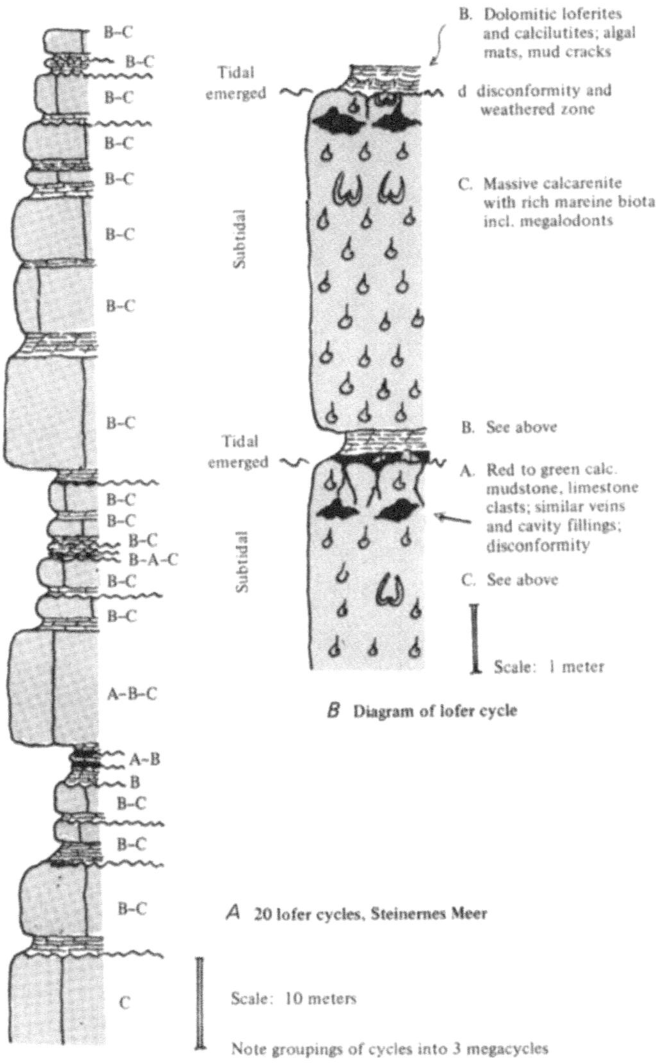

Figure 27-1. The Lofer cycle. *A*. Sequence of 20 cycles. *B*. The cycle as typically developed.

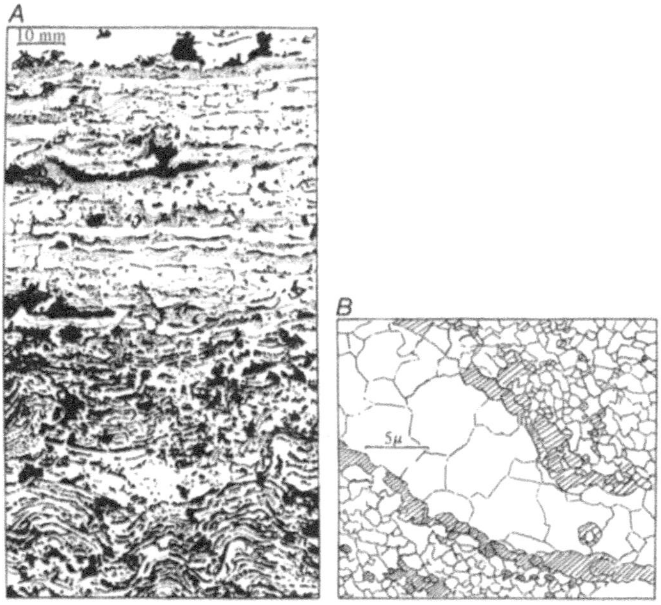

Figure 27-2. *A.* Algal-mat loferite: crinkled at base, flat-bedded above, showing fenestra and sheet cracks of varying geometry and dimensions. Golling quarry. *B.* Transmission electron micrograph of the same rock, showing a calcite-filled, dolomite-rimmed algal filament.

light limestone containing oncoids, dasycladacean and codiacean algae, Foraminifera, bryozoa, gastropods, megalodontid and other bivalves, and echinoderms. This lagoonal sediment grades laterally into coral-sponge reef facies.

Intertidal member B consists of a variety of rock types:

Loferites or birdseye limestone. Cream to light gray micritic carbonates, containing an abundance of pores in the millimeter range (birdseyes, shrinkage pores, or fenestra of other authors) that subsequently have become filled by geopetal mud and sparry cement. In the Dachstein limestone, these loferites normally contain much micritic dolomite, which produces a white bloom on weathered surfaces. *Laminated loferites* (L-F limestones of Tebbut, et al., 1965) are characterized by a millimeter-scale lamination induced by change of dolomite content, color, texture (homogeneous versus pelleted), and alignment of pores or their confluence into sheet cracks. Algal lamination may be flat or crinkled (Fig. 27–2*A*). Laminated loferites are not uncommonly mud-cracked (prism-cracked) (Fig. 27–3*C*). They grade into *massive loferites,* which generally show a clotted structure of poorly defined rounded bodies, some of which, in turn, are composed of vague pellets. These bodies seem comparable to the peds of soils (Fig. 27–3*A*, Fig. 27–3*B*), and are here named as such.

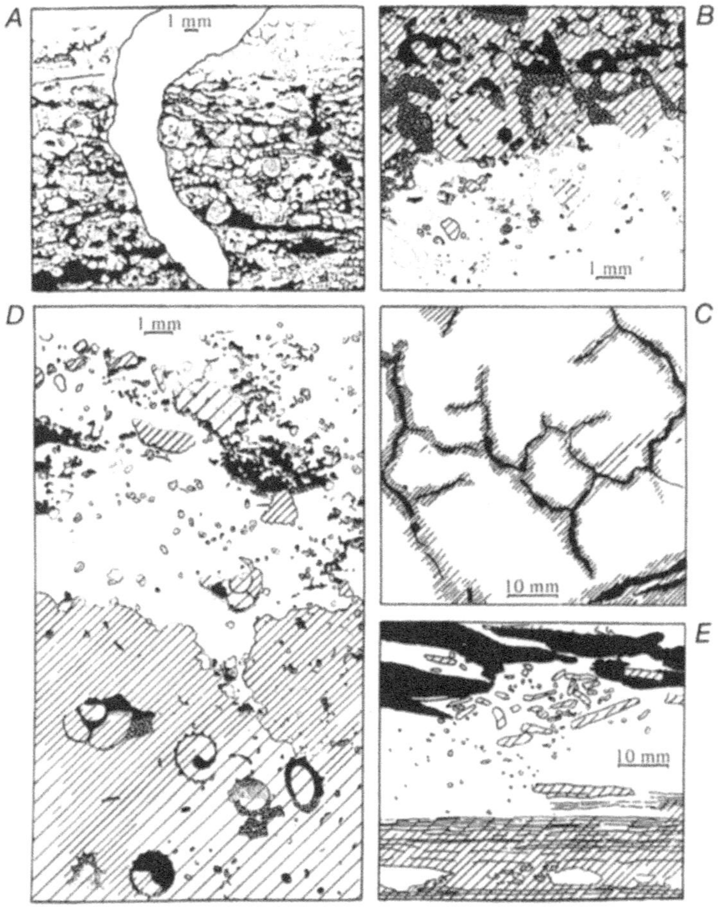

Figure 27-3. *A*. Calcite loferite showing development of rounded peds of varying dimensions, between fenestra confluent into sheet cracks. Cut by vein. Mt. Kehlstein. *B*. Loferitic dolomite crust overlying a calcilutite bearing dolomitic intraclasts. Geopetal muds in fenestra of crust are graded and have penetrated from below. Mt. Dachstein. *C*. Prism-cracked algal loferite, section parallel to bedding. Shallow, incomplete prism cracks, forming a second-order set within decimeter-scale prism (a bounding crack of which cuts lower-right corner of picture). Crack walls are dolomited. Loferer Steinberge. *D*. Somewhat dolomitic lutite bearing cerithid snail shells, disconformably overlain by calcilutite with intraclasts derived from substrate. Shells were dissolved, leaving molds, before deposition of calcilutite which furnished geopetal mud. Steinernes Meer. *E*. Laminated dololutite containing lenticular undolomitized remnants; disconformably overlain by calcilutite containing flat clasts of substrate, as well as mollusk shells which underwent solution and secondary filling. Mt. Kehlstein.

Nonloferitic lutites. Sediments lacking fenestra include laminated (Fig. 27–3*E*) and massive calcilutites and dololutites that range into pink and brown shades. The massive type commonly contains a restricted fauna of Foraminifera, ostracodes, cerithid snails, and small bivalves (Fig. 27–3*D*).

Intraclasts. Intraclasts are extremely common in either facies, consisting of reef-derived fossil debris (near reef belt), intraclasts of calcilutite, and, most commonly, angular clasts of dolomitic composition, frequently in calcilutite matrix. These dolomite clasts, derived from immediately underlying beds, range from flat-pebble derivatives of laminated dolomite (Fig. 27–3*E*) to irregular lumps (Fig. 27–3*D*). Some beds are largely composed of dolomitic intraclasts and intervening fenestra.

Evidence for Intertidal Origin

BIOTA. The laminated facies (crinkled and flat) represents mats of filamentous and coccoid algae (Fischer, 1964) characteristic of modern tidal flats. The fauna of the massive lutites is of tidal-flat type (Fig. 27–3*D*).

MUD CRACKS. (1) Prism cracks: Normal mud cracks, bounding prismatic columns of laminated loferites that have bowed to become concave upward; cracks may occur at several scales (see minor ones in Figure 27–3*C*). (2) Sheet cracks: In general, bedding-parallel planes of separation may be due to several causes (soft-sediment deformation, removal of soluble stratum, shrinkage). The loferites (Figs. 27–2*A*, 27–3*A*) show every gradation from aligned fenestra to confluent fenestra to sheet cracks, which here seem to have resulted from shrinkage of sediment due to desiccation, as discussed below.

PED AND FENESTRA STRUCTURE. The structure most typical of the nonlaminated loferites is one of rounded and commonly compoundly-pelletoid micritic bodies of sediment (peds), separated by irregular pore spaces (fenestra), which tend to be confluent into sheet cracks in the bedding dimension (Fig. 27–3*A*). This structure is common in soils, and is probably caused by repeated wetting and drying, with the associated cycles of expansion and contraction, transfer of dissolved and colloidal materials, and the deposition of such materials from thin, evaporating water films controlled by capillary forces. Many fenestra may owe their initial origin to other causes, such as burrowing or gas bubbles generated in the sediment or trapped in sediment during flooding. And some may be simple intergranular pores in intraclastic beds in which the clasts have lost their identity; but even these may be modified by desiccation.

DOLOMITE. All of the facies of member C may range from calcilutite to dololutite; dolomite content increases northward (in Hauptdolomit equivalents even geopetal mud and spar are dolomite). Dolomite micritic ($<1\ \mu$), anhedral (Fig. 27–2*B*); some of it formed as contemporaneous brittle surface crusts shown by intraclasts (Figs. 27–3*D*, 27–3*E*), and thus is of intertidal-supratidal type.

INTERNAL DISCONFORMITIES. Within member B, beds are commonly separated by discordant surfaces showing a sharp micro-relief, and accompanied by angular intraformational clasts (Figs. 27–3*D*, 27–3*E*). This is evidence of turbulent disruption of at least moderately well-cemented sediment, thus suggesting a tidal setting.

OTHER EVIDENCE OF LEACHING AND CEMENTATION. Preservation of penecontemporaneous fenestra in algal-mat facies (Fig. 27–2*A*) and of ped-and-fenestra fabric in massive loferites (Figs. 27–3*A*, 27–3*B*) are evidence of penecontemporaneous cementation. Preservation of snail shells (Fig. 27–3*D*) shows leaching of aragonite, leaving molds, before deposition of succeeding calcilutite (which invaded molds to form geopetal mud).

SEQUENCE. A theoretical, complete sequence of this kind of transgressive-regressive cycle might be expected to read *d* (disconformity)—*A* (soil)—*B* (intertidal, transgressive)—*C* (subtidal)—B (intertidal, regressive)—*d*. The Lofer sequence generally lacks the regressive phase of *B*, probably because of the succeeding erosion. When total emergence did not occur and the sequence passes gradationally from *C* to *B* and back into *C* (Fig. 27–1*A*), the *B* interval presumably contains both transgressive and regressive phases. *A* is also commonly not developed as a bed, presumably because of erosion at and beyond the margin of the tidal zone; however, remants of *A* are abundant in vein and cavity fillings.

Discussion

Absence of evaporites and gypsum pseudomorphs suggests the setting was not arid.

The Dachstein Limestone records periodic alternation between subtidal and tidal and even wholly emergent conditions. Any given sequence records about 300 such events. Assuming 15 million years for Norian-Rhaetian time, the Lofer cycles average 50,000 years, of which probably only some hundreds of years are represented by the intertidal member.

Three distinct models may be proposed as end-member possibilities for the sequence as a whole: a steady state tidal marsh model, a eustatic oscillation model, and a diastrophic oscillation model.

In a steady state tidal marsh model, we assume steady subsidence at the rate of about 100 Bubnoff units (millimeters per thousand years or meters per million years), and picture, behind the reef belt and a narrow lagoon, a development of tidal marshes and tidal creeks, tens of kilometers wide, presumably maintained by large tides. In this setting, the tidal creeks are the main progenitors of the subtidal sediments, each massive unit (member C) representing the laterally-accreted product (channel and point-bar sediment) of a meandering tidal creek, while the intertidal deposits (member B) accumulated on the tidal floodplain between creeks.

This model seems inapplicable to most of the sequence:

1. It does not explain the common occurrence of emergent horizons with features of weathering, and the soil-like characteristics of member A.
2. One would expect this setting to produce much more lenticular deposits.
3. One would not expect the normally rich biota of member C to have lived in tidal creeks, while, on the other hand, one would expect member C to contain many dolomitic intraclasts (which are absent).

While this model must therefore be rejected for the sequence as a whole, some of the beds within it were possibly formed in this way.

The other two models ascribe the sequence to regional oscillations in relative sea level, with an amplitude of perhaps 15 meters and a period of about 50,000 years. These oscillations could represent either interruptions of the regional subsidence pattern by episodes of uplift or worldwide changes in sea level. Either model would produce a broadening lagoon with landward-migrating tidal fringe during periods of transgression, and a narrowing lagoon with a seaward retreat of the tidal belt during periods of regression. Fischer (1964) prefers a eustatic model, and suggests that variations in rate of subsidence are a megacyclic overprint. The eustatic oscillation is tentatively attributed to distant glaciations under the influence of the Milankovitch climatic cycle.

References

FISCHER, A. G. 1964. The Lofer cyclothems of the Alpine Triassic. *In* Merriam, D. F., ed. Symposium on Cyclic Sedimentation. *Kansas Geol. Surv. Bull. 169*, 107–149.

TEBBUT, G. E., CONLEY, C. D., and BOYD, D. W. 1965. Lithogenesis of a distinctive carbonate rock fabric. *Wyoming Geol. Surv. Contr. Geol. 4*, 1.

28

Carbonate Tidal-Flat Deposits of the Early Devonian Manlius Formation of New York State

Léo F. Laporte

Occurrence

The Manlius Formation, 7 to 15 meters thick, is the basal unit of the Lower Devonian Helderberg Group in eastern and central New York State. The Helderberg Group is a 100-meter sequence of fossiliferous marine limestones deposited in the wake of a shallow transgressing sea (Fig. 28–1). The Manlius records the tidal-flat and shallow subtidal portion of this transgression. The Manlius is exposed in numerous outcrops, roadcuts, quarries, and streambeds west of the Hudson River from Kingston north to Albany, and north and south of Route 20 from Albany west to Syracuse (Fig. 28–2). Two localities where the tidal-flat deposits of the Manlius are particularly well exposed are on the abandoned east wall of a quarry, South Bethlehem, and on the abandoned east wall of a quarry, 1.6 km north-northwest of Perryville (Rickard's localities 53 and 144 [1962]).

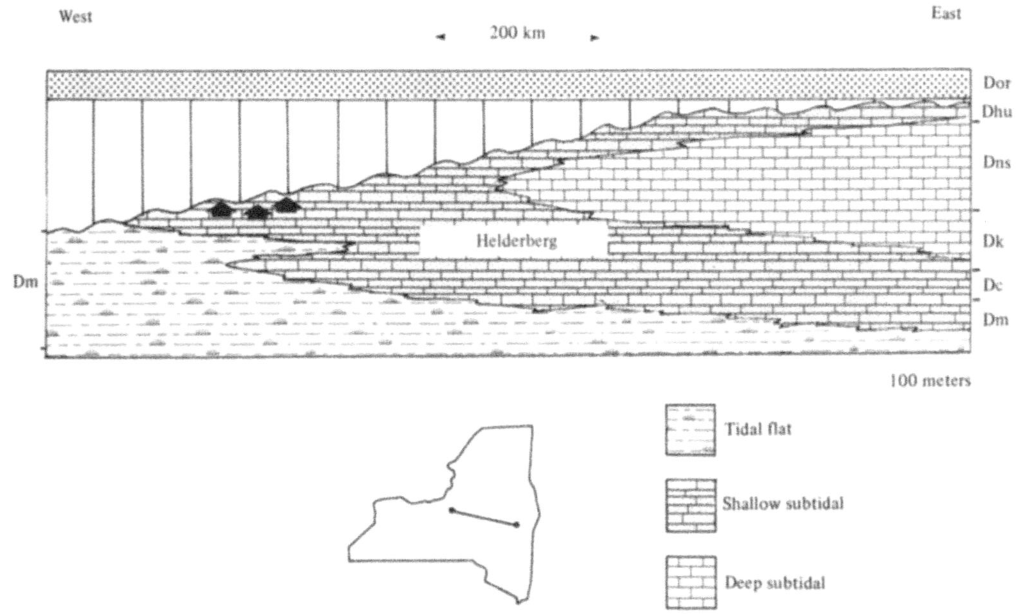

Figure 28-1. Restored section of the Helderberg Group, Lower Devonian, New York. The Helderberg carbonate sequence records a westward transgressing sea with tidal-flat deposits being successively overlain by shallow subtidal and deep subtidal facies (Laporte, 1969, 1971). *Dm* = Manlius, *Dc* = Coeymans, *Dk* = Kalkberg, *Dns* = New Scotland, *Dhu* = upper Helderberg, *Dor* = Oriskany Sandstone that lies disconformably on the Helderberg. Black arrowheads indicate position of coralline bioherms in the upper Coeymans in central New York. Note that formations within the Helderberg cross time lines, becoming increasingly younger in the west.

Figure 28-2. Helderberg and Manlius sections as typically found in the Hudson Valley, eastern New York. (Symbols conform to conventions used in this volume.)

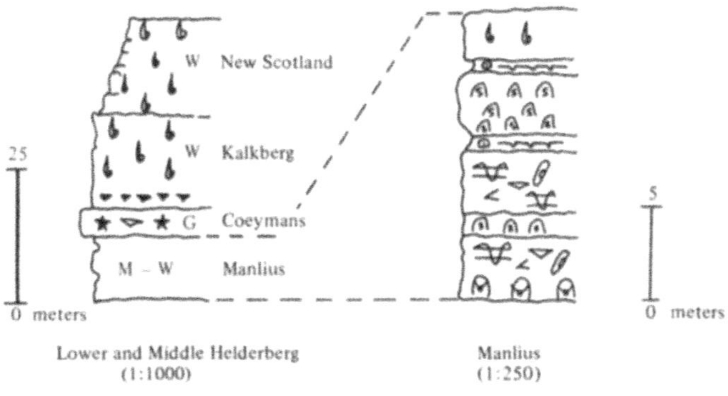

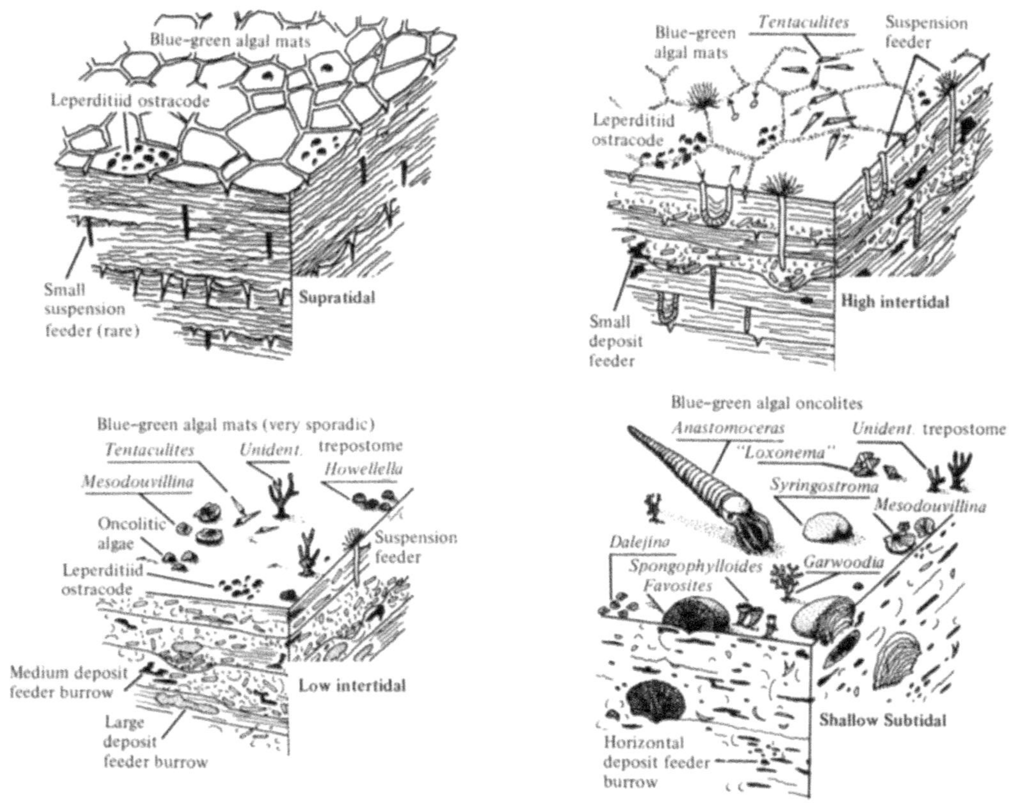

Figure 28-3. Inferred reconstructions of Manlius facies. (From Walker and Laporte, 1970.)

Facies

Three idealized end-member facies can be recognized in the Manlius: supratidal, intertidal, and subtidal.

The *supratidal facies* is characterized by irregular laminar stratification, mud cracks, birdseye, and scattered ostracod valves in a pelletal lime mudstone. Often the laminations, about 1 mm thick, are alternately calcite/dolomite, and are separated by thin, wispy, bituminous films (Figs. 28–4, 28–5).

The *intertidal facies* is typically thin-bedded, pelletal lime mudstones and skeletal grainstones with scour-and-fill (Fig. 28–6). Fossils include ostracods, tentaculitids, small spiriferid and strophomenid brachiopods, ramose ectoproct bryozoans, and small U-shaped spreiten burrows; organosedimentary structures interpreted as algal stromatolites and algal oncolites are also common.

The *subtidal facies* is a burrowed, pelletal lime wackestone with stromatoporoids, small solitary rugose corals, brachiopods and ectoproct bryozoans, codiacean algae, snails, an occasional favositid coral, and a rare orthocone cephalopod (Fig. 28–7).

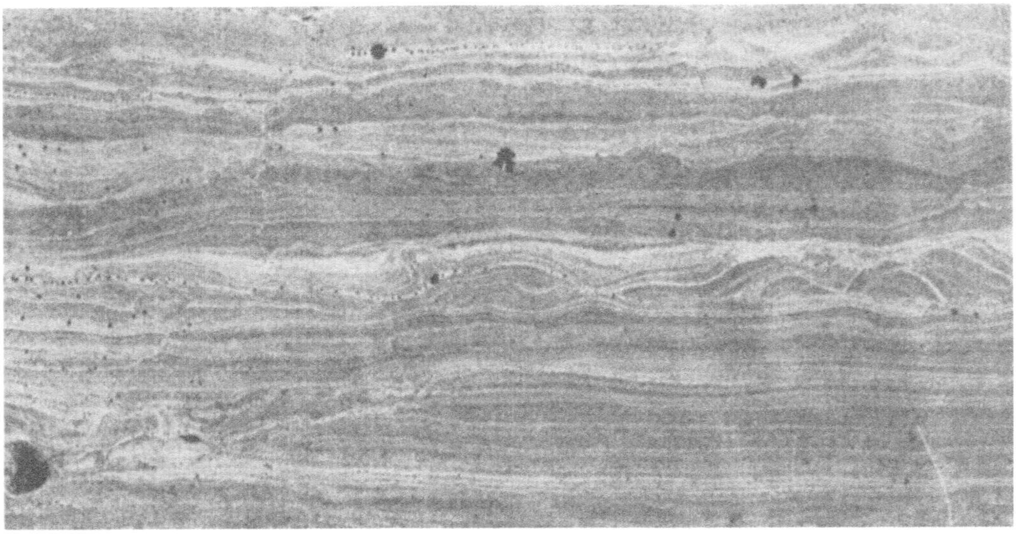

Figure 28-4. Irregularly laminated, fine-grain calcite (gray) with thin alternating laminations of dolomite (white). Individual laminations are disturbed in places by curling and disruption (of algal mats?) upon desiccation. A few skeletal grains are present. Supratidal facies, Manlius Formation, McConnell Corners, New York (loc. 128).

Figure 28-5. Laminated, fine-grain calcite, alternating with dolomitic laminations. Skeletal debris has accumulated in small depression. Note laminations overhanging the depression; the steep angle of these overhangs suggests binding by algal mats. Supratidal facies, Manlius Formation, Austin's Glen, New York (loc. 44).

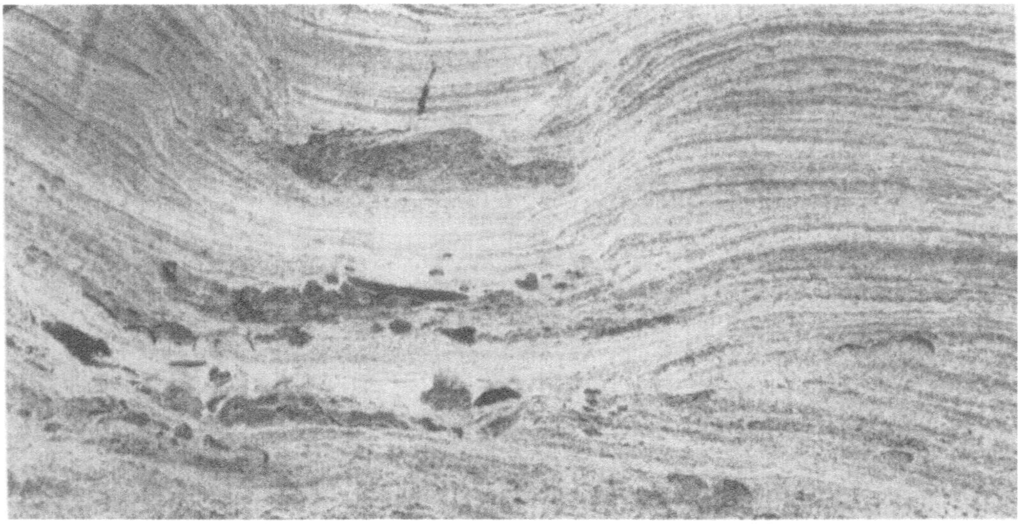

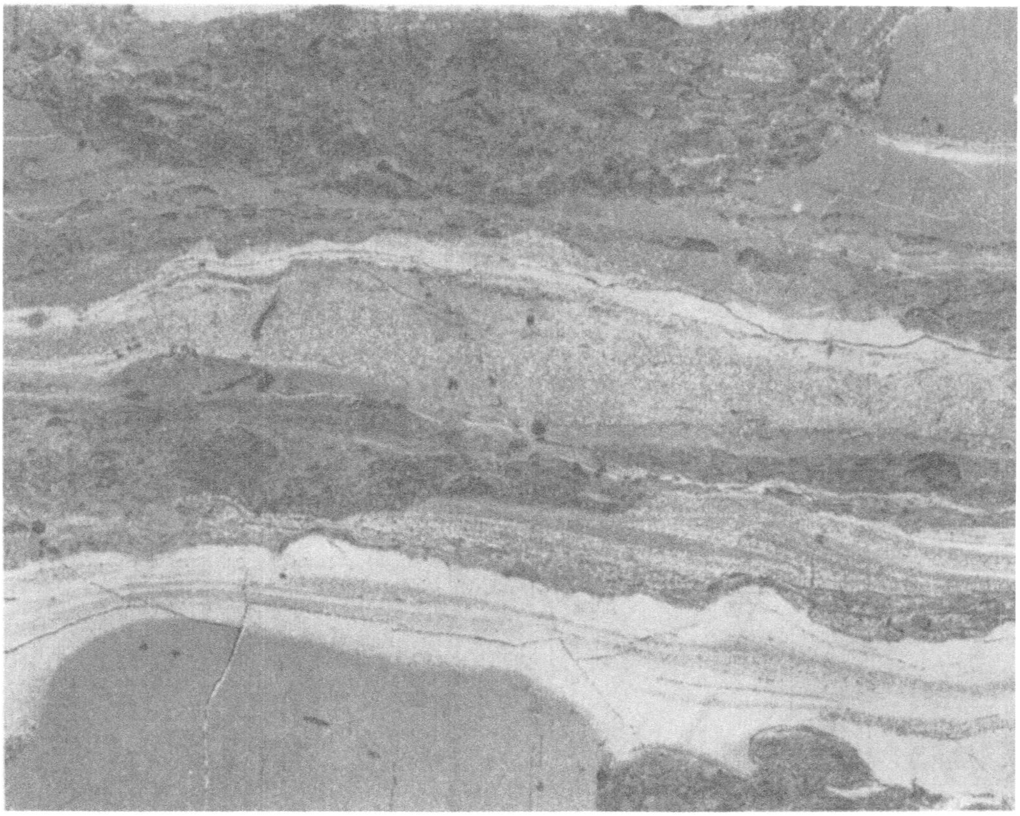

Figure 28-6. Lithified mudstones eroded and overlain by skeletal layers and thin, laminated dolomite. Typical example of scour-and-fill structures found throughout the Manlius. Skeletal debris includes tentaculitids, brachiopods, ostracods, and ectoproct bryozoans. Intertidal facies, Manlius Formation, South Bethlehem, New York (loc. 53).

Stromatoporoids often form tabular beds 2 to 3 meters thick and several tens of meters long.

In general, the supratidal and intertidal facies (collectively referred to as tidal-flat deposits) are best developed in the lower half of the Manlius, while the subtidal facies is better developed in the upper half of the unit. In detail, however, individual outcrops show repeated interbedding of one facies with another, suggesting lateral migration of tidal flats across shallow subtidal deposits during the initial phase of the Helderberg transgression. The Manlius is thus a complex facies mosaic; that is, a stratigraphic record of laterally shifting tidal-flat environments, with and opposite to the direction of the main advancing edge of the marine transgression.

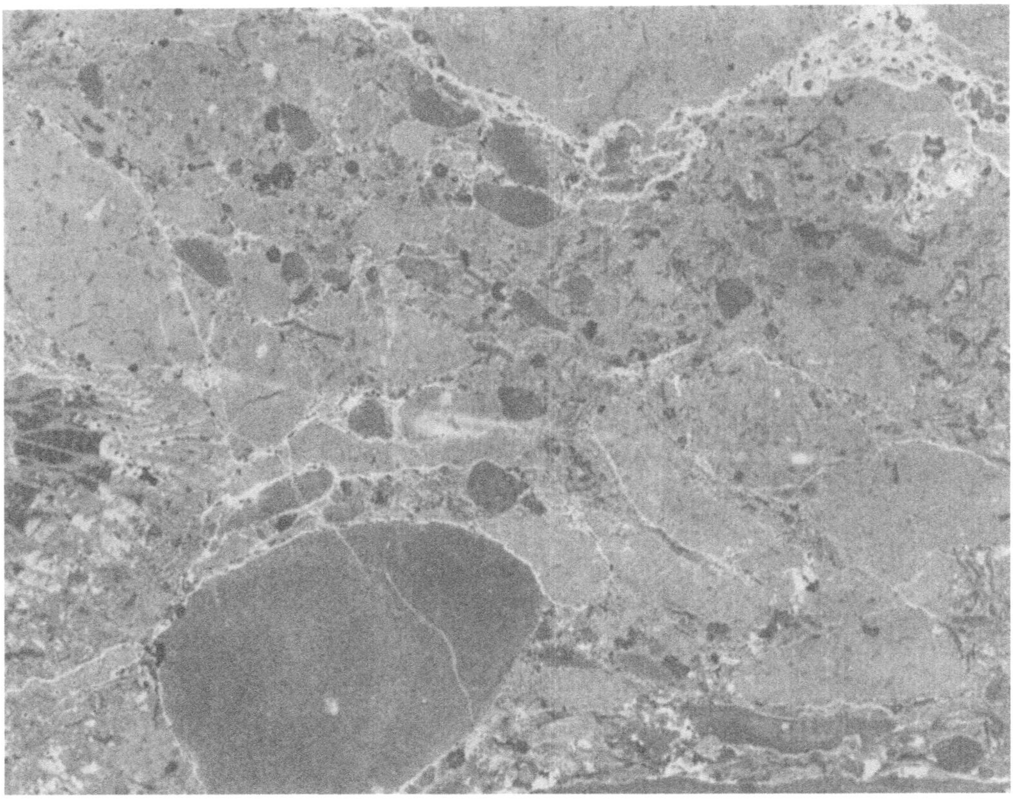

Figure 28-7. Burrowed wackestone with skeletal debris, including favositid, stromatoporoids, brachiopods, ostracods, and pelmatozoans. Subtidal facies, Manlius Formation, Perryville, New York (loc. 144).

Evidence for Tidal Origin

The abundance of mud cracks and birdseye is assumed to be the result of desiccation of Manlius sediments upon subaerial exposure. The beds containing these presumptive desiccation features (supratidal facies) are intimately and repeatedly associated with beds containing abundant marine fossils (subtidal facies). It is therefore concluded that during Manlius deposition there were numerous water level fluctuations.

Supporting this conclusion is the additional presence of abundant scour-and-fill structures that separate interbeds of fine-grained pelletal carbonate mudstones, 3 to 5 cm thick, from overlying, 1- to 3-cm thick, skeletal grainstones that contain intraclasts of the underlying fine-grained carbonate (intertidal facies, Fig. 28–8). Unless the fine-grained carbonates were lithified (cemented) under seawater, it might be supposed that the carbonate muds were lithified upon subaerial exposure

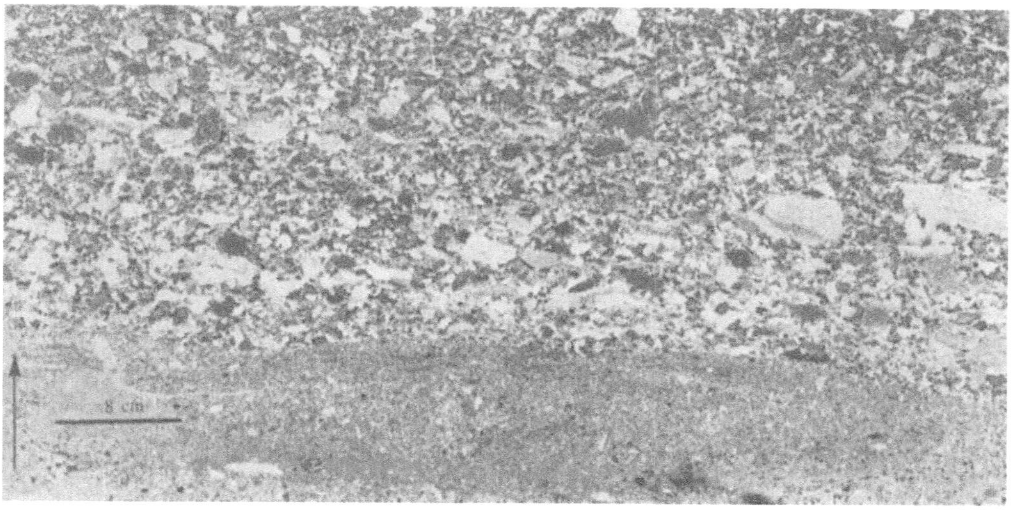

Figure 28-8. Unlaminated, vertically burrowed dolomite below; dolomitic intraclasts above; scour surface between. Supratidal facies, Manlius Formation, Jamesville, New York (loc. 151). This and Figures 28-5 to 28-8 are negative prints of acetate cellulose peels at about 2 × magnification. The peels are taken normal to bedding and the top of the peel is stratigraphically up. Locality numbers are from Rickard, 1962.

and later eroded. The intraclasts might be either reworked mud-cracked polygons or grains eroded from the tidal flats by high waters.

The faunal diversity is low in the Manlius, supporting the interpretation of a high stress tidal flat and restricted, shallow subtidal environment. The presence of U-shaped spreiten burrows is also consistent with a tidal-flat interpretation. The bituminous films, present virtually everywhere in the supratidal facies (Figs. 28–4, 28–5), are further interpreted as the residues of thin algal mats.

Finally, the occurrence of dolomitic laminations (in which the dolomite rhombs are 5 to 10 μ in diameter) alternating with fine-grained calcite laminations is similar to occurrences of penecontemporaneous dolomite in Holocene carbonate tidal-flat deposits.

The phrase "tidal flat" refers to depositional environments judged to have been subaerially exposed for indeterminate lengths of time, whether hours, days, months, or years. In a similar vein, the terms "supratidal," "intertidal," and "subtidal" refer to degrees of subaerial exposure, whether occasionally wetted, regularly wetted, or continuously wetted by marine waters.

Clearly the tidal-flat deposits accumulated in a zone of fluctuating water level, but I do not know—and it may remain forever unknown—if the water fluctuations were the result of astronomic tides, wind tides, storms, monsoon climates, or whatever, or how long the intervals were between wettings.

Discussion

Given the regional paleogeographic and stratigraphic framework of the Helderberg Group carbonates, a tidal-flat and restricted shallow subtidal origin for the Manlius is reasonable. Higher Helderberg units record more offshore, high and low energy, open marine environments that succeeded the inshore, tidal flat, and restricted subtidal Manlius. The shoreline facies of the Helderberg, as recorded within the Manlius, did not, however, migrate in a simple supratidal-intertidal-subtidal pattern. Rather, the complex mosaic of facies within the Manlius indicates local and repeated shifts in shoreline environments.

Tidal-flat deposits at times prograded over the subtidal deposits; at other times the subtidal facies overrode the tidal flats. Variations in carbonate production, subsidence, and hydrography undoubtedly controlled whether any one place along the Manlius shore was just above, awash, or just below "mean sea level." Integrated through time, however, there was a gradual net change in sea level with tidal-flat environments dominating in early Manlius deposition and shallow, restricted subtidal dominating in late Manlius deposition.

Recognition of tidal-flat deposits within the Manlius provides a valuable environmental datum for interpreting laterally equivalent Helderberg facies that lack such environmentally diagnostic carbonate facies. As with oolitic and reefy carbonates, tidal-flat carbonates are an important key to environmental stratigraphic interpretation of more ambiguous associated facies.

References

LAPORTE, L. 1967. Carbonate deposition near mean sea-level and resultant facies mosaic: Manlius Formation (Lower Devonian) of New York State. *Am. Assoc. Petrol. Geol. Bull. 51*, 73–101.

——— 1969. Recognition of a transgressive carbonate sequence within an epeiric sea: Helderberg Group (Lower Devonian) of New York State. *Soc. Econ. Paleontol. Mineralo. Spec. Publ. 14*, 98–119.

——— 1971. Paleozoic carbonate facies of the central Appalachian shelf. *J. Sed. Petrol. 41*, 724–740.

RICKARD, L. 1962. Late Cayugan (Upper Silurian) and Helderbergian (Lower Devonian) stratigraphy in New York. *N.Y. State Mus. Sci. Serv. Bull. 386*, 157 pp.

WALKER, K., and LAPORTE, L. 1970. Congruent fossil communities from Ordovician and Devonian Carbonates of New York. *J. Paleontol. 44*, 928–944.

29

Tidal-Flat Facies in Carbonate Cycles, Pillara Formation (Devonian), Canning Basin, Western Australia

J. F. Read

Occurrence

Cyclic carbonate units consisting of stromatoporoid biostromes overlain by pellet limestones with cryptalgal and fenestral fabrics occur in the Pillara Formation (Devonian), Canning Basin, Western Australia (Read, 1973). About 70 cyclic units, 2 to 10 meters thick, occur at the type section in the Pillara Range. Part of the sequence at the type section is shown in Figure 29–1*A*; an idealized cyclic unit is shown in Figure 29–1*B*.

Sediments

Stromatoporoid biostromes

SUBSPHERICAL-STROMATOPOROID LIMESTONE. Massive limestones less than 3 meters thick, composed mainly of subspherical-stromatoporoid colonies (up to 0.6 meters in diameter),

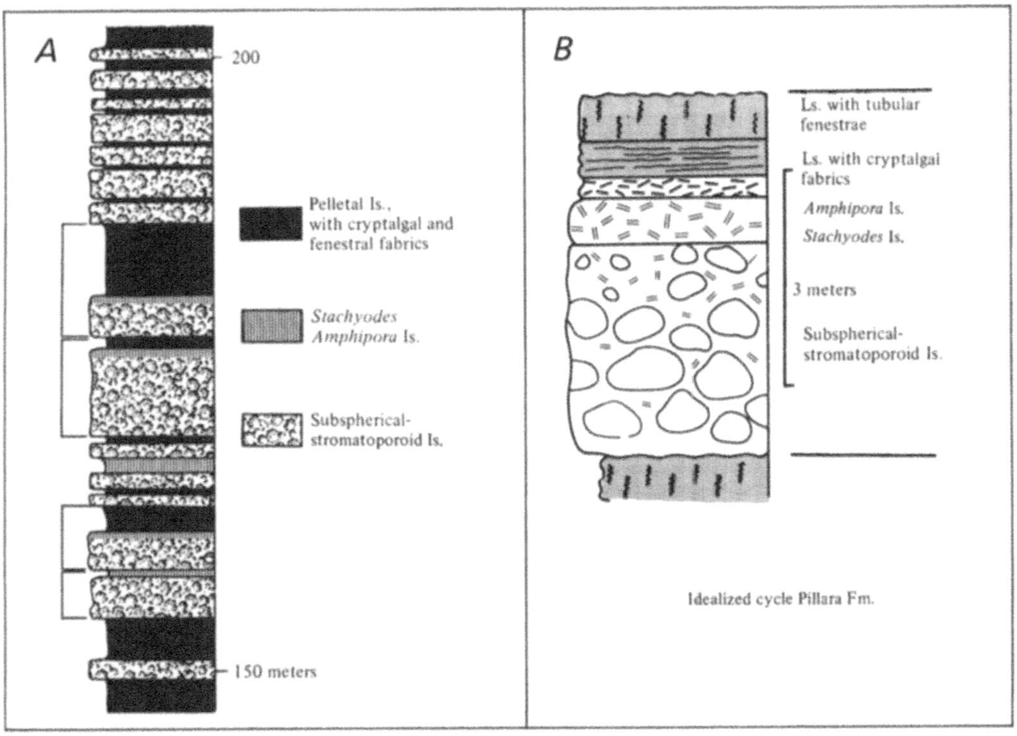

Figure 29-1. *A.* Part of stratigraphic section, Pillara Formation, type section. Typical cycles are in brackets. *B.* Idealized cycle, Pillara Formation.

flat-lying *Stachyodes, Amphipora,* and minor cylindrical corals in a matrix of skeletal packstone and wackestone. Sand-sized grains include angular stromatoporoid fragments, dasyclad algal fragments, calcispheres, small gastropods, and ostracods, together with pellets (in packstones) and lime mud (in wackestones).

STACHYODES LIMESTONE. Medium-bedded to massive wackestones and packstones less than 1 meter thick, containing abundant flat-lying *Stachyodes, Amphipora,* and small subspherical stromatoporoids (less than 10 cm in diameter) together with common, sand-sized grains (calcispheres, dasyclad algal fragments, ostracods, spicules, and minor mollusks), and variable amounts of pellets, intraclasts, and lime mud.

AMPHIPORA LIMESTONE. Thin-bedded packstones and wackestones composed of flat-lying *Amphipora,* sand-sized skeletal grains (*Amphipora* fragments, calcispheres, dasyclad algal fragments, and rare mollusks), variable quantities of pellets, intraclasts, and lime mud.

Pellet limestone and lime mudstones

SEDIMENTS. Pellet limestones are thin- to medium-bedded sediments (1 to 10 meters thick) that commonly have cryptalgal fabrics or fenestral fabrics. Most are packstones or wackestones composed of pellets, intraclasts of pellet packstone, small amounts of skeletal grains, and variable amounts of lime mud.

FABRICS. Cryptalgal and fenestral fabrics in pellet limestones tend to occur in a definite vertical succession (Fig. 29–2). The sequence outlined in Figure 29–2 is common in thin units (less than 2 meters) of pellet limestone at tops of cycles, although some elements are absent locally. Thick sequences (2 to 10 meters) of pellet limestone and lime mudstone tend to have repeated successions or minor reversals.

Fine, laminoid-fenestral fabric consists of subparallel close-spaced, flattened spar-filled voids (fenestrae) up to 1 mm high (Fig. 29–3*C*). Sediments are generally flat-laminated pellet packstones with lamination marked by grain-size differences; some sediments with this fabric are unlaminated, with a poorly defined layering marked by alignment of flattened voids.

Medium, irregular fenestral fabric consists of irregular to subspherical equidimensional fenestrae from 1 to 5 mm in size (Fig. 29–3*B*); some voids tend to be flattened horizontally.

Figure 29-2. Sequence of cryptalgal and fenestral fabrics.

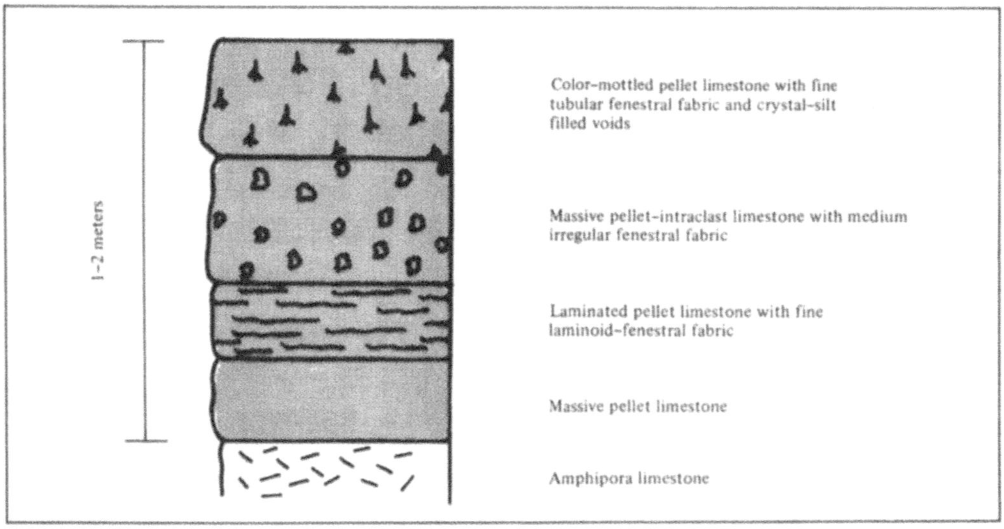

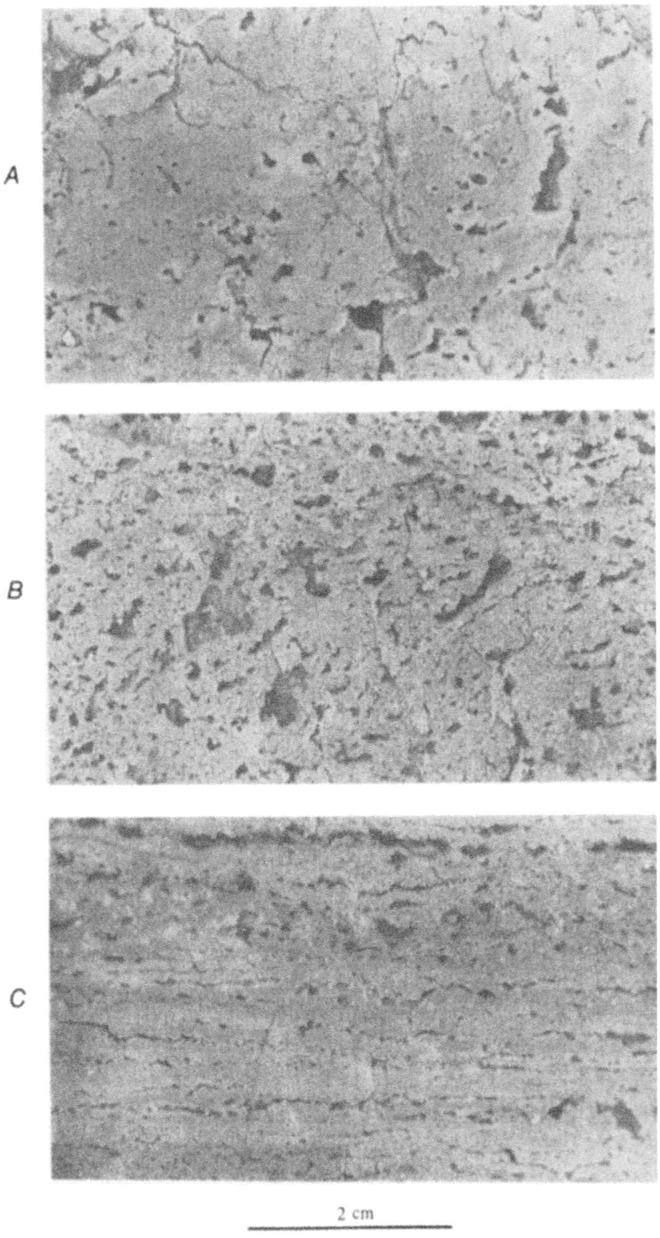

Figure 29-3. Cryptalgal and fenestral fabrics. Polished slabs. *A*. Lime mudstone with tubular fenestrae. *B*. Pelletal limestone with irregular fenestrae. *C*. Flat, algal-laminated limestone with laminoid fenestrae.

Sediments with this fabric are mainly massive, unlaminated, intraclast-pellet packstones.

Fine, tubular fenestral fabric consists of randomly oriented to subvertical tubular fenestrae 0.5 to 1 mm in diameter that commonly bifurcate (Fig. 29–3*A*). Rare, larger irregular fenestrae reach 1 cm or more in diameter. Sediments are typically color-mottled lime mudstone and rarely pellet packstone. Mottles consist of bleached cream-colored halos around fenestrae, passing outward through orange or tan (iron-stained) sediment into light tan-colored sediment. Many fenestrae are partly or completely filled with crystal silt.

Environmental Analysis

Pellet limestones with cryptalgal and fenestral fabrics are tidal-flat facies: they are similar both in appearance and vertical succession to those described from tidal flats in Shark Bay, Western Australia (Hagan and Logan, 1974; Logan et al., 1974; Read, 1974). Fine, laminoid-fenestral fabrics are formed beneath smooth mat communities in the lower intertidal zone; irregular fenestral fabrics are developed beneath pustular mat communities in middle and upper intertidal zones; tubular fenestral fabrics are associated with salt-tolerant land plants inhabiting upper intertidal to supratidal areas.

Stromatoporoid biostromes formed under shallow subtidal conditions as evidenced by abundant marine biota. The cycles reflect rapid submergence followed by shoaling by sedimentation under stable conditions. Rapid submergence is suggested by the general absence of reversed cycles and the presence of the most diverse assemblages of marine organisms in lower parts of cycles. Limited thickness of biostromes indicates growth under stable sea-level conditions rather than under subsiding conditions (which would result in thick biostromes). Tidal-flat facies were deposited following shoaling by sedimentation to intertidal levels.

The first occurrence of cryptalgal and fenestral limestones capping cycles defines the approximate position of sea level during cycle deposition and represents a sea-level datum. The distance of a lithofacies below this datum closely approximates the water depth where sedimentation occurred under stable conditions; where sedimentation occurred under gradually subsiding conditions, the distance represents an overestimated value.

The use of cryptalgal and fenestral fabric limestones as approximate sea-level datums for estimation of water depths of cyclic lithofacies provides valuable information on paleoenvironments and paleoecology of the Pillara Formation.

References

HAGAN, G. M., and LOGAN, B. W. 1974. History of Hutchinson Embayment tidal-flat, Shark Bay, Western Australia. *Amer. Assoc. Petrol. Geol. Memoir 22*, 283–315.

LOGAN, B. W., HOFFMAN, P., and GEBELEIN, C. D. 1974. Algal mats, cryptalgal fabrics and structures, Hamelin Pool, Western Australia. *Amer. Assoc. Petrol. Geol. Memoir 22*, 140–194.

READ, J. F. 1974. Carbonate bank and wave-built platform sedimentation, Edel Province, Shark Bay, Western Australia. *Amer. Assoc. Petrol. Geol. Memoir 22*, 1–60.

——— 1973. Carbonate cycles, Pillara Formation (Devonian), Canning Basin, Western Australia. *Bull. Can. Petrol. Geol. 21*, 38–51.

30

Shoaling-Upward Shale-to-Dolomite Cycles in the Rocknest Formation (Lower Proterozoic), Northwest Territories, Canada

Paul Hoffman

Occurrence

The Rocknest Formation (Fraser and Tremblay, 1969) is exposed in an area 200 km long and 100 km wide in the northwest corner of the Canadian Shield. It is part of the Epworth Group, which comprises the supracrustal rocks in the foreland fold and thrust belt of the Coronation geosyncline (Hoffman, 1973a). These rocks were deposited between 2,200 and 1,800 million years ago.

The Rocknest Formation constitutes the upper part of a westward-facing continental shelf sequence. It consists of cyclically interstratified shale and dolomite, and thickens from less than 500 meters in the inner shelf area to 1,200 meters at the outer edge of the shelf. There is an abrupt facies change at the shelf edge, west of which there is only 110 meters of off-shelf shale with beds of dolomite megabreccia.

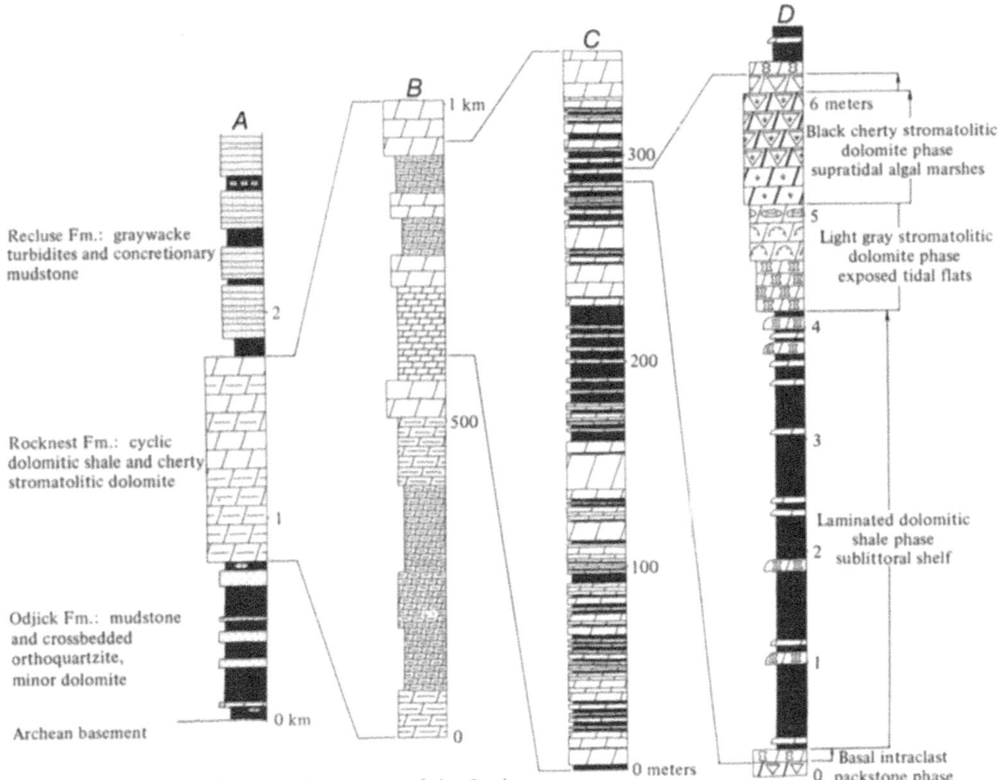

Figure 30-1. Column A. Stratigraphic setting of the Rocknest Formation. Column B. Subdivision of the formation into 12 members. Column C. Alternation of shale and dolomite in the upper part of the formation. Column D. Representative cycle (that is, one major shale-dolomite couplet) with four depositional phases.

Figure 30-2. Aerial photograph of the Rocknest Formation in the Epworth fold belt south of Coronation Gulf, Northwest Territories. The folds have a wavelength of 1 km. The bedding surfaces are expressed by recessive shale half-cycles between resistant dolomite. The linear transverse depressions are diabse dikes. (Reproduced with permission of the Geological Survey of Canada.)

Figure 30-3. Ground view of six shale-dolomite cycles. Top of the formation is to the right, and approximately 65 meters of section are shown in the foreground. (Reproduced with permission of the Geological Survey of Canada.)

Cycles

The Rocknest Formation can be subdivided into 12 members on the basis of differences in shale-dolomite ratio (Fig. 30–1, column B). More importantly, within each member are many alternations of shale and dolomite (Fig. 30–1, column C; Fig. 30–2), most apparent where their proportions are nearly equal (Fig. 30–3). The alternations are asymmetric inasmuch as the shale invariably lies sharply on the dolomite (Fig. 30–4),

Figure 30-4. Sharp contact of laminated dolomitic shale overlying nonshaly dolomite. At the top of the dolomite is a 12-cm-thick bed of intraclast packstone. The scale is 5 ft (1.5 meters) long. (Reproduced with permission of the Geological Survey of Canada.)

Figure 30-5. Intraclast packstone bed overlying stromatolitic dolomite at the base of a cycle. The packstone is itself overlain by dark recessive dolomitic shale at the top of the photograph. The scale is graduated in 3-cm intervals. (Reproduced with permission of the Geological Survey of Canada.)

whereas the upward transition from shale to dolomite is generally more gradational. Each shale-dolomite couplet constitutes one cycle. The cycles are 2 to 20 meters thick and the formation as a whole contains almost 200 cycles.

The cycles are made up of four consistently ordered phases (Fig. 30–1, column D), of which the third from the bottom may in some cycles be absent.

BASAL INTRACLAST PACKSTONE. At the base of nearly every cycle is a laterally persistent bed, normally less than 20 cm thick, of dolarenite or dolorudite that sharply overlies the stromatolitic dolomite of the underlying cycle (Fig. 30–5). It consists of well-rounded dolomite intraclasts packed in a matrix of dololutite or terrigenous shale. The intraclasts are normally well-sorted, grain-supported, and imbricated, and, rarely, there are thin, discontinuous stromatolitic layers.

Figure 30-6 (left). Dark reddish-brown dolomitic shale with poorly graded intervals of yellowish-brown dololutite. The laminations are cut by soft-sediment syneresis dikes. (Reproduced with permission of the Geological Survey of Canada.)

Figure 30-7 (right). Oolitic dolarenite bed with scattered intraclasts from the transition in a cycle between laminated shale and cryptalgal dolomite. Many of the ooids are selectively silicified. (Reproduced with permission of the Geological Survey of Canada.)

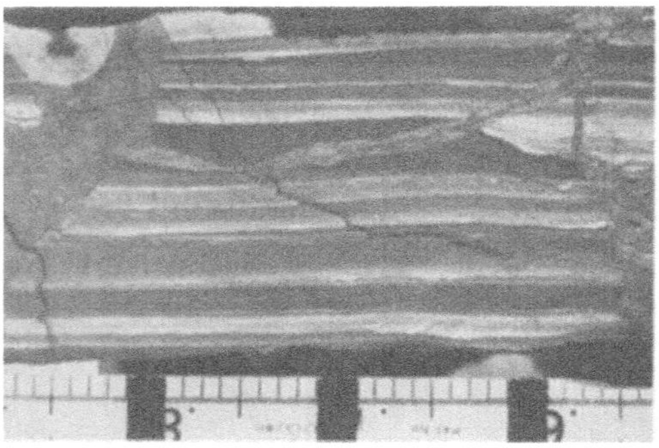

LAMINATED DOLOMITIC SHALE. Recessive reddish-brown to black shale sharply overlies the basalt intraclast packstone bed. The shale has conspicuous variously spaced laminations of yellowish-brown dololutite (Fig. 30–6). Also present are graded layers of dololutite or dolorenite, 1 to 3 cm thick, some of which have load-casted bottoms and ripple-laminated tops. Commonly there are complex networks of sedimentary dikes and microbreccias. The shale becomes more dolomitic and the dolarenite layers much more numerous upward. At the top are beds of intraclastic and oolitic dolarenite (Fig. 30–7) separated only by thin shale partings. These beds commonly have truncated ripple marks and thin discontinuous stromatolitic layers.

LIGHT GRAY CRYPTALGAL DOLOMITE. Gradationally overlying the shale is light yellowish-gray nonshaly dolomite. The commonest inorganic structure is edgewise conglomerate (Fig. 30–8), but the dolomite is dominantly cryptalgal. There is a general decrease in relief of the cryptalgal structures upward. At the base are laterally linked domes (Fig. 30–9), elliptical and preferentially oriented in plan view. In the middle, in some cycles, are discrete, actively branched columns (Fig. 30–10). At the top are stratiform sheets, commonly with discoidal oncolites (Fig. 30–11). On the basis of internal texture, independent of form, three intergradational types of cryptalgal structures are distinguished: (1) stromatolites, the commonest, which are well laminated; (2) thrombolites (Aitken, 1967), which are unlaminated and have a clotted texture; and (3) loferites (Fischer, 1964), which are laminated or unlaminated

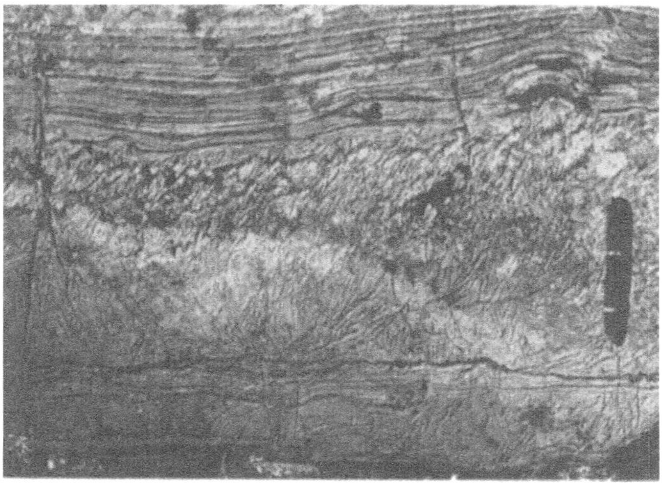

Figure 30-8. Edgewise conglomerate in light gray stromatolitic dolomite. (Reproduced with permission of the Geological Survey of Canada.)

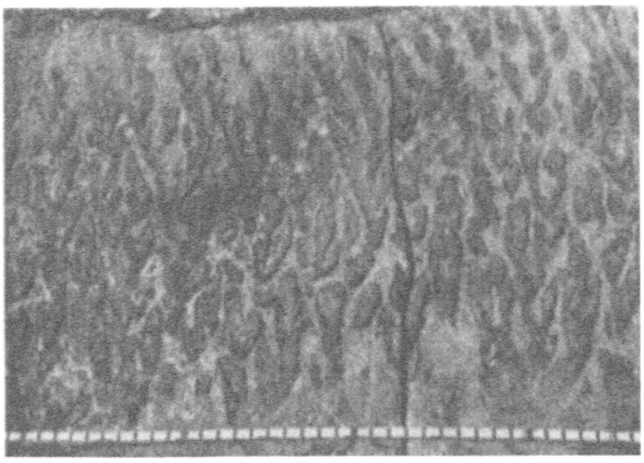

Figure 30-9 (left). Laterally linked domal stromatolites in cherty light gray dolomite. There is a hammer in the right center for scale. In the lower left is laminated dolomitic shale beneath the stromatolites. In plan view, the stromatolites are elliptical and preferentially oriented. (Reproduced with permission of the Geological Survey of Canada.)

Figure 30-10 (right). Bed of discrete actively branching columnar stromatolites. Such stromatolites have no preferential elongation in plan view. (Repreduced with permission of the Geological Survey of Canada.)

and have a birdseye texture. In general, stromatolitic textures occur both below and above the thrombolites and loferites. White chert, habitually replacing the lateral margins of the cryptalgal structures, is ubiquitous.

BLACK CRYPTALGAL DOLOMITE. Within the light gray dolomite, mostly near the top, are units of black cherty dolomite with tiny arborescent stromatolites (Fig. 30–12) that resemble structures in modern algal tufa. Where silicified, microscopic filament molds are preserved in the stromatolites. The stromatolites are separated by light gray dololutite or void-filling sparry dolomite. Within the black dolomite of a few cycles are layers with the distinctive conically laminated columnar stromatolite *Conophyton* Maslov (Fig. 30–13). The columns are unusually slender, less than 1 cm, commonly aggregate to form domal mounds, and tend to be selectively silicified.

Figure 30-11. Discoidal oncolite in light gray stromatolitic dolomite (Reproduced with permission of the Geological Survey of Canada.)

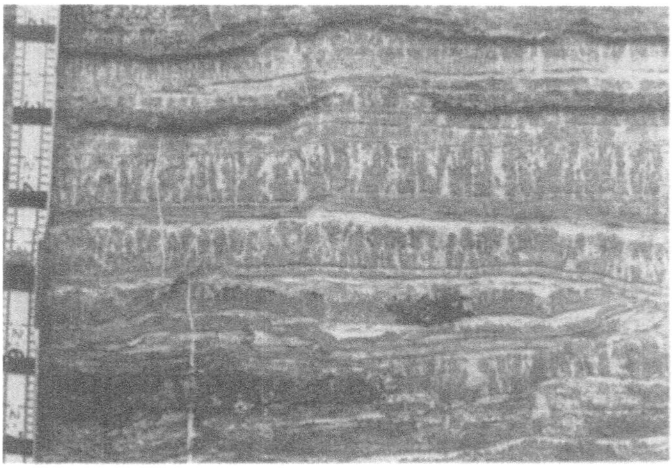

Figure 30-12. Tiny arborescent stromatolites in black cherty dolomite. This important tufalike structure occurs in nearly all the thicker cycles. (Reproduced with permission of the Geological Survey of Canada.)

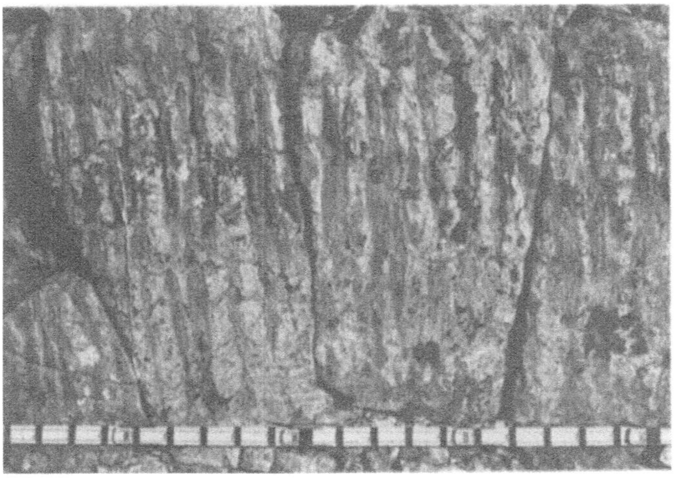

Figure 30-13. Slender unbranched columns of the conically laminated stromatolite *Conophyton* Maslov. Such stromatolites are associated with the arborescent forms in the black cherty dolomite of member 5 only. (Reproduced with permission of the Geological Survey of Canada.)

Paleobathymetric Interpretation

This analysis follows Coogan (1969), among many others, in applying Walther's law to the interpretation of cyclicity in shelf carbonates. Systematically superposed lithotopes must have been deposited in laterally adjacent depositional environments; nondepositional environments are represented by diastems. Each cycle begins with a condensed record of marine transgression across the shelf and ends with a more complete shoaling-upward sequence resulting from progradation of carbonate tidal flats. This paleoenvironmental interpretation of the cycles is independent of whether their fundamental cause is autocyclic or allocyclic (Beerbower, 1964).

During the transgression, the basal intraclast packstone phase was deposited in the surf zone of the retreating shoreline. The intraclasts were eroded from the top of the underlying cycle and rounded in the surf. The fine-grained matrix filtered down into the intraclastic sand as the water deepened. Below prevailing wave base, the laminated dolomitic shale phase accumulated from wind-blown dust, subaqueously suspended sediment, and periodic quickly waning seaward-directed bottom currents capable of moving silt and sand. The sedimentary dikes and microbreccias result from syneresis, and can be distinguished from subaerial mud cracks by the lack of systematic relationship to individual bedding surfaces and by the absence of mud-pebble conglomerate.

Shoaling of the bottom during progradation is indicated in the laminated dolomitic shale phase by the great increase upward in intraclastic and oolitic dolarenite. The lower part of the light gray cryptalgal dolomite phase was deposited in the shallow sublittoral and lower intertidal zones, as evidenced by flat-pebble conglomerate, a surf indicator, and by elongate domal stromatolites, which accrete by sediment adhesion shaped by wave scour (Logan et al., 1974). The upward trend to stratiform cryptalgal sheets with recumbent rather than edgewise pebbles indicates buildup into the upper intertidal, landward of the zone of persistent wave action.

The black cryptalgal dolomite was deposited in protected coastal embayments, stagnant tidal ponds, and seasonally flooded supratidal marshes landward of the main tidal flats. Here, carbonate was precipitated directly in thick algal carpets to produce the tiny arborescent stromatolites and related cryptalgal structures.

Paleoclimate and Origin of the Dolomite

The best paleoclimatic indicator is the supratidal facies, located at the top of the shoaling-upward sequence (Hoffman, 1973b, Fig. 1). In arid climates, this is the well-known sabkha facies with its gypsum or anhydrite. In humid climates, there are brackish algal marshes, such as those in the Bahamas (Shinn, et al., 1969), in which crusts of calcareous tufa are precipitated. On this basis, the Rocknest climate was clearly humid.

Although completely dolomitized, sedimentary structures and textures in the Rocknest carbonates are well preserved. The usual sabkha model of early dolomitization is unacceptable in light of the paleoclimatic interpretation. The dorag model of dolomitization (Badiozamani, 1973; Folk and Land, 1975), related to mixing of meteoric and marine groundwaters, is consistent with a humid paleoclimate and would operate during each progradation of tidal flats and marshes across the shelf.

References

AITKEN, J. D. 1967. Classification and environmental significance of cryptalgal limestones and dolomites, with illustrations from the Cambrian and Ordovician of southwestern Alberta. *J. Sed. Petrol. 37*, 1163–1178.

BADIOZAMANI, K. 1973. The dorag dolomitization model—application to the Middle Ordovician of Wisconsin. *J. Sed. Petrol. 43*, 965–984.

BEERBOWER, J. R. 1964. Cyclothems and cyclic depositional mechanisms in alluvial plain sedimentation. *In* Merriam, D. F., ed., Symposium on Cyclic Sedimentation. *State Geol. Surv. Kansas Bull. 169*, 31–42.

COOGAN, A. H. 1969. Recent and ancient carbonate cyclic sequences. *In* Elam, J. G. and Chuber, S., eds., Cyclic Sedimentation in the Permian Basin. *West Texas Geol. Soc. Publ. No. 69-56*, 5–16.

FISCHER, A. G. 1964. The Lofer cyclothems of the Alpine Triassic. *In* Merriam, D. F., ed., Symposium on Cyclic Sedimentation. *State Geol. Surv. Kansas Bull. 169*, 107–149.

FOLK, R. L., and LAND, L. S. 1975. Mg/Ca vs salinity: two controls over crystallization of dolomite. *Am. Assoc. Petrol. Geol. Bull. 59*, 60–68.

FRASER, J. A., and TREMBLAY, L. P. 1969. Correlation of Proterozoic strata in the northwestern Canadian Shield. *Can. J. Earth Sci. 6*, 1–9.

HOFFMAN, P. 1973a. Evolution of an early Proterozoic continental margin—The Coronation geosyncline and associated aulacogens of the northwestern Canadian Shield. *Phil. Trans. Roy. Soc. London Ser. A 273*, 547–581.

——— 1973b. Recent and ancient algal stromatolites—Seventy years of pedagogic cross-pollination. *In* Ginsburg, R. N., ed., Evolving Concepts in Sedimentology. *Johns Hopkins Univ. Stud. Geol. No. 21*, 178–191.

LOGAN, B. W., HOFFMAN, P., and GEBELEIN, C. D. 1974. Algal mats, cryptalgal fabrics and structures, Hamelin Pool, Western Australia. *In* Logan, B. W., ed., Evolution and Diagenesis of Quarternary Carbonate Sequences, Shark Bay, Western Australia. *Am. Assoc. Petrol. Geol. Memoir 22*, 140–194.

SHINN, E. A., LLOYD, R. M., and GINSBURG, R. N. 1969. Anatomy of a modern carbonate tidal-flat. *J. Sed. Petrol. 39*, 1202–1228.

Section V

Ancient Carbonate Examples

Laminated, Thin-Bedded, and Stromatolitic

Section V has seven Paleozoic examples, each of which presents convincing evidence of accumulation in the zone of short-term fluctuations of sea level. In all examples a significant element of this evidence is the occurrence of distinct stromatoids, crinkled laminations, or mud-cracked laminations—which are frequently interrupted abruptly and irregularly by cross-cutting lenses, channels or beds of coarse-grained sediments; intraformational conglomerate, ooid, or skeletal sand. It is significant that the six examples of Lower Paleozoic age do not show repeated, small-scale vertical sequences of lithofacies and structures like the examples in Section IV. This absence of distinct vertical sequences, together with the prominence of fine-grained laminated beds interrupted by coarse-grain lenses, and the vast extent of these early Paleozoic deposits, indicate a distinctive kind of tidal-flat deposit. There were two levels of short-term fluctuations in sea level: the smaller and more frequent fluctuations produced the laminations and thin beds; the larger and less frequent fluctuations are probably responsible for the coarser interbeds. Most of the deposition was in the intertidal zone, and beds with clear evidence of permanent submergence or prolonged exposure are less frequent.

31

Carbonate Tidal Flats of the Grand Canyon Cambrian

Harold R. Wanless

For the past 27 years the Cambrian sedimentary sequence exposed in the walls of Grand Canyon, Arizona, has served as a classic illustration of time-transgressive sedimentation and as a nearly ideal example of the textbook concept of the deepening offshore succession of marine environments—from the shallow, nearshore, high energy Tapeats Sandstone to the deeper, quieter water Bright Angel Shale to the yet deeper marine areas of Muav Limestone accumulation (McKee, 1945, 1969; Shelton, 1966).

During reexamination of the Grand Canyon Cambrian in 1969 and 1970, the writer encountered numerous lithologic features and sequences that suggest quite a different paleoenvironment and depositional pattern. In the Bright Angel Shale, unburrowed, channeled, flaser-bedded sandstone units record tidal-flat sedimentation, and herringbone cross-laminated glauconitic sandstones capped by hematite oolite beds record shoaling to an exposure surface. In the Muav Limestone, dolomitized

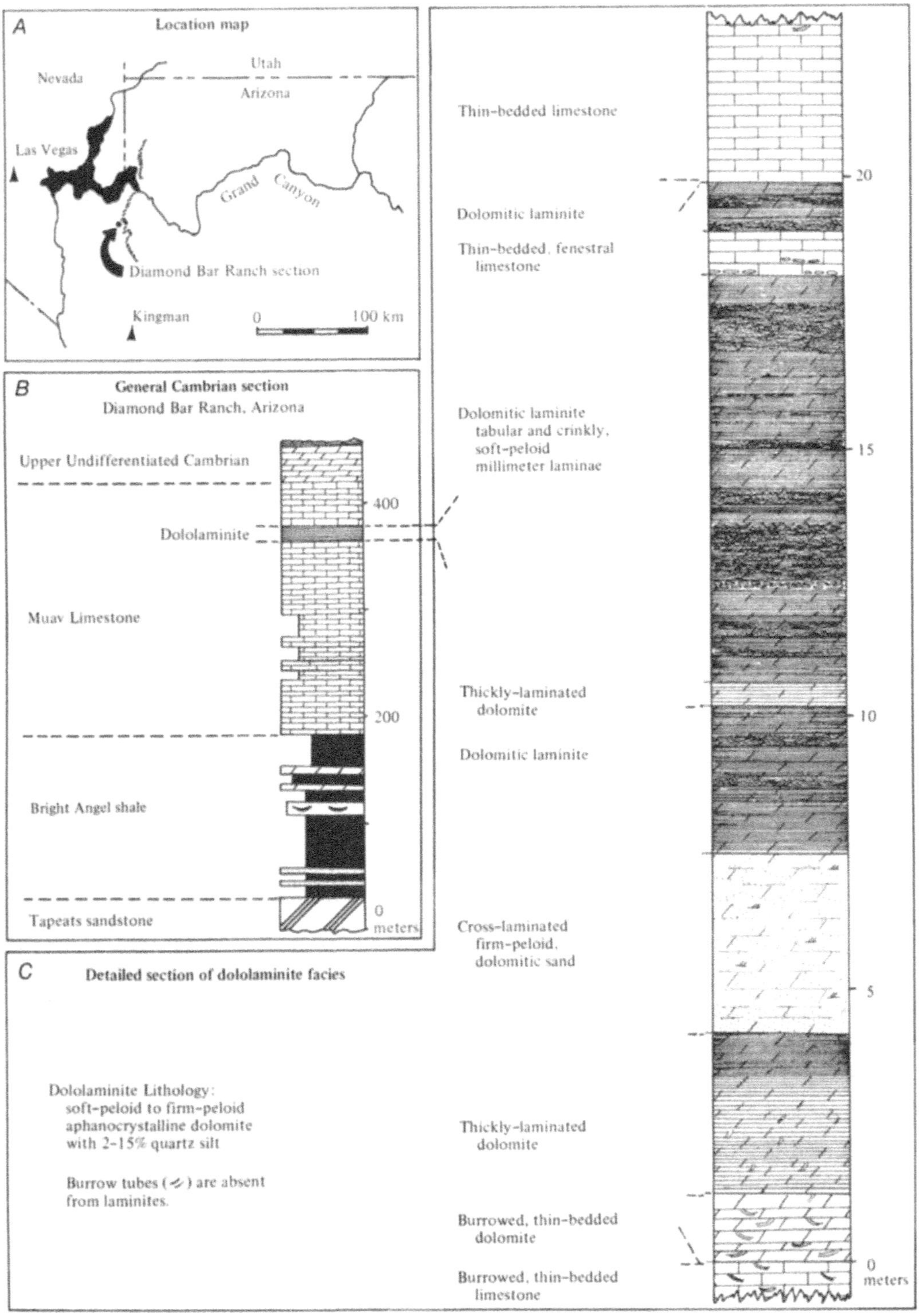
A
Location map
Nevada
Utah
Arizona
Las Vegas
Grand Canyon
Diamond Bar Ranch section
Kingman
0
100 km
B
General Cambrian section
Diamond Bar Ranch, Arizona
Upper Undifferentiated Cambrian
400
Dololaminite
Muav Limestone
200
Bright Angel shale
Tapeats sandstone
0
meters
C
Detailed section of dololaminite facies
Dololaminite Lithology:
soft-peloid to firm-peloid
aphanocrystalline dolomite
with 2-15% quartz silt
Burrow tubes ( ) are absent
from laminites.
Thin-bedded limestone
20
Dolomitic laminite
Thin-bedded, fenestral
limestone
Dolomitic laminite
tabular and crinkly,
soft-peloid
millimeter laminae
15
Thickly-laminated
dolomite
10
Dolomitic laminite
Cross-laminated
firm-peloid,
dolomitic sand
5
Thickly-laminated
dolomite
Burrowed, thin-bedded
dolomite
0
meters
Burrowed, thin-bedded
limestone

eocrinoidal biocalcarenites, algal-ball limestones, and rare stromatolites similarly suggest very shoal-water sedimentation.

The most striking evidence for shallow-water paleoenvironment is found in the Muav Limestone at the western (most seaward and supposedly deepest marine) end of Grand Canyon (Fig. 31–1). Here, the characteristic burrowed, thin-bedded limestone is interrupted by a 20-meter thick sequence composed dominantly of very thinly laminated, soft-pellet dolomicrites. This dololaminite facies contains textures, sedimentary structures, and associated beds that are strikingly similar to those found in the modern carbonate tidal-flat sediments accumulating along the western coast of Andros Island, Bahamas. The following comparative description summarizes the attributes of the Cambrian and modern laminates.

Comparative Description

Setting

Cambrian. Dololaminate facies locally interrupts very thin-bedded, burrowed soft pelmicrite (nodular limestone) in western Grand Canyon (Fig. 31–1). Total facies unit persistent east-west (across depositional strike) for more than 2 km but less than 15 km.

Modern. Tidal-flat wedge, 15 km x 60 km x 3 meters or less, on sheltered west side of Andros Island, Bahamas; borders shallow, pelleted lime muds of Great Bahama Bank; climate subtropical, humid, wet; prevailing east winds with strong north and northwest winds during winter cold fronts; laminites on broad (10 meters to 1 km) supratidal beach and channel levees near tidal-flat margins.

Laminite composition

Cambrian. (1) Aphanocrystalline dolomite; (2) soft-pellet, aphanocrystalline dolomite containing 2 to 15 percent angular quartz and feldspar silt; ovoid pellets 120 to 200 μ in long diameter (Fig. 31–2).

Modern. (1) Aragonitic mud; (2) pelleted aragonitic mud; ovoid pellets 130 to 220 μ long in diameter; rare local patches of penecontemporaneous protodolomite.

Figure 31-1 (opposite). *A.* Location map of measured Cambrian section at western end of Grand Canyon. *B.* General Cambrian section. *C.* Detailed section of dololaminite facies.

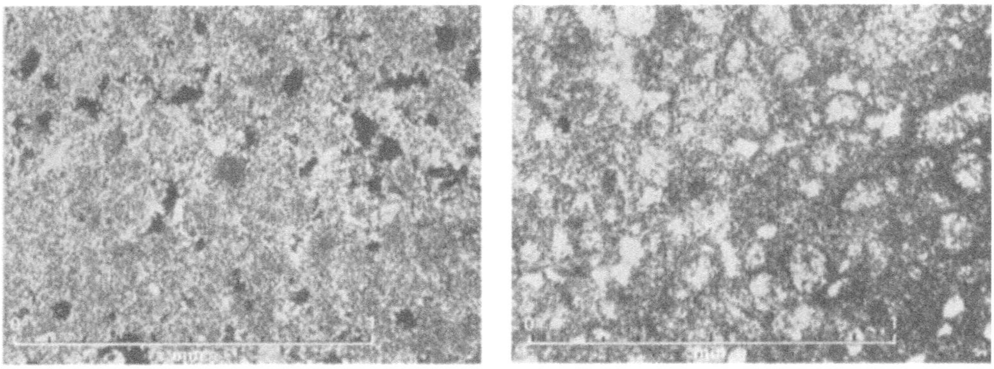

A B

Figure 31-2. *A.* Photomicrograph of soft-pellet aphanocrystalline dolomite and quartz silt. *B.* Photomicrograph of firm-pellet; fine crystalline dolomite. (Scale in millimeters.)

Laminations

One mm or less in thickness; either stacks of thin, single mud laminae that are continuous with even thickness or couplets of (1) thin, continuous mud lamina and (2) thin, discontinuous pellet-sand lamina; continuous mud laminae have uniform thickness and drape irregularities, and are commonly oversteepened. Discontinuous pellet-sand laminae are quite well sorted, form thin starved ripples and depression fillings, and do not drape. See Figure 31–3.

Laminite morphology

TUBULAR CONTINUOUS. *Cambrian.* Fig. 31–4*B*, stacks of poorly pelleted, continuous mud laminae; no desiccation but local, sharp, stepped erosional surfaces; no burrowing.

Modern. Same on levee crest (exposure index >98); smooth, firm, featureless surface bound by filamentous blue-green algal mat (*Schizothrix*); sharp erosional channel and shoreline edges; no burrowing.

UNDULATORY CONTINUOUS. *Cambrian* (Fig. 31–5). Couplets of continuous mud lamina and discontinuous pellet-sand lamina; no desiccation; no burrowing.

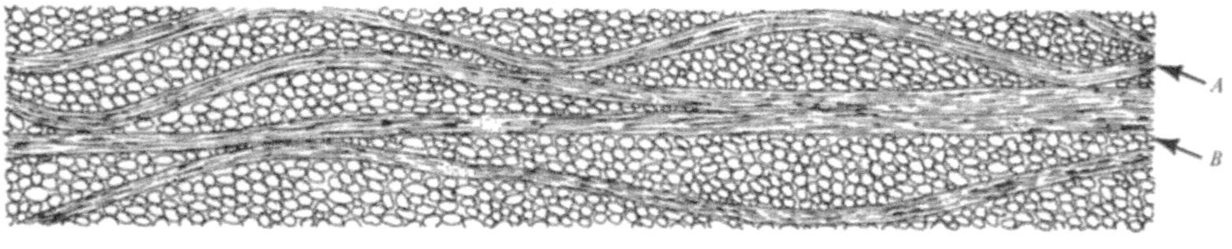

Figure 31-3. Sketch of couplets of (*A*) thin, continuous mud lamina and (*B*) thin, discontinuous pellet-sand lamina.

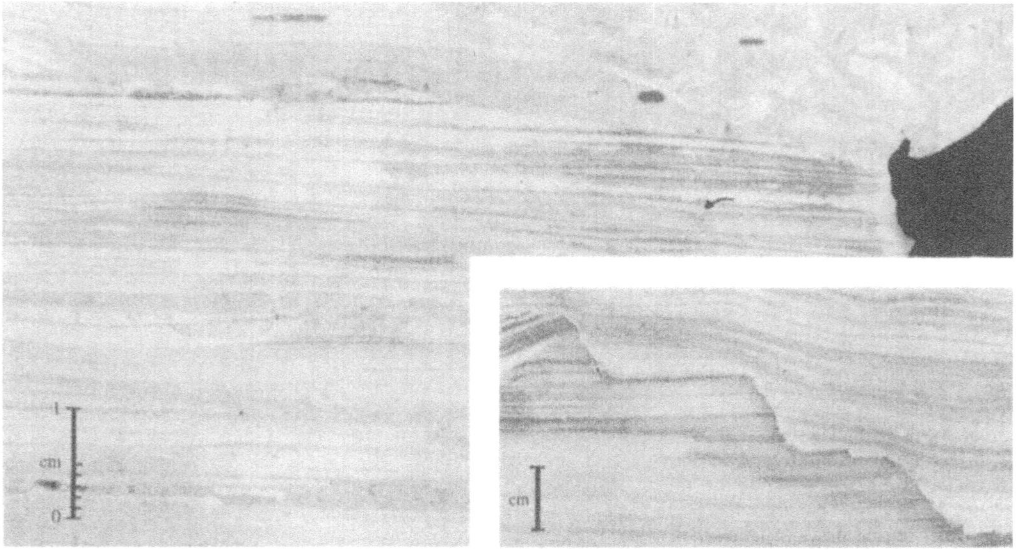

A *B*

Figure 31-4. *a.* Tabular disrupted dololaminite; Cambrian, polished slab. *b.* Erosional surface in tabular, continuous dololaminite; Cambrian, polished slab.

Figure 31-5. Undulatory, continuous dololaminite; Cambrian, polished slab.

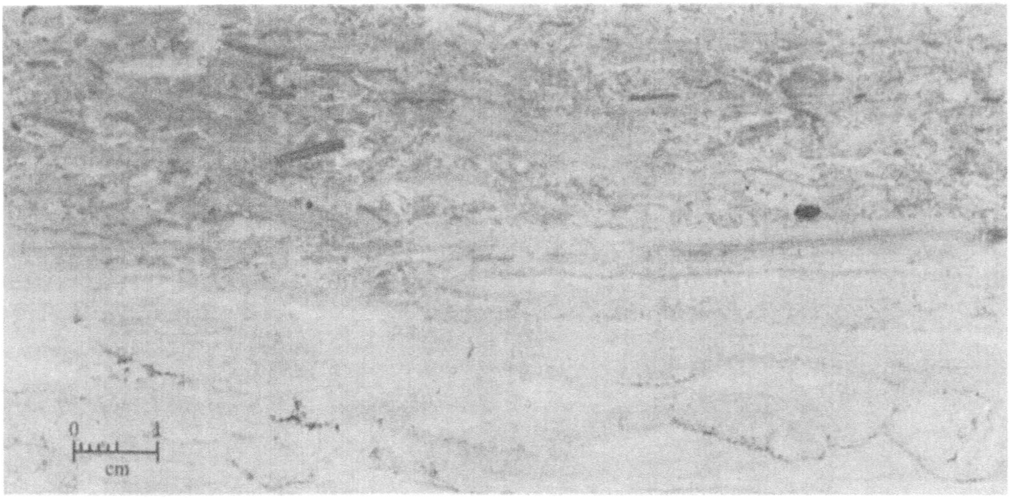

Figure 31-6. Flat-pebble intraclast bed overlying crinkly and tabular; disrupted dololaminite; Cambrian, polished slab.

Modern. Same on and near levee crest (EI 93 to 98); firm *Schizothrix* mat surface with small starved ripples and current lineations of pelleted sand; rare, incomplete desiccation of upper lamina only, no burrowing.

TABULAR DISRUPTED. *Cambrian* (Figs. 31–4*A* and 31–6). Stacks of continuous mud laminae disrupted by thin, filled (arrow) and unfilled desiccation cracks that only penetrate one or two laminae; irregularities and small scour pockets filled with pellet sand and small intraclast chips; no burrowing.

Modern. Same on upper levee backslope (EI 87 to 93); firm, finely desiccated surface covered with *Schizothrix* mat; intraclast chips concentrated in small scour pockets (Fig. 31–6); oligocheate burrows.

Figure 31-7. Crinkly, disrupted (microstromatolitic) dololaminite; Cambrian, polished slab.

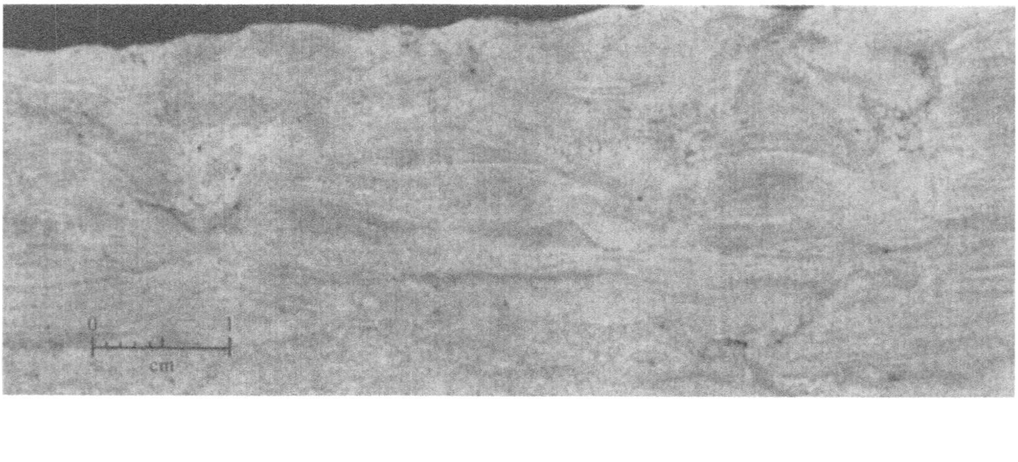

CRINKLY DISRUPTED. *Cambrian* (Figs. 31–6 and 31–7). Small stromatolitic domes and accentuated ripples; oversteepened continuous mud laminae draping domes; pellet-sand laminae discontinuous, pinching out on domes; thin irregular desiccation cracks separate domes and penetrate 1 or 2 cm; no burrowing.

Modern. Same on upper levee backslope (EI 87 to 93); small ripples and other surface irregularities accentuated by soft *Schizothrix* or mixed *Schizothrix Scytonema* (tufted) algal mat; mud laminae drape surface; thin, irregular desiccation cracks penetrate 1 cm or more; scattered fiddler crab burrows.

Associated beds

THICKLY LAMINATED DOLOMITE. *Cambrian.* Continuous layers 1 to 15 mm thick with minor small scour pockets; grade upward to dololaminites; small vertical burrows.

Modern. Similar continuous laminations of irregular thickness in intertidal to subtidal ponds behind levees; intensely churned by grazing and burrowing fauna.

CROSS-LAMINATED, DOLOMITIZED, FIRM PELLET SAND. *Cambrian* (Fig. 31–8). Dominantly small (<3 cm thick) trough cross-laminations; also planar high-angle and low-angle cross laminations in upper part of sequence; no burrows.

Modern. Similar small, trough cross-laminated skeletal and firm-pellet sands in channels that deeply penetrate tidal flat.

THIN-BEDDED, FENESTRAL LIMESTONE. *Cambrian.* Very thinly bedded, firm-pellet microsparite (Fig. 21–2*G*); bedding

Figure 31-8. Trough cross-laminated, dolomitized, firm-pellet sand; Cambrian, negative print from peel.

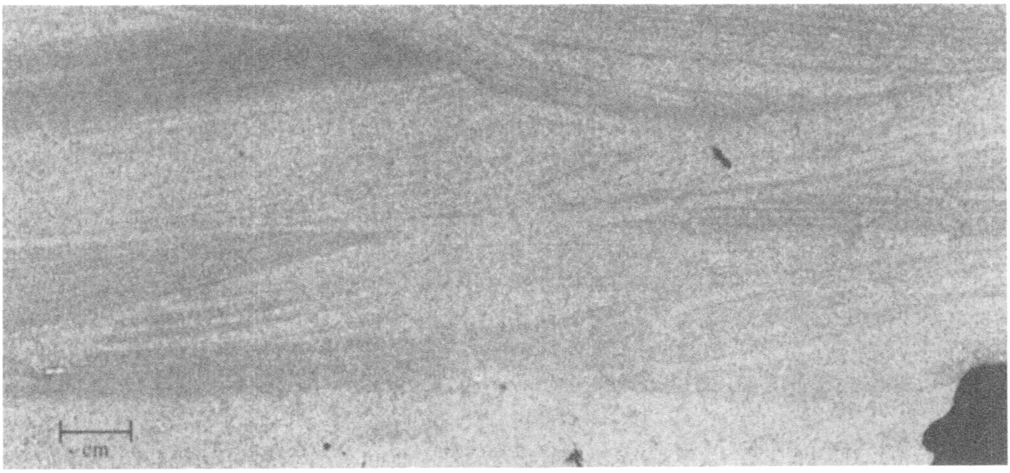

disrupted; numerous planar discontinuous fenestrae zones; flat-pebble limestone intraclasts fill small scour depressions; no burrowing.

Modern. Possibly similar to thinly to thickly laminated crusts on lower levee backslopes in which early induration preserves pellet texture and fenestral fabric.

Discussion

From this comparison the Cambrian dolominite facies is interpreted to record an extended period of tidal-flat sedimentation. The abundance of quartz silt and pelleted carbonate grains in these Cambrian laminites demonstrates that much, if not all, of the laminite sediment was detrital, derived from an adjacent area where quartz silt was available and where faunal activity was producing pelleted sediment. The abundance of detrital grains, current ripples, cross laminations, and intraclasts in the dololaminite facies indicates that the laminites accumulated in association with strong currents.

Oversteepened crinkly mud laminae (Fig. 31–7) are clear evidence of an algal-bound sediment surface. Comparison with the modern laminates suggests that *all* of the dololaminite morphologies were the result of an algal-bound sediment surface.

Laminites form only on the supratidal levees that receive sediment by overbank flooding of sediment-laden waters during sporadic onshore storms. Both the mud and pellet-sand laminae are deposited during these brief periods of flooding. The continuous, uniform, often draping mud laminae appear to be an algal stick-on deposit during the early stages of overbank flooding. The mucilaginous filament sheaths of the *Schizothrix* mat act as a flypaper-like surface to the mud in the flooding storm waters. Filament length limits the effectiveness of this process to less than a millimeter in thickness. The discontinuous pellet-sand laminae form as a traction load ripple, lineation, or depression filling over the mud laminae.

Thick laminae and beds are totally absent from the modern and Cambrian laminites because the positive levees are not a site of poststorm settle-out sedimentation, and the firm, algal-bound surface prevents reworking of the levee sediment during overbank flooding. The total spectrum of sporadic storms of variable intensity and duration has produced a very uniform record of millimeter laminations.

The rubbery algal-bound surface also strongly inhibits desiccation. Lamina-thick cracks form in fresh storm laminations before algal stabilization.

The dololaminite tidal-flat sequence is a prolonged record of storm tide sedimentation at and just above the level of high water. Although all of the bedding attributes of the modern Andros tidal flat are recognized in the dololaminite facies (cross-laminated channel, thickly laminated and burrowed pond, levee laminites, and layers of early induration), organization of these attributes into a series of small cycles, such as that described by Ginsburg and Hardie (Chapter 23), is not apparent.

Tidal-Flat Lamination

The above comparative analysis suggests that the fine-scale attributes of the laminae may give indications of a wind-tidal-flat origin for finely laminated carbonate deposits. The significant attributes are uniformity of laminae with at least some mud-sand laminae couplets and microstromatolitic oversteepening and draping of mud laminae.

The diagnostic morphology and fabric of laminations deposited by alternating of flooding and exposure will be a valuable aid in recognizing the tidal origin of fossil carbonates in which other criteria—mud cracks, desiccation fenestra, linked stromatolites—are rare or absent.

References

McKee, E. D. 1945. *Cambrian History of the Grand Canyon Region.* Pt. 1. Stratigraphy and ecology of the Grand Canyon Cambrian, Carnegie Inst. Washington Publ. *653*, 1–168.

——— 1969. Paleozoic rock of Grand Canyon. In *Geology and Natural History of the Fifth Field Conference, Powell Centennial River Expedition*, Four Corners Geol. Soc. pp. 78–90.

Shelton, J. S. 1966. *Geology Illustrated.* Freeman, San Francisco, 434 pp.

32

Peritidal Lithologies of Cambrian Carbonate Islands, Carrara Formation, Southern Great Basin

Robert B. Halley

Occurrence

The Carrara Formation is a heterogeneous sequence of quartzites, siltstones, shales, limestones, dolostones, and mixed terrigenous-carbonate rocks. It is Early and Middle Cambrian in age (Stewart, 1970; Palmer, 1971). Figure 32–1 illustrates the general distribution of Carrara lithologies along a transect approximately normal to depositional strike (Fig. 32–2). The formation contains three "grand cycles" (Aitken, 1966; Palmer, 1971), which terminate at the top of massive limestone members. A fourth cycle begins with the uppermost shale of section 9 and is not illustrated in sections 3 through 8. This fourth cycle grades into the overlying Bonanza King Formation.

Sections 3 through 8 are within the Carrara Formation and are correlated eastward with the Tapeats Sandstone, Bright Angel Shale, Lyndon Limestone, and Chisholm Shale (section 9). Westward, the Carrara can be correlated with the Mule Spring Limestone and Emigrant Formation (section 1) and the

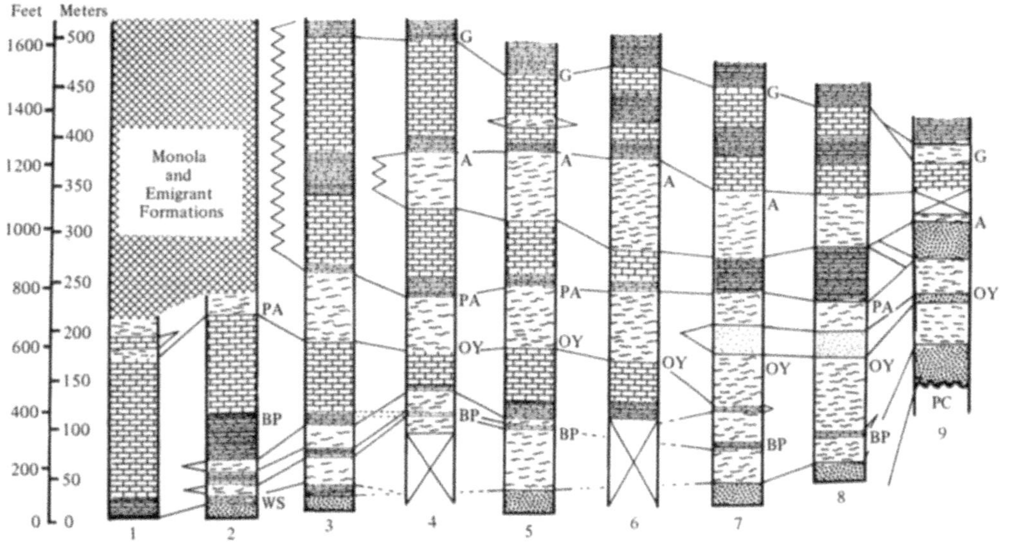

Figure 32-1. Generalized lithologic distribution and correlation of the Carrara Formation. Position of faunules within the stratigraphic section indicated as follows: WS—*Wanneria, Salterella*; BP—*Bristolia, Peachella*; OY—Youngest *Olenellus*; PA—pre-*Albertella*; A—*Albertella*; G—*Glossopleura*.

Figure 32-2. Locality index for Figures 32-1 and 32-3.

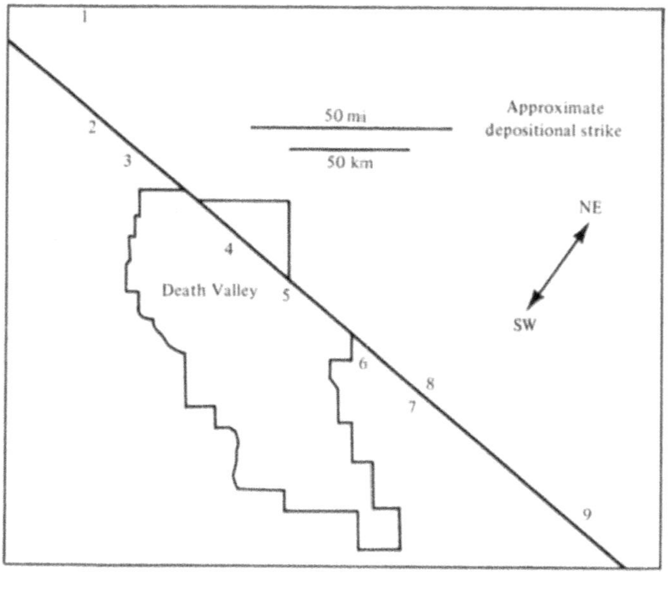

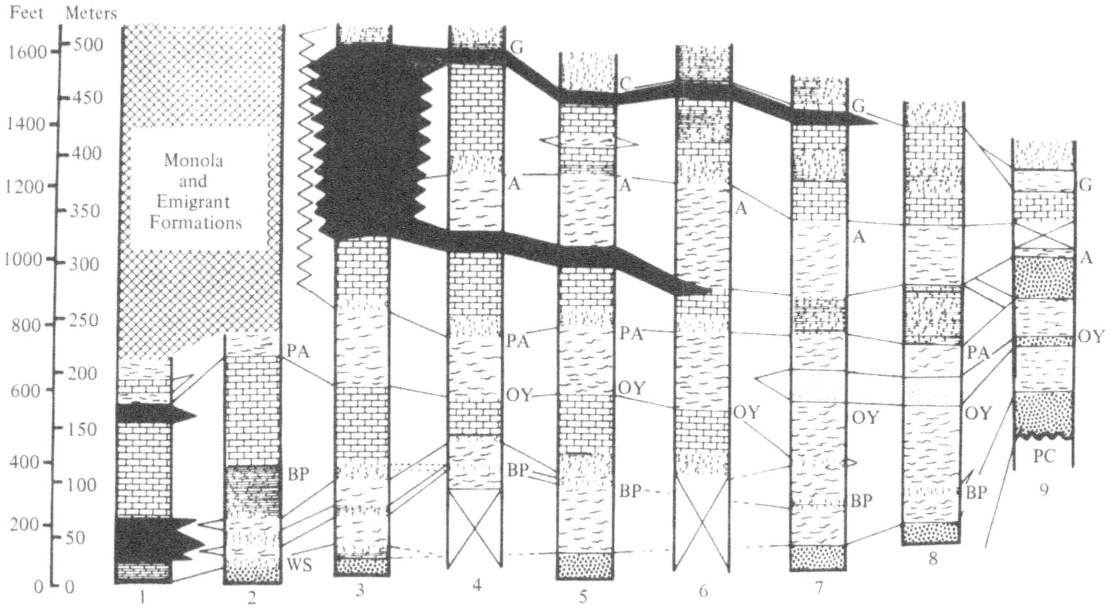

Figure 32-3. Generalized distribution of Carrara peritidal deposits (in black) along a transect approximately normal to depositional strike. No peritidal lithologies are present in the Monola or Emigrant Formations overlying sections 1 and 2.

Figure 32-4. Middle limestone member (section 4) of the Carrara capped by light colored peritidal limestones (*arrow*); × 1/500.

Saline Valley Formation, Mule Spring Limestone, and Monola Formation (section 2). The Monola and Emigrant Formations are, in part, equivalent to the upper two-thirds of the Carrara, but lack the shallow water depositional features of the Carrara. The distribution of intertidal-supratidal deposits within the Carrara is diagrammed in Figure 32–3. These rocks are distinctive light gray and white limestones and subordinate, brown siltstones (Fig. 32–4, *arrow*).

Evidence for Tidal Origin

The recognition that these lithologies are of intertidal and supratidal origin is based on comparison of rock fabrics, textures, structures, and composition with primary depositional and early diagenetic features of Recent intertidal and supratidal sediments. Although Laporte (1967), Roehl (1967), and a host of others have presented more detailed comparisons of Recent and ancient peritidal deposits than space allows here, brief descriptions of the features of the Carrara peritidal lithologies follow, with allusions to similar features in Holocene sediments. Any one of these features may not be diagnostic, but their cooccurrence in some Carrara intervals indicates deposition within a few feet of sea level.

ONCOLITES, STROMATOLITES, AND CRYPTALGALAMINITES. Algal mats trap and bind sediment by acting as a baffle, by filaments growing over and around grains, and by having sticky mucilaginous sheaths to which grains adhere (Gebelein, 1969). Fluctuations in the sediment supply or growth rate of the algae cause the deposits produced by mats to be laminated. These laminations are destroyed or never produced in the subtidal due to the burrowing and grazing activities of invertebrates (Garrett, 1970).

The morphology of the laminated structure is largely controlled by the physical parameters of the environment. Logan et al. (1974) found that decreasing relief of intertidal algal stromatolites reflects decreasing wave energy in Shark Bay, Western Australia. Logan et al. (1964) suggest that oncolites form in agitated lower intertidal conditions. Buchanan et al. (1972) indicate that oncolite formation may extend into the shallow subtidal. Oncolites occur throughout Carrara intertidal and subtidal limestones. Most are well laminated and range in diameter from 0.1 to 5 cm. (Fig. 32–5). In areas of good preservation these nodules retain algal filament molds similar to *Girvanella*. The intimate association of oncolites with intertidal

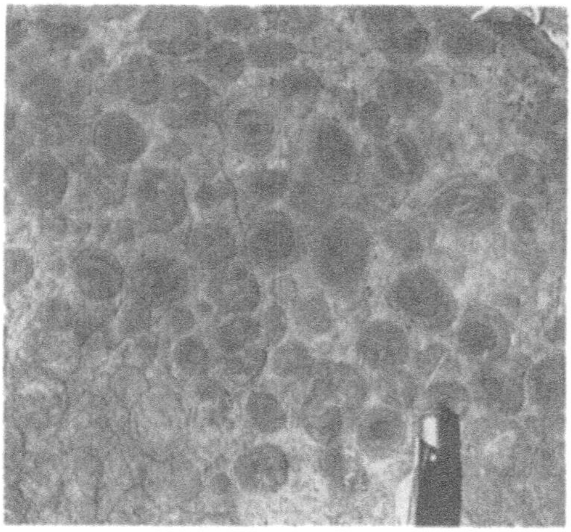

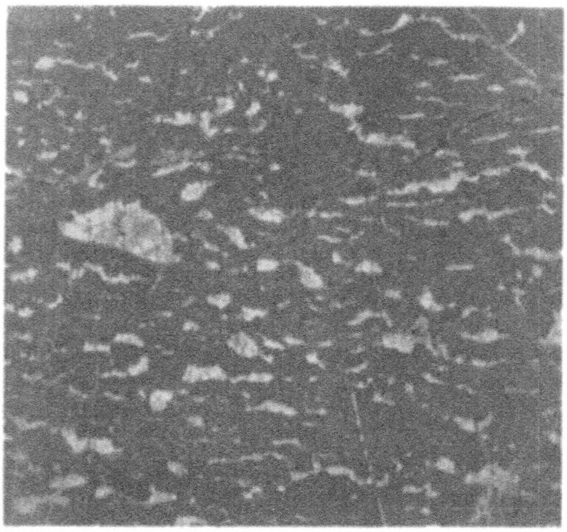

Figure 32-5 (left). Oncolites from lower limestone member; × 1/6.

Figure 32-6 (right). Birdseyes developed in micrite; × 2/3.

lithologies as well as subtidal lithologies suggests deposition in both the lower intertidal and shallow subtidal. Stromatolites are not common in the Carrara, but small (up to 20 cm wide and 30 cm high) domes and columns are found within laminites and loferites and are draped over the edges of upturned mud-crack polygons. Cryptalgalaminites are 0.1 to 1 mm laminated limestones, limestone-dolomite, and dolomite rocks. Where preserved, the primary depositional texture of stromatolitic and cryptalgalaminitic laminations is a vertical alternation of grain size.

BIRDSEYE LIMESTONES, LOFERITES, AND LAMINOID-FENESTRAL FABRICS. Vugs form in supratidal carbonate muds during desiccation and gas bubble migration (Shinn, 1968). These vugs later become rimmed with carbonate cement (Shinn et al., 1969) and filled with sparry calcite that produces the birdseye structure. Birdseye limestones as described by Laporte (1967), loferites of Fischer (1964), and laminoid-fenestral fabrics of Tebbutt and others (1965) are considered partly synonymous, and intergradational fabrics occur in the Carrara supratidal limestones.

In general, horizontally elongate, discrete, spar-filled voids with matching upper and lower surfaces are considered shrinkage pores and the rocks containing them are birdseye limestones (Fig. 32–6). Laminoid-fenestral fabrics and loferites are anastomosing, horizontally and vertically elongate or irregular voids now filled with calcite spar. They occur in well-laminated and mud-cracked pelmicrite. Many of these early voids were

Figure 32-7. Flat pebbles in intraclastic grainstone overlying cryptalgalaminite; $\times$ 1/4.

initially filled with an internal sediment similar to the M_2 sediment of Fischer (1964). Rim cements overlie the Carrara internal sediment and encrust walls of the voids. Reverse grading, a feature reported from the laminoid-fenestral fabrics of Tebbutt et al. (1965), is not present in the Carrara supratidal limestones.

ROUND- AND FLAT-PEBBLE CONGLOMERATES: Ginsburg (1957) illustrated rounded lime mud clasts accumulating along the shore of Nest Key, Florida. These clasts were being eroded from supratidal sun-dried mud and were cohesive enough to be rounded by wave action. Shinn et al. (1969) illustrate flat, laminated, redeposited clasts of algally laminated beach sediment. Rounded and flat-pebble conglomerates of the Carrara intertidal and associated subtidal limestones are found in deposits differing in geometry and associated lithologies. Round-pebble conglomerates flank birdseye limestones and are themselves flanked by subtidal lithologies. The horizontal extent of beds of rounded pebbles is normally less than 10 meters. In contrast, beds containing flat pebbles are several tens of meters wide. The flat-pebble conglomerates are clearly associated with cryptalgalaminites (Fig. 32–7). In several localities where these millimeter-laminated dolomitic platy pebbles overlie cryptalgalaminites, pebbles are preserved in all stages of separation from the laminite upper surface.

EARLY DOLOMITE. Gebelein and Hoffman (1971) have suggested that magnesium concentration in algal filaments provides a source of magnesium for the dolomite in millimeter-laminated limestone-dolomite cryptalgalaminites. Dolomite-rich laminations of the Carrara cryptalgalaminites contain small (5 to 50 μ) rhombs of dolomite; the larger rhombs are often stained red, brown, or yellow with iron oxides. The crystal size of this dolomite and its well-ordered crystal structure are features that differ from Recent early dolomite. However, its distribution with respect to primary depositional features and associated lithologies suggest this dolomite is of early diagenetic origin. Later diagenetic changes could account for the larger crystal size and better ordering of ancient supratidal dolomite.

MUD CRACKS. Well-separated, polygonal mud cracks from the interior of Crane Key, Florida Bay, have been described by Ginsburg (1957). They occur in the supratidal zone in areas of seasonally ponded waters of variable salinity. Mud cracks similar to these occur in the Carrara supratidal limestones (Fig. 32–8), and where seen in vertical section, appear to have been

Figure 32-8. Vertical traces of upturned edges of polygonal mud cracks in dolomitic limestone; × 1/6.

self-propagating, that is, the trace of a crack may extend as much as a meter vertically. Mud cracks also occur in the dolomitic siltstones overlying supratidal carbonate deposits; these are not self-propagating, and individual cracks reach a vertical maximum of 2 cm.

SALT CRYSTAL CASTS. Salt crystal casts can form in intermittently flooded areas even in regions of abundant rainfall (Pettijohn, 1957). Salt crystal casts up to 5 mm sq are associated with the mud cracks in dolomitic siltstones. These salt crystals are believed to have formed from evaporation of stranded pools of seawater in the supratidal; they are usually ephemeral, and their preservation considered fortuitous. Salt crystal casts in the Carrara are rare.

ALGAL FILAMENT MOLDS. Monty (1967) and Shinn et al. (1969) have reported preservation of algal filament molds in areas of seasonally ponded brackish and fresh water. These filament molds are produced when calcium carbonate is precipitated around the algae, presumably as a result of algal photosynthesis. Algal filament molds are found sporadically in the Carrara supratidal limestones (Fig. 32–9). They are associated with loferites, and are similar to the preserved filaments described and illustrated by Fischer (1964).

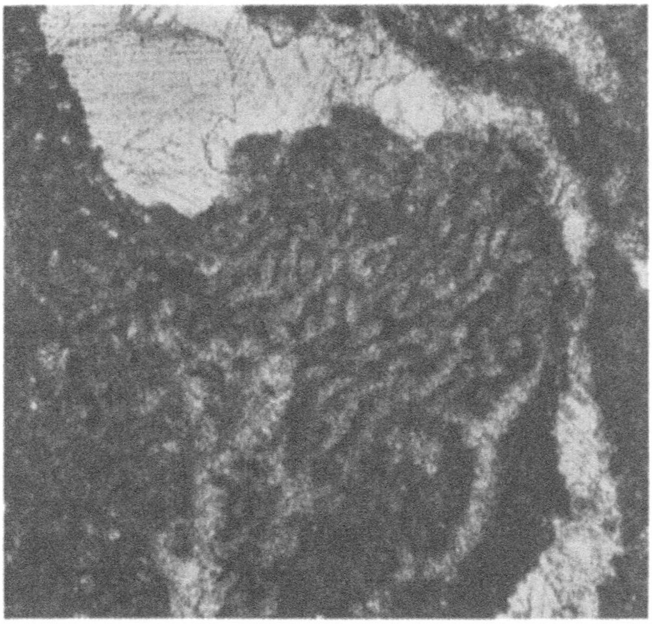

Figure 32-9. Molds of algal filaments preserved within loferite-like limestones; × 10.

Discussion

Most peritidal lithologies of the Carrara form thin to medium beds. Only the laminoid-fenestral fabrics and loferites are massively bedded. At any one locality the vertical distribution of lithologies within the Carrara tidal carbonates is more or less random. This distribution suggests that the peritidal rocks were deposited as a complex mosaic of subtidal, intertidal, and supratidal facies—much like the Manlius Formation of New York State (Laporte, 1967). The lateral distribution of lithologies, however, is not random. Supratidal limestones are more frequent in the west.

The intertidal and supratidal deposits of the Mule Spring and Carrara Formations display a consistent pattern of development (Fig. 32–3). First, prior to the deposition of tidal sediments, 50 to 100 meters of subtidal sediment accumulated as a large shallow platform. Second, intertidal and supratidal sediments accumulated on the western half of these banks, and periodically extended eastward to form the tongues in Figure 32–3. Because the intertidal and supratidal deposits are localized, they are interpreted as offshore carbonate islands. To the east of these islands was an epeiric sea extending 300 to 500 km to the cratonic shoreline. The western half of this epeiric sea was a shallow, carbonate-producing platform that was a major source of carbonate sediment for the construction of the westward islands, and was probably the only source for the eastward extension of these islands.

References

Aitken, J. D. 1966. Middle Cambrian to Middle Ordovician cyclic sedimentation, Southern Rocky Mountains of Alberta. *Bull. Can. Petrol. Geol. 14*, 405–442.

Buchanan, H., Streeter, S., and Gebelein, C. D. 1972. Possible living algal-foraminiferal consortia in nodules from modern carbonate sediments of the Great Bahama Bank. Am. Assoc. Petrol. Geol. Ann. Meeting, *56*, No. 3, 606.

Fischer, A. G. 1964. The Lofer cyclothems of the Alpine Triassic. *In* Merriam, D. F., ed., Symposium on Cyclic Sedimentation. *State Geol. Surv. Kansas Bull. 169*, 107–149.

Garrett, P. 1970. Phanerozoic stromatolites: Noncompetitive ecologic restriction by grazing and burrowing animals. *Science 169*, 171–173.

Gebelein, C. D. 1969. Distribution, morphology, and accretion rate of Recent subtidal algal stromatolites, Bermuda. *J. Sed. Petrol. 39*, 49–69.

Gebelein, C. D., and Hoffman, P. 1971. Algal origin of dolomite in interlaminated limestone-dolomite sedimentary rocks. *In* Bricker,

O. P., ed., *Carbonate Cements*. Baltimore, Johns Hopkins Univ. Press, pp. 319–326.

GINSBURG, R. N. 1957. Early diagenesis and lithification of shallow water carbonate sediments in South Florida. *In* Le Blanc, R. J., and Breeding, J. G., eds., Regional Aspects of Carbonate Deposition. *Soc. Econ. Paleontol. Mineral. Spec. Publ. No. 5*, 80–100.

LAPORTE, L. F. 1967. Carbonate deposition near mean sea-level and resultant facies mosaic: Manlius Formation (Lower Devonian) of New York State. *Am. Assoc. Petrol. Geol. Bull. 51*, 73–101.

LOGAN, B. W., REZAK, R., and GINSBURG, R. N. 1964. Classification and environmental significance of algal stromatolites. *J. Geol. 72*, 68–83.

LOGAN, B. W., HOFFMAN, P. and GEBELEIN, C. D. 1974. Algal mats, cryptalgal fabrics and structures, Hamelin Pool, Western Australia. *In* Logan, B. W. ed., Evolution and diagenesis of Quaternary carbonate sequences, Shark Bay, Western Australia, *Am. Assoc. Petrol. Geol. Memoir 22*, 140–194.

MONTY, G. L. V. 1967. Distribution and structure of Recent stromatolitic algal mats, Eastern Andros Island, Bahamas. *Soc. Geol. Belg. Ann. 90*, 55–102.

PALMER, A. R. 1971. The Cambrian of the Great Basin and adjacent areas, Western United States. *In* Holland, C. H., ed., *Cambrian of the New World*. New York, Wiley-Interscience, pp. 1–78.

PETTIJOHN, F. J. 1957. *Sedimentary Rocks*. New York, Harper & Row, 718 pp.

ROEHL, P. O. 1967. Stony Mountain (Ordovician) and Interlake (Silurian) facies analogs of Recent low energy marine and subaerial carbonates, Bahamas. *Am. Assoc. Petrol. Geol. Bull. 51*, 1979–2032.

SHINN, E. A. 1968. Practical significance of birdseye structures in carbonate rocks. *J. Sed. Petrol. 38*, 215–223.

SHINN, E. A., GINSBURG, R. N., and LLOYD, R. 1969. Anatomy of a modern carbonate tidal-flat, Andros Island, Bahamas. *J. Sed. Petrol. 39*, 1201–1228.

STEWART, J. H. 1970. Upper Precambrian and Lower Cambrian strata in the Southern Great Basin, California and Nevada. *U. S. Geol. Surv. Prof. Paper 620*, 206 pp.

TEBBUTT, G. E., CONLEY, C. D., and BOYD, D. W. 1965. Lithogenesis of a distinctive carbonate rock fabric. *Univ. Wyoming Contr. Geol. 4*, 1–13.

33

Peritidal Origin of Cambrian Carbonates in Northwest Spain

Isabel Zamarreño

The Láncara Formation, a unit composed up of carbonates, ranging in age from Early to Middle Cambrian, is well exposed in the Cantabrian zone (northwest Spain). It generally forms the sole of several thrust sheets emplaced by a general décollement of the Paleozoic stratigraphic sequence. Where a complete Cambrian succession exists, the Láncara Formation overlies a sequence of coarse-grained sandstones and interbedded shales (Herrería Formation) of Early Cambrian age, and is overlain by another sandstone-shale formation ranging in age from Middle to Late Cambrian (Oville Formation).

The Láncara Formation has been divided into two informal members based on lithology (Zamarreño, 1972). The upper one (Middle Cambrian) consists of abundant skeletal grains and displays a nodular texture. The lower member (Lower Cambrian) is made up of dolostones or dolostones and limestones; it is devoid of fauna and shows several features (birdseyes, mud cracks, algal mats, oncolites, and so on) similar to Recent carbonate tidal-flat deposits. For the purposes of this

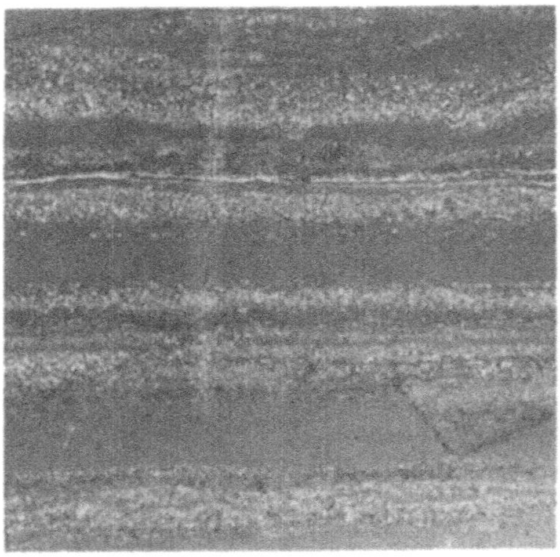

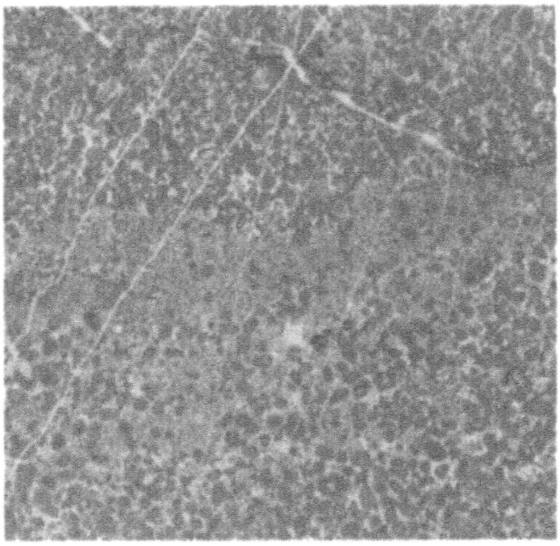

Figure 33-1 (left). Laminated dolopelmicrite of lithofacies 1; Tarna (C-52); × 10.

Figure 33-2 (right). Pelsparite of lithofacies 1 with euhedral dolomite distributed in patches; Carangas (C-3); × 10.

compilation, we deal only with the lithofacies and sedimentary structures peculiar to the lower member of the Láncara Formation.

Facies

Five major lithofacies associations can be distinguished. Each is characterized by distinctive textures, sedimentary structures, and stratigraphic position in the sequence: (1) laminated dolopelmicrite and pelsparite lithofacies; (2) algal-laminated dolostone and laminated dolomicrite lithofacies; (3) birdseye limestone lithofacies; (4) grapestone lithofacies; and (5) oosparite and quartzarenite lithofacies. In the field, lithofacies 1 and 2 appear as thin-bedded, tan-colored, laminated dolostone, while lithofacies 3 and 4 appear as thick-bedded to massive gray limestones. Lithofacies 5 is less widespread and the brown color and cross-bedded structures are its distinctive features.

LAMINATED DOLOPELMICRITE AND PELSPARITE LITHOFACIES. The abundance of peloids and the laminations are the distinctive features of this lithofacies (Fig. 33–1). It has two main textures: laminated dolopelmicrites and pelsparites. The former consists of laminae composed of graded peloids and detrital quartz alternating with laminae of dolomite. The laminations are of several types: parallel, distorted, and wavy laminations, as well as scour-and-fill structures.

The second predominant texture consists of peloids cemented

by sparry calcite (Fig. 33–2) and includes pelsparites and intrapelsparites devoid of laminations. Euhedral dolomite crystals are distributed in patches in the interstices between peloids, but when the degree of dolomitization increases, dolomite crystals also occur within the peloids. Less commonly, algae referable to the genus *Nuia* or some trilobite debris are also associated with the peloids. Interbedded with these predominant textures, homogeneous dolomicrites are minor constituents in some localities.

ALGAL-LAMINATED DOLOSTONE AND LAMINATED DOLOMICRITE LITHOFACIES. The laminated character of the sediments in this lithofacies is due to the trapping and binding of carbonate sediment onto an algal mat, in contrast to the preceding lithofacies, where the laminations are inorganic in origin. Gray to black planar laminations alternate with clear carbonates containing in places peloids (Figs. 33–3 and 33–4). In general, the calcitic molds of algae are poorly preserved but the texture is distinctive. Birdseye structures (fenestral fabric) are well developed in the algal-laminated dolostones and the cavities are filled with sparry calcite, megaquartz, or chert. Flat-pebble breccias as well as intrapelmicrudites and intrapelsparudites are also present, interbedded with the algal-laminated dolostones that constitute the predominant lithology.

These products of fragmentation are probably due to the desiccation or to the erosion of the algal mats. In a few localities the intrapelmicrudite and intrapelsparudite exhibit large

Figure 33-3 (left). Algal-laminated dolostone of lithofacies 2; Boñar (C-992); × 10.

Figure 33-4 (right). Algal-laminated dolostone of lithofacies 2. Note a storm deposit at top; Barrios de Luna (C-134); × 10.

nodulelike cavities filled with megaquartz and length-slow chalcedony, suggesting a sulfate-rich environment (Folk et al., 1971).

Laminated dolomicrites are also a common type of texture beneath algal-laminated dolostones. They consist of rather straight, parallel laminae alternating and differing in texture or composition. It is difficult to attribute an algal origin to these deposits, due to the lack of distinctive algal-mat texture. Homogeneous and burrowed dolomicrites are subordinate constituents in this lithofacies.

BIRDSEYE LIMESTONE LITHOFACIES. The framework of this lithofacies consists mainly of algal debris (Cyanophyta) set in a matrix of micrite, with planar and spherical birdseye structures (fenestral fabric) filled with sparry calcite (Figs. 33–5 to 33–7). Silt-sized detrital quartz, dolomite rhombs, and idiomorphic quartz are also scattered throughout the framework (Fig. 33–8). The quartz suggests hypersaline conditions (Grimm, 1962). Desiccation structures such as polygonal mud cracks (Fig. 33–9) and sheet cracks (Fig. 33–10) can be recognized in several localities. Some beds containing oncolites are also present in a few localities (Fig. 33–11).

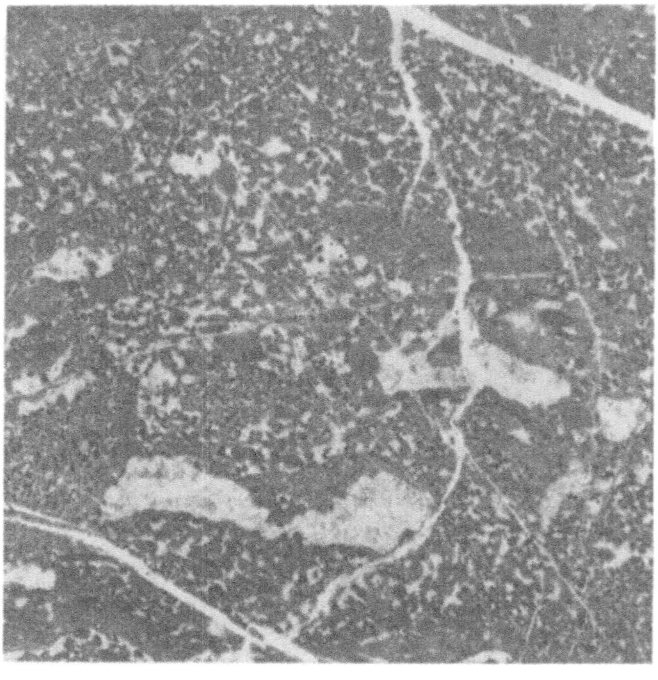

Figure 33-5. Birdseye limestone of lithofacies 3 with abundant algal remains; Corias de Arriba (C-462); × 10.

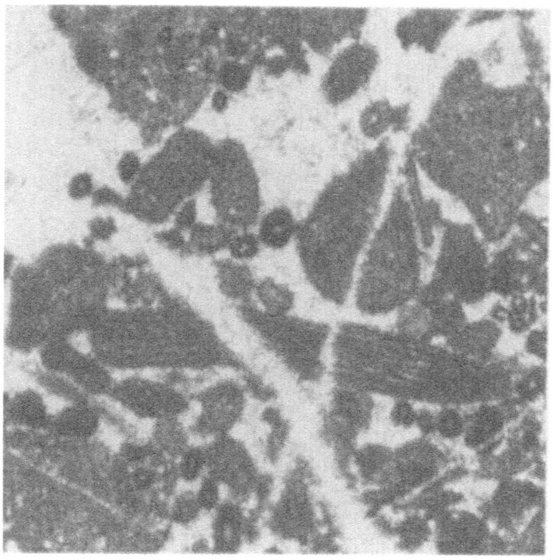

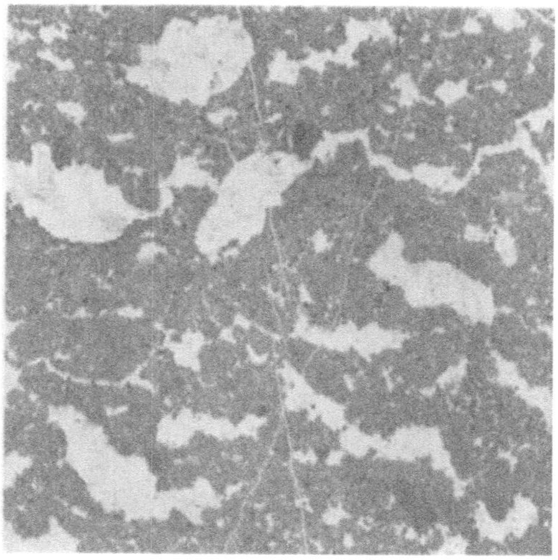

Figure 33-6 (left). A detail of the specimen in Figure 32-5 showing the algae; × 40.

Figure 33-7 (right). Birdseye limestone of lithofacies 3 with tiny dolomite crystals and pyrite scattered in the framework; Torre de Babia (C-1177); × 10.

GRAPESTONE LITHOFACIES. This lithofacies consists of intrapelsparudites in which the intraclasts closely resemble grapestones.

OOSPARITE AND QUARTZARENITE LITHOFACIES. The oosparites consist of ooids displaying concentric and radial structures around a nucleus of micrite, although some nuclei are fossil debris or quartz (Fig. 33–12). Detrital quartz, glauconite, and/or skeletal debris are also associated with the ooids. The quartzarenites are composed of fine-grained, angular, detrital quartz

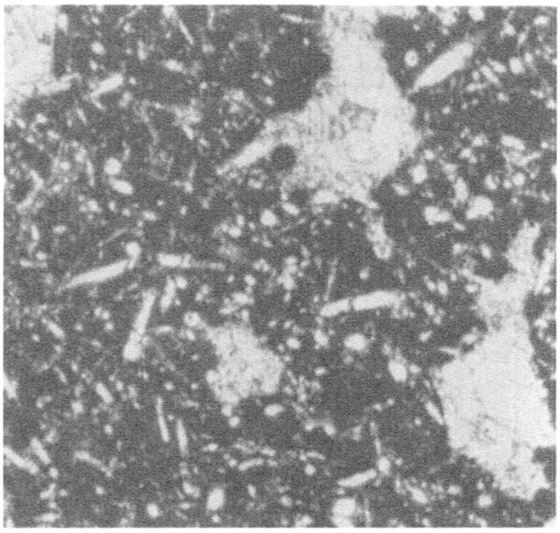

Figure 33-8. Idiomorphic quartz in the framework of a birdseye limestone; Torre de Babia (C-1190); × 40.

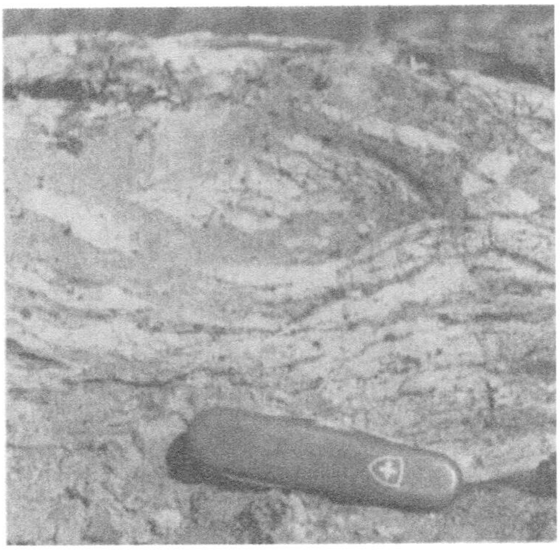

Figure 33-9 (left). Mud cracks in the birdseye limestone lithofacies; Valdoré.

Figure 33-10 (above right). Sheet cracks in birdseye limestone lithofacies; Valdoré.

along with glauconite and occasional ooids; all these components are cemented by sparry calcite.

Environment of Deposition and Distribution of Facies

In the lower member of the Láncara Formation a combination of criteria has been used as evidence of an environment analogous to Recent carbonate tidal flats: textures, sedimentary structures, scarcity of fauna due to facies, and regional distribution of lithofacies. It is rather difficult, however, to distinguish

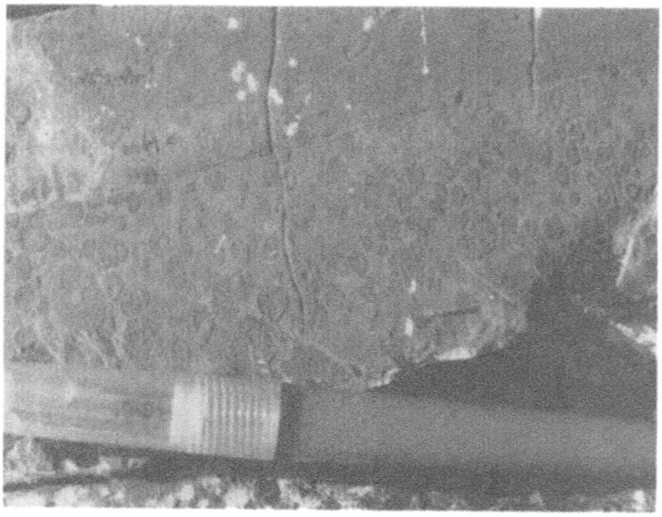

Figure 33-11 (right). Oncolites interbedded in the birdseye limestone lithofacies; Barrios de Luna.

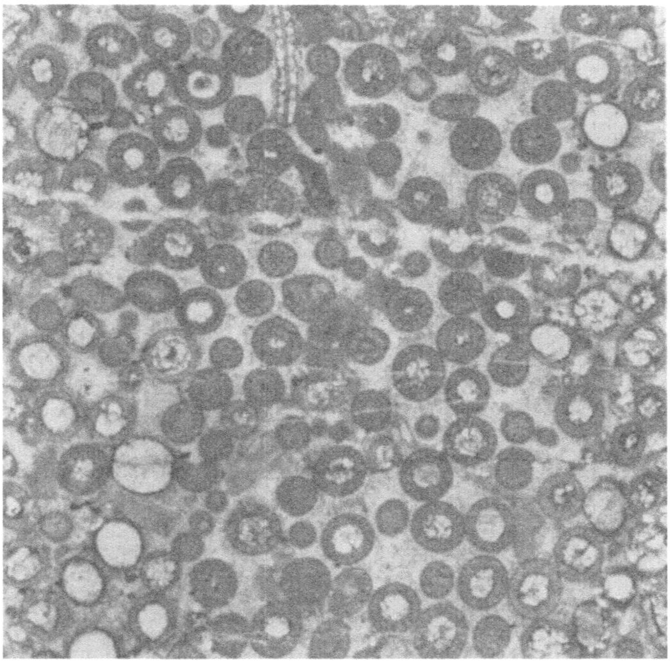

Figure 33-12. Oosparite lithofacies. Note some calcitized ooids; Primajas (C-701); × 10.

between the intertidal and supratidal paleoenvironments in certain circumstances; for that reason the term "peritidal" is used in this paper.

The laminated dolopelmicrites and pelsparites (lithofacies 1) are analogous to the levee and beach-ridge sediments described by Shinn et al. (1969) from the supratidal environment of Andros Island, Bahamas. Some of the minor textural types (pelsparites with fossiliferous remains, oosparites, and homogeneous dolomicrites) that occur interbedded in lithofacies 1 and that are only present in certain localities may represent sediments laid down in ponds or channels within the supratidal zone. Lithofacies 1 is found throughout the Ponga Nappe Province, suggesting that in this province supratidal conditions prevailed during deposition.

The algal-laminated dolostones are the predominant textural type in lithofacies 2. Algal mats are characteristic of the supratidal zone in Florida and the Bahamas, whereas in the Persian Gulf and Shark Bay they are more prevalent in the protected intertidal environment. It should be emphasized that in the Persian Gulf the intertidal algal mats beneath the sabkha are dolomitized (Kinsman et al., 1971). By comparison with the algal-laminated sediments and their associated sediment types in Shark Bay (Davies, 1970) and the Persian Gulf (Kendall

Figure 33-13. Lithostratigraphy of the lower member of the Láncara Formation and lithofacies distribution.

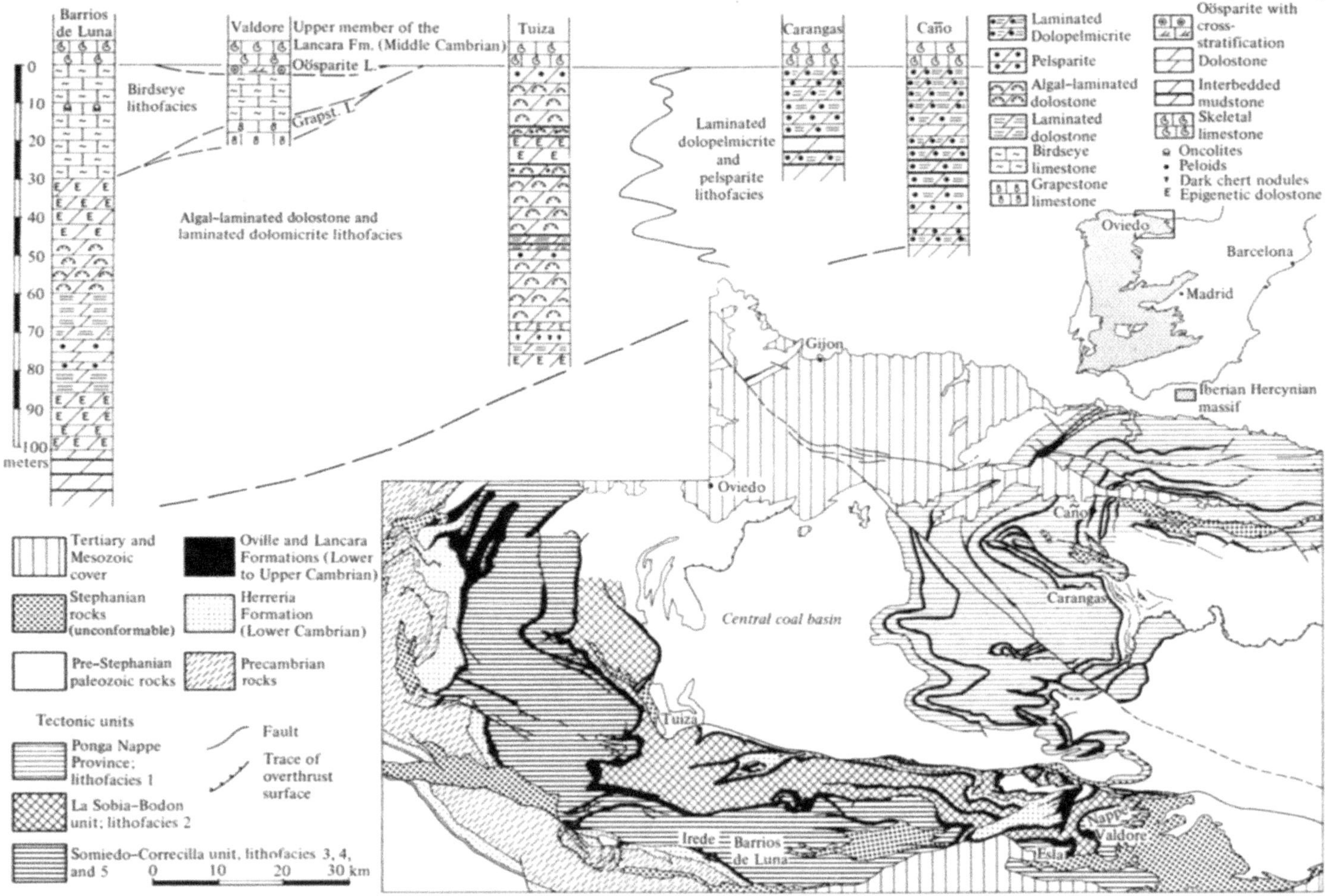

et al., 1968; Kinsman et al., 1971), it is inferred that this lithofacies in the Láncara Formation may represent the intertidal environment, although a low supratidal setting is not excluded. Lithofacies 2 occurs in the La Sobia-Bodón unit.

The birdseye limestones are typical of lithofacies 3. Desiccation features (polygonal mud cracks and sheet cracks) have been observed in certain horizons in the sequence. In these horizons the birdseye limestones contain abundant fine-grained dolomite. The birdseyes cannot be considered diagnostic solely of the supratidal environment. In the Láncara Formation birdseyes are believed to be due mainly to the release of gases by organic decay, which is in accordance with the abundance of algae in the limestones. Lithofacies 3 presumably represents a low intertidal to subtidal environment, although some periods of long subaerial exposure also occurred, as suggested by the desiccation features. This lithofacies is peculiar to the Somiedo-Correcilla unit and the Esla Nappe Province.

The grapestone lithofacies is believed to represent a subtidal deposit. It is present only in the Esla Nappe Province and in one locality (Irede) in the Somiedo-Correcilla unit. In both regions it occurs beneath the birdseye limestone lithofacies.

Lithofacies 5 (oosparites and quartzarenites) is also restricted in its distribution, and represents a high energy event. This lithofacies occurs in the Esla Nappe but is absent from localities to the west.

In northwest Spain there is a close relationship between the lithofacies distribution and the main tectonic units that can be recognized in the Cantabrian zone (Fig. 33–13).

Carbonate peritidal deposition appears to be an important feature of the Cambrian seas in northwest Spain, as well as elsewhere. Several similar sequences have been reported from North America (Kepper, 1972; Aitken, 1967). As noted by Kepper (1972), such cyclic sequences may permit the recognition of eustatic and/or epeirogenic events on a worldwide scale. The peritidal lithofacies in northwest Spain are of Early Cambrian age, thus being older than those in North America, where their age ranges from Middle to Late Cambrian.

References

AITKEN, J. D. 1967. Classification and environmental significance of cryptalgal limestones and dolomites, with illustrations from the Cambrian and Ordovician of southwestern Alberta. *J. Sed. Petrol. 37*, 1163–1178.

DAVIES, G. R. 1970. Algal-laminated sediments, Gladstone Embayment, Shark Bay, Western Australia. *In* Logan, B. W., Davies, G. R., Read, J. F., and Cebulski, D. E., eds., Carbonate sedimentation and environments, Shark Bay, Western Australia. *Am. Assoc. Petrol. Geol. Memoir. 13*, 169–205.

FOLK, R. L., and PITTMAN, J. S. 1971. Length-slow chalcedony: A new testament for vanished evaporites. *J. Sed. Petrol. 41*, 1045–1058.

GRIMM, W. D. 1962. Idiomorphe Quarze als Leitmineralien für Salinare Fazies. *Erdöl Kohle Erdgas. Petrochem. 15*, 880–887.

KENDALL, C. G. ST. C., and SKIPWITH, P. A. D'E. 1968. Recent algal mats of a Persian Gulf Lagoon. *J. Sed. Petrol. 38*, 1040–1058.

KEPPER, J. C. 1972. Paleoenvironmental patterns in Middle to Lower Upper Cambrian interval in Eastern Great Basin. *Am. Assoc. Petrol. Geol. Bull. 56*, 503–527.

KINSMAN, D. J. J., PARK, R. K., and PATTERSON, R. J. 1971. Sabkhas: Studies in recent carbonate sedimentation and diagenesis, Persian Gulf. *Geol. Soc. Am. Abs. 3*, 772–774.

SHINN, E. A., LLOYD, R. M., and GINSBURG, R. N. 1969. Anatomy of a modern carbonate tidal-flat, Andros Island, Bahamas. *J. Sed. Petrol. 39*, 1202–1228.

ZAMARREÑO, I. 1972. Las litofacies carbonatadas del Cámbrico de la Zona cantábrica (NW España) y su distribución paleogeográfica. *Trabajos Geol. 5*, 1–118.

34

Intertidal and Associated Deposits of the Prairie du Chien Group (Lower Ordovician) in the Upper Mississippi Valley

Richard A. Davis, Jr.

Occurrence

The Prairie du Chien Group represents the lowermost Ordovician sequence in the Upper Mississippi Valley. Although rarely are fossils preserved, conodonts in this sequence have been identified as being Tremodocian in age. This sequence of dolomites represents the earliest Paleozoic carbonate deposits in the upper midwest area of the United States. Exposures of the Prairie du Chien form the prominent bluffs along the Mississippi River and adjacent coulees in Minnesota, Iowa, and Wisconsin (Fig. 34–1). Numerous quarries and roadcuts also provide access to these strata.

Pre-St. Peter erosion on the upper Prairie du Chien surface produced wide variations in thickness ranging from 0 to 85

This paper was based on a study supported by the Wisconsin Alumni Research Foundation, a Sigma Xi-RESA Grant-in-Aid, the Wisconsin Geological and Natural History Survey, and the Western Michigan University Faculty Research Fund. The author is grateful to M. E. Ostrom and the late Lewis M. Cline for their help.

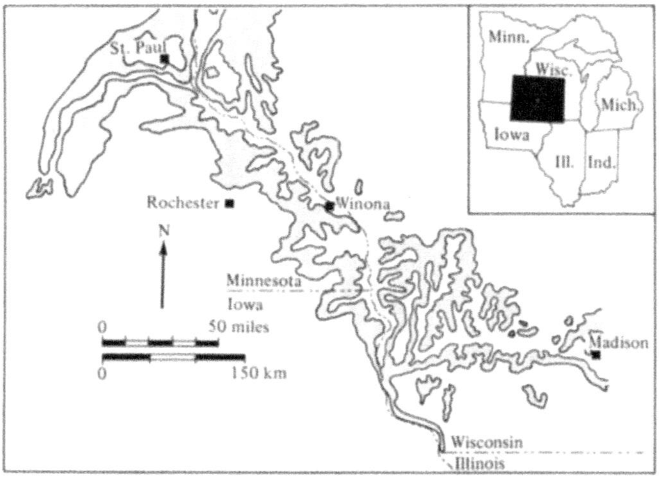

Figure 34-1. Distribution of Prairie du Chien outcrop belt in the Upper Mississippi Valley.

meters. The dominantly carbonate Prairie du Chien Group is underlain by the Jordan Sandstone and overlain by the St. Peter Sandstone, both of which are rather pure quartz sandstone units.

Prairie du Chien strata contain two formations, each of which has two members. Dolomite, quartzitic dolomite, and dolomitic quartz sandstone make up the bulk of the sequence with thin, quartz sandstone layers (Fig. 34–2). Minor occurrences of gray-green shale and nodular chert horizons are also present. Algal stromatolites are the only abundant organic remains that are preserved, although molds and fragments of invertebrates are present locally.

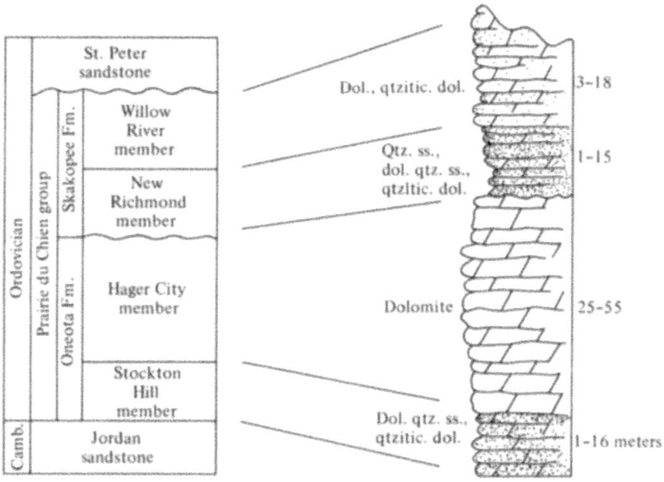

Figure 34-2. Stratigraphic units and gross lithology of Prairie du Chien strata.

Facies

The Prairie du Chien Group contains a wide variety of lithic facies, even though it is dominantly dolomitized carbonate. The depositional character of these strata can, however, be recognized throughout the Prairie du Chien. The lithofacies generally occur in thin layers, and form repeated sequences. The major predolomitization lithofacies include grain sparite, intrasparite, oosparite, algal biolithite, and quartz sandstone.

GRAIN SPARITE. Because dolomitization has destroyed the internal character of biogenic detrital grains, the term "grain sparite" (Davis, 1966a) is applied to spar-cemented, grain-supported rocks. This rock type is the most prevalent in the Prairie du Chien. The upper half or so of the Hager City and Willow River members is nearly pure grain sparite. The grains are nearly equidimensional with at least 50 percent "grain bulk." Grains are sorted and rounded, and only a few thin sections contain small amounts of micrite. Although dominantly fine to medium calcarenite, a few grains extend into the calcirudite category. At some horizons, oolites, intraclasts, and quartz grains are minor percentages of the total grain composition in this lithic facies.

INTRASPARITE. Aphanocrystalline to very finely crystalline clasts of dolomite show little or no textural effect of dolomitization. The intraclasts range from medium sand to diameters of a few centimeters (Fig. 34–3*D*). Large clasts tend to be disc-shaped but all sizes are fairly well rounded. Pure intrasparite is rare, with "grains," quartz, and oolites generally present. Although most clasts are pure micrite, a few contain floating quartz or carbonate grains.

Most intrasparite zones in the Prairie du Chien are associated with algal stromatolites, although they also occur within grain sparite sequences. At many locations the clasts are actually "algalclasts" in that they show laminations and were constructed by algae. These clasts were torn up while semilithified, representing the mode of origin in Folk's (1959) definition of the term.

OOSPARITE. Prairie du Chien oolites may be dolomitized, silicified, or partially silicified dolomite. Virtually all oolite types—including elongate, superficial, compound, reworked, and quiet water oolites (Freeman, 1962; Davis, 1966b)—are present (Fig. 34–3*A*). Although typically both rounded and sorted, there are occurrences of moderate to poorly sorted

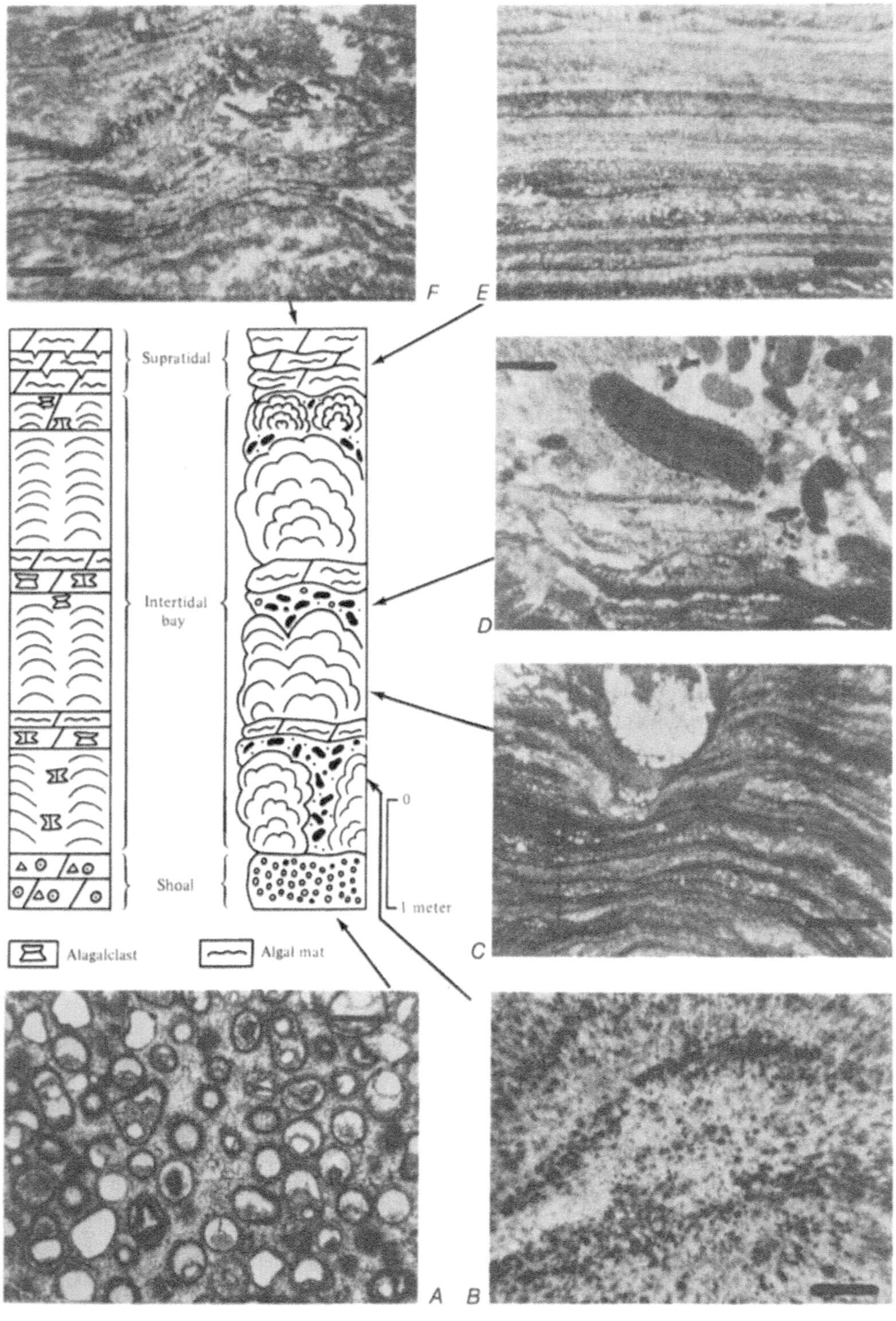

F
E
Supratidal
Intertidal bay
Shoal
0
1 meter
Alagalclast
Algal mat
D
C
A
B

Figure 34-3 (opposite). Generalized stratigraphic section of Prairie du Chien intertidalites showing lithic types (*A–F*) associated with various environments of deposition. (Bar scale in all photomicrographs 1 mm long.)

oosparites. Prairie du Chien oolites have nuclei composed of quartz grains, small intraclasts, and what apparently are biogenic grains.

ALGAL BIOLITHITE. Dolomitization has partially obscured stromatolites at some horizons (Fig. 34–3*B*, 3*C*); however, they are readily recognizable in the field. In thin section, the algal structure may be invisible, although at many horizons the delicate algal laminations are preserved. The thickness of the laminae ranges from less than one to several millimeters. Quartz, biogenic grains, and oolites may be incorporated in these laminations.

Although all of the varities of stromatolites described by Logan et al. (1964) are present in the Prairie du Chien Group, most are of the laterally linked, hemispheriod (LLH) type (Fig. 34–4*A*). Stacked hemispheroids (SH) (Fig. 4*B*) are rather common in the Hager City member, where they occur in the digitate form described by Howe (1966). Spherical algal stromatolites (SS) are rare with essentially planar algal mats fairly common. The LLH type may reach a meter in height.

QUARTZ SANDSTONE. The only terrigenous lithic type in the Prairie du Chien is well sorted and rounded, fine- to medium-grain quartz sandstone. It occurs as thin beds in the carbonates of the Stockton Hill and Willow River members, and as thick accumulations in the New Richmond Member of the Shakopee Formation. There are three types of quartz sandstone: (1) friable with practically no cement; (2) cemented by secondary growth of the quartz grains; and (3) cemented by carbonate. Although the quartz sandstones are generally pure, minor amounts of reworked silicified oolites, rock fragments, intraclasts, "grains," and feldspar occur. Cross-bedding is rare, but well-formed ripples are rather common, particularly in the areas of friable quartz sandstone.

Idealized Stromatolitic Sequence

Although the variety of fossils is extremely limited, a number of sedimentary features plus the algal stromatolites permits a fairly detailed interpretation of the depositional environments.

A particular stratigraphic sequence appears at numerous locations and various horizons throughout the Prairie du Chien Group in the upper Mississippi Valley. Detailed environmental analysis of this sequence (Fig. 34–3) provides a description of the intertidal portion of the Prairie du Chien deposits.

At many locations, the base of the intertidal sequence consists of a thin (<0.5 meter) bed of oosparite (Fig. 34–3*A*), which is locally silicified. These beds display no diagnostic structures; however, the oolites themselves are indicative of shallow to intertidal shoaling areas. The fact that the ooids are sorted and exhibit rather thick coatings at most locations supports this hypothesis. These shoals are interpreted as being located along the mouths of intertidal embayments.

The bulk of this sequence is composed of various types of algal stromatolite heads and associated intrasparites (Fig. 34–3). There are a number of cyclic stromatolite units within the strata that may represent an intertidal bay environment. Each of these cycles contains large LLH-C type stromatolites at the base. The individual heads reach nearly a meter in height and diameter. The algal structure is well preserved (Fig. 34–3*C*) with sharply defined laminae throughout. The nature of the detrital material accumulated by the algae has been obliterated in most cases; however, pellets (Fig. 34–3*B*) and biogenic debris were undoubtedly the major constituents. Most stromatolites have a few small intraclasts, ooids, fossil fragments, or quartz grains incorporated between laminations (Fig. 34–3*C*). Small cut-and-fill structures (Fig. 34–3*C*) are also common in the algal heads. Intraclasts and algalclasts (Fig. 34–3*D*) are commonly present also.

The algal stromatolite heads and associated intraclastic facies

Figure 34-4. *A.* Well-preserved LLH-C stromatolites. *B.* Poorly preserved digitate SH stromatolites.

A

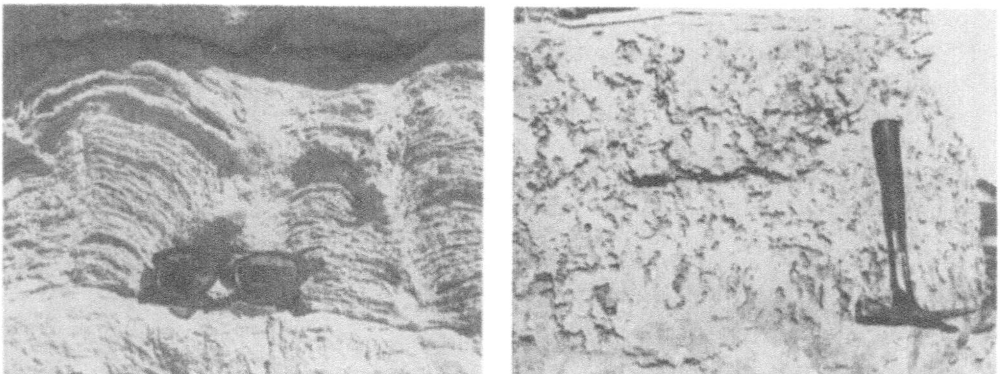

B

A

B

Figure 34-5. *A.* View of quarry floor, showing the tops of stromatolites; compare with Figure 34-5*B*. *B.* View of modern algal stromatolites in the intertidal zone of Shark Bay, Western Australia (courtesy of Brian Logan); compare with Figure 34-5*A*.

represent an intertidal depositional environment, probably much like that of the modern Shark Bay, Western Australia (Fig. 34–5*A*, 5*B*). The large heads were probably formed in a zone with a tidal range of 1 meter if stromatolite height equals tidal range (Logan, 1961). Current and wave energy were moderate to low as evidenced by the lack of large-scale disruptions of heads and lack of spacing at most places. Extensive exposures are not available so that lateral changes cannot be determined. Clasts were derived from algal heads and mats as well as from lime mud that was accumulating between and adjacent to algal heads.

Immediately above the large algal heads at some horizons are small stromatolites of about 25 cm diameter. These are of the LLH-S type and probably represent deposition near the intertidal bay margin.

The sequence is capped by a rather thin (<0.5 meter) algal-mat unit. The microscopic nature of these mats is much like

Figure 34.6. *A.* Mud-cracked micrite. *B.* Wrinkled algal mats.

A

B

that of heads; however, sand-size debris such as quartz, ooids, and so on is noticeably absent. Desiccation features are generally abundant on both the hand specimen and microscopic scales (Figs. 34–3*F* and 34–6*A*). Wrinkled mats (Fig. 34–6*B*) are interpreted as being representative of the supratidal environment on the bay margin.

The sequence described above is rather typical of the intertidalites (Klein, 1970) in the Prairie du Chien. This sequence occurs at various horizons and locations and may underly or overly a grain sparite facies, which represents the open, shallow, carbonate bank environment.

References

DAVIS, R. A. 1966a. Willow River dolomite; Ordovician analog of modern algal stromatolite environments. *J. Geol. 74,* 908–923.

——— 1966b. Quiet water oolites from the Ordovician of Minnesota. *J. Sed. Petrol. 36,* 813–818.

FOLK, R. L. 1959. Practical petrographic classification of limestones. *Bull. Am. Assoc. Petrol. Geol. 43,* 1–38.

FREEMEN, T. 1962. Quiet water oolites from Laguna Madre, Texas. *J. Sed. Petrol. 32,* 475–483.

HOWE, W. B. 1966. Digitate algal stromatolite structures from the Cambrian and Ordovician of Missouri. *J. Paleo. 40,* 64–77.

KLEIN, G. DEV. 1970. Paleotidal sedimentation. *Geol. Soc. Am. Abstr. 2,* 598.

LOGAN, B. W. 1961. Cryptozoan and associate stromatolites from the Recent, Shark Bay, Australia. *J. Geol. 69,* 517–533.

LOGAN, B. W., REZAK, R., and GINSBURG, R. N. 1964. Classification and environmental significance of algal stromatolites. *J. Geol. 72,* 68–83.

35

Shoaling and Tidal Deposits that Accumulated Marginal to the Proto-Atlantic Ocean: The Tribes Hill Formation (Lower Ordovician) of the Mohawk Valley, New York

Gerald M. Friedman and
Moshe Braun

Occurrence

The carbonates of the Tribes Hill Formation (Canadian, Lower Ordovician) crop out long the Mohawk Valley in east-central New York (Fig. 35–1). Descriptions of these rocks have been published by Braun and Friedman (1969) and by Friedman (1972).

Lithofacies

Ten lithofacies are recognized:

10. Biointramicrite and biomicrite
9. Oobiosparite and biointraoosparite
8. Intrasparite and biointrasparite
7. Mottled dolomitic micrite and biomicrite
6. Dolostone
5. Shale stringers and layers
4. Pebble conglomerate
3. Feldspathic pelmicrite
2. Laminated feldspathic dolomite
1. Mottled feldspathic dolomite

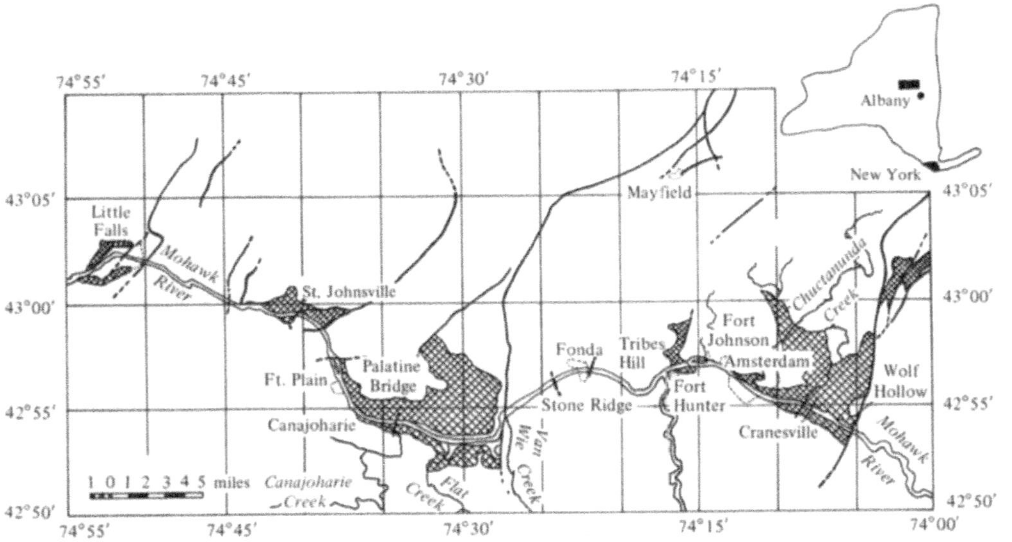

Figure 35-1. Index map showing distribution of Lower Ordovician rocks in Mohawk Valley, New York. Outcrop areas are shown in quadrille ruling. Heavy solid lines indicate faults. (After Fisher, 1954.)

Figure 35–2 relates the 10 lithofacies to four members defined by Fisher (1954).

Lithofacies 1: Mottled feldspathic dolomite

This facies occurs as thin dolomite beds, 2 to 25 cm but locally more than 50 cm thick, with a few thin interbeds of black argillaceous dolomite that are up to 5 cm thick. In the field, the dolomite shows gray-black mottling and in places birdseye structures. In one sample, the infilling of the birdseyes shows a black bituminous rim that may be anthraxolite. Trace fossils are present, but actual fossil remains are rare. Authigenic alkali feldspar (microcline) is ubiquitous throughout. The feldspar is dispersed through the dolomite and occurs in patches; it thus creates a mottled effect.

Lithofacies 2: Laminated feldspathic dolomite

Lithofacies 2 is mineralogically identical to lithofacies 1 but differs from it texturally and structurally. Lithofacies 2 is irregularly bedded and contains abundant undulating stromatolitic structures ("pseudoripples") as well as disturbed and discontinuous laminae. In places a few thin interbeds of black argillaceous dolomite are present. The thickness of the laminae of this facies ranges from 0.5 to 2 or 3 mm; on freshly broken surfaces the thinner laminae are black and the thicker laminae are gray.

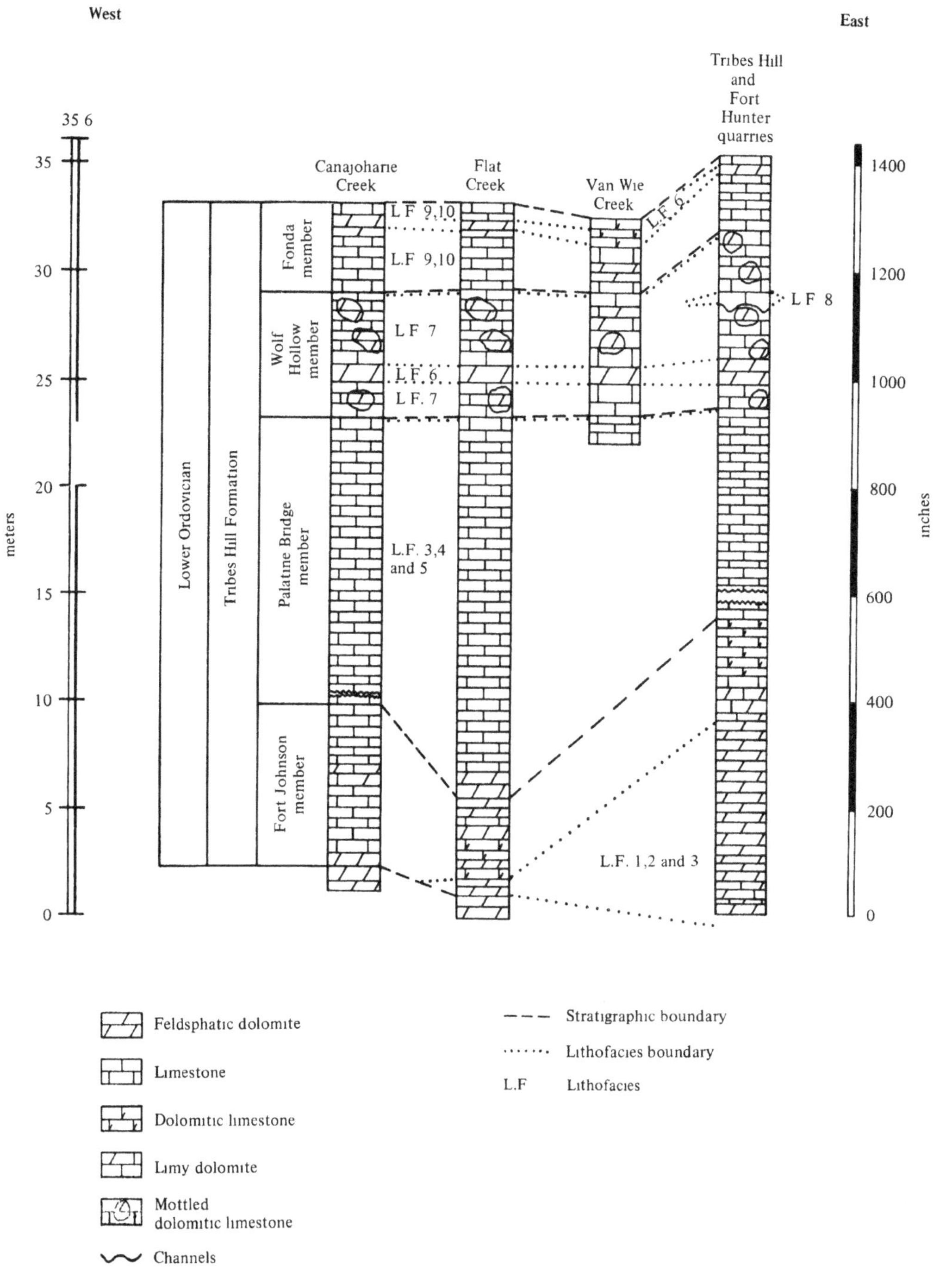

Figure 35-2. Columnar section showing the relationship of 10 lithofacies to four members in Tribes Hill Formation (Lower Ordovician). (After Braun and Friedman, 1969.)

Lithofacies 3, 4, and 5: Feldspathic pelmicrite, pebble conglomerate, and shale stringers and layers

These three facies are closely interbedded and form a cyclic pattern. The pebbles of lithofacies 4, a pebble conglomerate, are composed of lithofacies 3 lithology, a feldspathic pelmicrite. Because of their cyclical nature these three facies are considered together.

LITHOFACIES 3: FELDSPATHIC PELMICRITE. Lithofacies 3 is a feldspathic limestone, in contrast with lithofacies 1 and 2, which are feldspathic dolomites. The rocks of lithofacies 3 occur as thin interbeds 1 to 10 cm thick. Their upper surfaces commonly show erosional features and are overlain by thin films of lithofacies 5, in the form of shale stringers. Calcitic and feldspathic layers are finely interlaminated, commonly cross-bedded; the couplets are like those in lithofacies 2, except that calcite has taken the place of dolomite. Pellets are distinctive depositional features of lithofacies 3.

LITHOFACIES 4: PEBBLE CONGLOMERATE. Lithofacies 4 occurs as thin limestone interbeds 5 to 20 cm thick. Because of differential weathering between pebbles and matrix on weathered surfaces, the pebbles stand out prominently. Some of the pebbles occur as irregular, rounded, or angular fragments, a structure known as lumpy (Matter, 1967). Mud cracks are found, which pass laterally into flat pebbles, and small channels with cross-bedded layers form scour-and-fill structures.

LITHOFACIES 5: SHALE STRINGERS AND LAYERS. Lithofacies 5 occurs in thin, dark-colored laminae 2 to 3 mm thick, on irregular erosion surfaces, and as thin, dark-colored layers of shaly material, 2 to 10 cm thick, with enclosed boundinagelike (lumpy) limestone lenses 2 to 3 cm long.

Lithofacies 6: Dolostone

Lithofacies 6, which makes ledges up to 1 meter thick, is thin bedded—10 to 20 cm thick. In places the uppermost part of a ledge may be mottled as in lithofacies 7. Lithofacies 6 differs from lithofacies 1 and 2 in that it has no feldspars.

Lithofacies 7: Mottled dolomitic micrite and biomicrite

Lithofacies 7 consists of compact, well-bedded, mottled limestone in which the mottles are composed of irregular patches of dolomite. In places on weathered bedding planes the outlines of large gastropods and cephalopods stand out. Birdseye structures are present in some beds as are pyrite patches. The limestone contains abundant dolomite-filled burrows.

Petrographically this limestone is a mottled dolomitic micrite and biomicrite.

Lithofacies 8: Intrasparite and biointrasparite (channel fill)

Lithofacies 8 is confined to channels that have been cut into lithofacies 7. The lower contacts between lithofacies 7 and 8 are truncation surfaces typical of channels (Fig. 35–3).

Blocks of lithofacies 7 lithology are lodged in the channels of lithofacies 8 (Fig. 35–4). These blocks vary in shape and size, with diameters ranging from 0.25 to 1 meter. They resemble similar blocks in tidal channels of the Bahamas, which are derived from the banks of the channels by undercutting.

Rocks of the channel fill are composed of two microlithologies: intrasparite and biointrasparite. The grains in the intrasparite are sand- to pebble-sized intraclasts of lithofacies 7 (micrite and biomicrite).

Lithofacies 9: Oobiosparite and biointraoosparite

The allochems making up lithofacies 9 are ooids, skeletal fragments, and intraclasts. The ooids are the diagnostic grains of lithofacies 9.

Lithofacies 10: Biointramicrite and biomicrite

The megascopic appearances of lithofacies 10 and 9 are identical; in the field these two lithofacies cannot be distinguished.

The allochthonous constituents of lithofacies 10 are skeletal grains and intraclasts; locally ooids are present but they are

Figure 35-3. Truncation at base of tidal channel. Rocks in channel consist of lithofacies 8 (intrasparite and biointrasparite), Tribes Hill Formation(Lower Ordovician). North Tribes Hill quarry.

Figure 35-4. Block of lithofacies 7 (mottled dolomitic micrite and biomicrite) foundered in tidal channel (lithofacies 8), Tribes Hill Formation (Lower Ordovician). North Tribes Hill quarry.

uncommon. The matrix between the grains and pebbles has been thoroughly altered diagenetically. However, where relicts are present, they are micrite.

Discussion

During the Cambrian and Ordovician periods, a shallow epeiric sea covered most of the North American continent. The area resembled the present-day Bahama Bank. At the eastern edge of this bank, that is, at the then-eastern edge of the continent, the limestones and dolostones of the Tribes Hill Formation accumulated. The steep paleoslope, which marked transition from the continent into the deep sea, lay about 35 mi east of the present outcrops of the Tribes Hill Formation. Carbonate sediments moved down this steep paleoslope by slides, slumps, turbidity currents, mud flows, and sand flows to oceanic depths where terrigenous mud was accumulating (Sanders and Friedman, 1967, pp. 240–248; Friedman, 1972). The paleoslope probably was an active hinge line. The ocean lying east of the North American continental block has been interpreted as the Proto-Atlantic Ocean (Bird and Dewey, 1970).

The carbonate rocks of the Tribes Hill Formation show many features that suggest that they were subjected to repeated shoaling and intermittently were exposed subaerially. These features include mud cracks, birdseye textures, undulating stromatolitic structures, mottles, lumpy structures, scour-and-fill structures, flat pebbles, cross-beds, and, as a lithology, syngenetic dolostone (Friedman and Sanders, 1967). Features identical to these are known from most Paleozoic shallow-water carbonates that underlie much of North America. The site of accumulation of the Tribes Hill carbonates, however, differed markedly from that of most other Paleozoic carbonates that stretch across North America. The Tribes Hill carbonates were deposited close to the edge of the continent. Hence diurnal or semidiurnal fluctuations of the waters of the deep ocean should have left their mark on the Tribes Hill deposits. If so, such deposits can be classed as tidal.

In modern tidal sediments perhaps the most obvious of the morphologic features are tidal channels. In the rocks of the Tribes Hill Formation, what may be ancient tidal channels can be observed. To our knowledge such channels have not been reported from the Cambro-Ordovician carbonate-rock sequences in other parts of North America.

The sizes of the channels in the Tribes Hill Formation are comparable to the sizes of modern tidal channels. Sharp basal truncations are typical (Fig. 35–3). The material filling the channels consists of lithofacies 8 (intrasparite and biointrasparite), a high energy facies. These channels cut into lithofacies 7 (a mottled dolomitic micrite and biomicrite), a low energy facies. Large blocks of lithofacies 7, up to 1 meter in diameter, which are lodged in the fills within the channels, are thought to have been derived by undercutting of the banks (Fig. 35–4). Hence, to accomplish such undercutting, the currents in these channels must have flowed fast. The contrast between the high energy facies filling the channels and the low energy facies in the flats adjacent to the channels likewise suggests that currents in the channels flowed swiftly.

Although in Paleozoic limestones the products of shoal waters are ubiquitous, tidal deposits may have been restricted to the margins of the continents where the epeiric shelf faced the deep ocean. The carbonate rocks of the Tribes Hill Formation may be an example of such a tidal sequence.

Authigenic feldspar is an essential constituent of the carbonate rocks of the Tribes Hill Formation, especially in lithofacies 1 and 2, and in the shale stringers and layers of lithofacies 5. The high concentration of feldspars causes the laminae in lithofacies 2 to weather in positive relief. Such feldspars commonly are the end products of the alteration of zeolites. However, zeolites are unknown from sedimentary rocks as old as Early Ordovician. In rocks older than mid-Paleozoic, any original zeolites probably have changed to feldspars. In volcaniclastic rocks of Cenozoic age, authigenic feldspar is known to be the end product of volcanic glass whose initial alteration product was a zeolite (Sheppard and Gude, 1969; Goodwin, 1973).

The feldspars in the Tribes Hill Formation are interpreted as wind-transported tephra that accumulated at the margin of the Proto-Atlantic Ocean. The active volcanoes responsible for such tephra may have been parts of ancient island arcs.

References

BIRD, J. M., and DEWEY, J. F. 1970. Lithosphere plate-continental-margin tectonics and the evolution of the Appalachian orogen. *Geol. Soc. Am. Bull. 81*, 1031–1060.

BRAUN, MOSHE, and FRIEDMAN, G. M. 1969. Carbonate lithofacies and environments of the Tribes Hill Formation (Lower Ordovician) of the Mohawk Valley, New York. *J. Sed. Petrol. 39*, 113–135.

FISHER, D. W. 1954. Lower Ordovician stratigraphy of the Mohawk Valley, N. Y. *Geol. Soc. Am. Bull. 65*, 71–96.

FRIEDMAN, G. M. 1972. "Sedimentary facies": Products of sedimentary environments in Catskill Mountains, Mohawk Valley, and Taconic sequence, eastern New York State. *Guidebook*. Soc. Econ. Paleontol. Mineral. Eastern Sect., 48 pp.

FRIEDMAN, G. M., and SANDERS, J. E. 1967. Origin and occurrence of dolostones. *In* Chilingar, G. V., Bissell, H. J., and Fairbridge, R. W., eds., *Carbonate Rocks*. Amsterdam, Elsevier, pp. 267–348.

GOODWIN, J. H. 1973. Analcime and K-feldspars in tuffs of the Green River Formation, Wyoming. *Am. Mineral. 58*, 93–105.

MATTER, ALBERT. 1967. Tidal flat deposits in the Ordovician of Western Maryland. *J. Sed. Petrol. 37*, 601–609.

SANDERS, J. E., and FRIEDMAN, G. M. 1967. Origin and occurrence of limestones. *In* Chilingar, G. V., Bissell, H. J., and Fairbridge, R. W., eds., *Carbonate Rocks*. Amsterdam, Elsevier, pp. 169–265.

SHEPPARD, R. A., and GUDE, A. J. III. 1969. Diagenesis of tuffs in the Barstow Formation, Mud Hills, San Bernardino County, California. *U.S. Geol. Surv., Prof. Paper 634*.

36

Ordovician Tidalites in the Unmetamorphosed Sedimentary Fill of the Brent Meteorite Crater, Ontario

F. W. Beales and G. P. Lozej

The term "tidal zone" as used here includes areas of the wind-tidal flat that are intermittently, perhaps only rarely, flooded by marine water. It is one of the most precisely definable bathymetric zones in a stratigraphic sequence, and we argue that this contention is further supported by our studies of the sedimentary fill of a terrestrial meteorite crater, the Brent Crater, Ontario (lat. 46°04′ N; long. 78°29′ W).

The Crater Fill

Surrounded by a low rim and underlain by 650 meters of shocked, impact breccia (Dence, 1968), 260 meters of lower Middle Ordovician tidalites and continental margin sediments

We are most grateful to C. S. Beals, who initiated the Dominion Observatory of Canada meteorite crater investigations; to M. R. Dence, who has substantiated the meteoritic origin of a series of craters, including Brent; and to B. A. Johnson, who survived the tedium of counting successive laminae for us. G. V. Middleton and R. G. Walker kindly reviewed the manuscript but the authors are solely responsible for its final form.

have been accommodated in the center of the crater cavity and thereby preserved as an isolated outlier on the southern Canadian Shield (Figs. 36–1 and 36–2).

LOWER PART. In units 1 to 4, thin beds and laminae of dolostone and micaceous mudstone recur conspicuously. Lesser, thin, gypsiferous bands occur (now selenite of satin-spar variety, oriented perpendicularly to the bedding or the faces of fractures). In these units remarkably numerous laminations characterize the restricted crater sequence. About 125,000 laminae, related to variation in dolostone, mudstone, and gypsum composition, are present in the 100 meters of section, comprising some 14,000 more pronounced lithic changes (laminated beds). The abundance of dolomite and gypsum with scattered grains and thin interbeds of fresh feldspathic silt is particularly noteworthy. Gypsum is common, but was presumably much more abundant because evaporite-solution collapse-breccias and intrasediment slumps are frequent.

It is difficult to envisage the deposition of such a thick predominantly dolostone sequence without considering a proximal marine origin, since the abundance of calcium, magnesium, carbonate, and sulfate constituents would otherwise have had to come from crater wall weathering, of which there is no evidence. The restricted mineralogy does not correspond with likely compositions for a source from groundwater derived from the surrounding Grenville terrain. A partially restricted marine-marginal supratidal environment is further supported by the presence of abundant desiccation cracks in dolostone beds, the intercalated evaporite-solution breccias, the minor bioturbation

Figure 36-1. Brent Meteorite Crater. (Modified after Dence, 1968.)

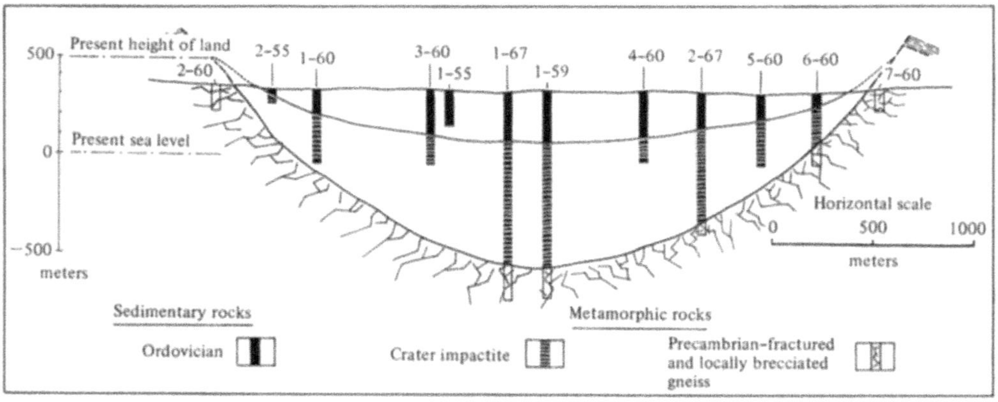

Brent Crater.

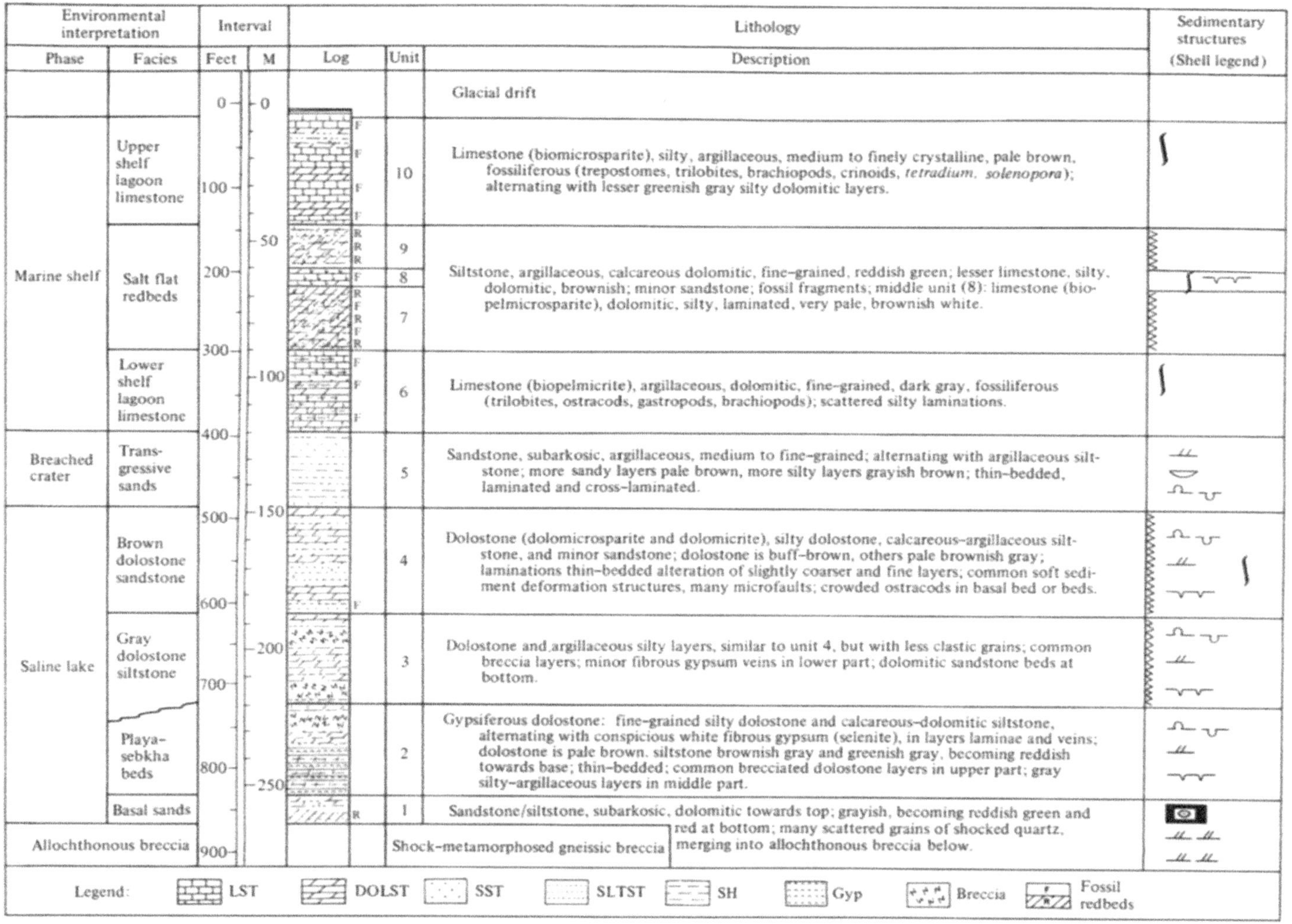

restricted to some layers only, the lack of fossils other than sporadic crinoid ossicles that survived dolomitization, possible blue-green algal remains, and rare ostracod-bearing green mudstone.

TRANSITION. A radical environment change has to be inferred from the variation in the lithology of unit 5, marked by a coarsening in grain size, disappearance of dolostone laminae, alternation of sandstone-mudstone beds, presence of current structures (cross-laminations, truncations, and channeling), and the appearance of calcispheres near the top. Unit 5 was probably deposited rapidly within a mobile, transgressive, intertidal to subtidal setting.

UPPER PART. In units 6 to 10, the crater geometry no longer influenced sedimentation. The upper section begins with a clearly subtidal marine-lagoon phase. Units 7 to 9 preserve a splendid record of sea-level fluctuations in the intertidal to supratidal range. Mud cracks, birdseye structures, secondary dolomitization, extensive episodic oxidation, bioturbations, hurricane-deposited (?) biosparite, and biosparrudite laminae and beds are common. Unit 10 records mostly the upper subtidal zone of sedimentation, with minor sea-level oscillations, suggested by groups of coarser and finer layers and variation in fossil content.

Discussion

The evidence used in the assignment of a tidal influence to the Brent Crater sediments are not individually diagnostic but, collectively, they assist in the formulation of a likely sedimentary model. The presumed wind-tidal influence is much more obvious in units 2 to 5 during the crater phase and in units 7 to 9 during the marine shelf phase.

Compositionally, the dominant dolomicrite is presumed to be primary, or very early diagenetic, throughout much of the core (units 2, 3, 4, and, in part, 8). It is quite different from the porphyrotopic scattered rhombs found in the upper dominantly limestone units. Also, in the lower beds, associated gypsum, intercalated red beds, and the lack of fossils, other than rare algal traces and rare ostracod and crinoid-bearing layers, are all consistent with a supratidal setting. Structurally, the repeated layering and lamination (Figs. 36–3 and 36–4) are preeminent, desiccation cracks are common, and minor bioturbation is re-

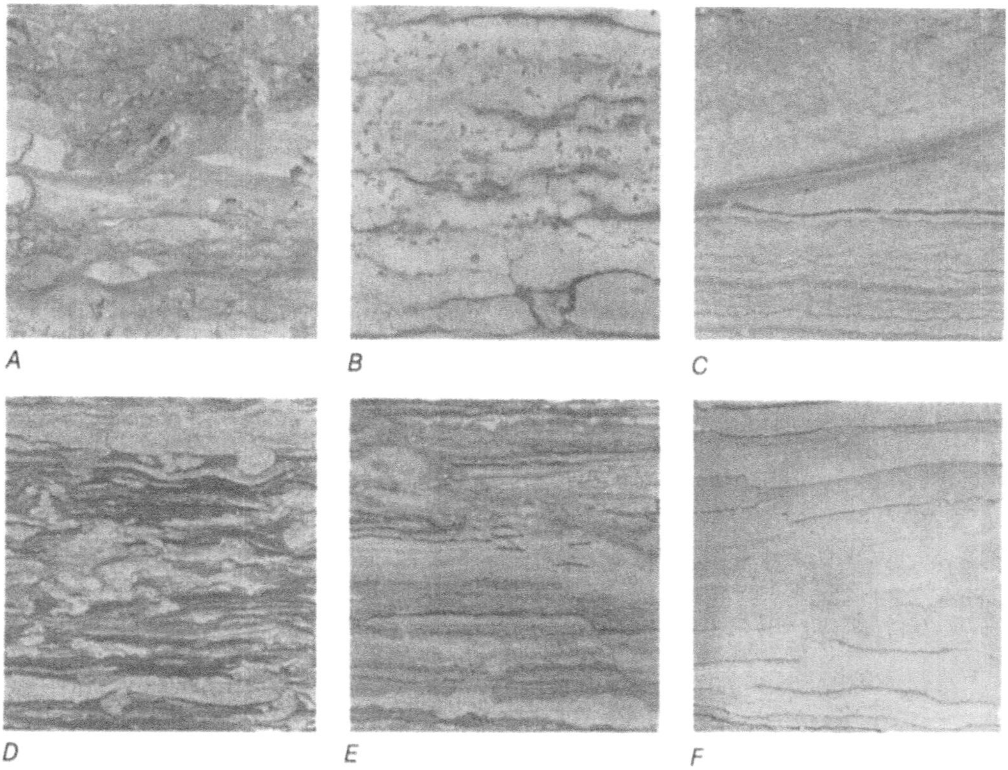

Figure 36-3. Numbered sedimentary units are as described in Figure 36-2; × 1.5.)

A. Unit 10 (−22 meters). Pale buff-gray, repeated beds of fine ooliths, skeletal materials, mostly trilobite and brachiopod; fine, shaly, microstylolitic seams. Scour of micritic bed by oosparite with skeletal lag in the center? Lower intertidal/upper subtidal?

B. Unit 8 (−59 meters). White, micritic limestone beds; exposure cracks and birdseye structures; pyrite and organic matter concentration along seams and stylolites; rare skeletal debris. Lagoonal, intertidal?

C. Unit 7 (−70 meters). Greenish-gray microundulatory shaly laminae and whitish buff dolomitic limestone; part reddish, silty, pelletal, with fine debris. Upper intertidal/supratidal (oblique streak is artifact).

D. Unit 5 (−132 meters). Burrowed, dolomitic, arkosic siltstone. Tidal flat?

E. Unit 4 (−177 meters). Brown-gray, laminated dolomicrite, showing irregular microconcretionary lenses and soft-sediment deformation structures; both tensional and compressional effects in different groups of layers. Semirestricted, supratidal lagoon?

F. Unit 3 (−184 meters). Micritic dolostone with repeated fine organic-residue laminae; microfaults and soft-sediment deformation, in part compressional. Semirestricted supratidal lagoon?

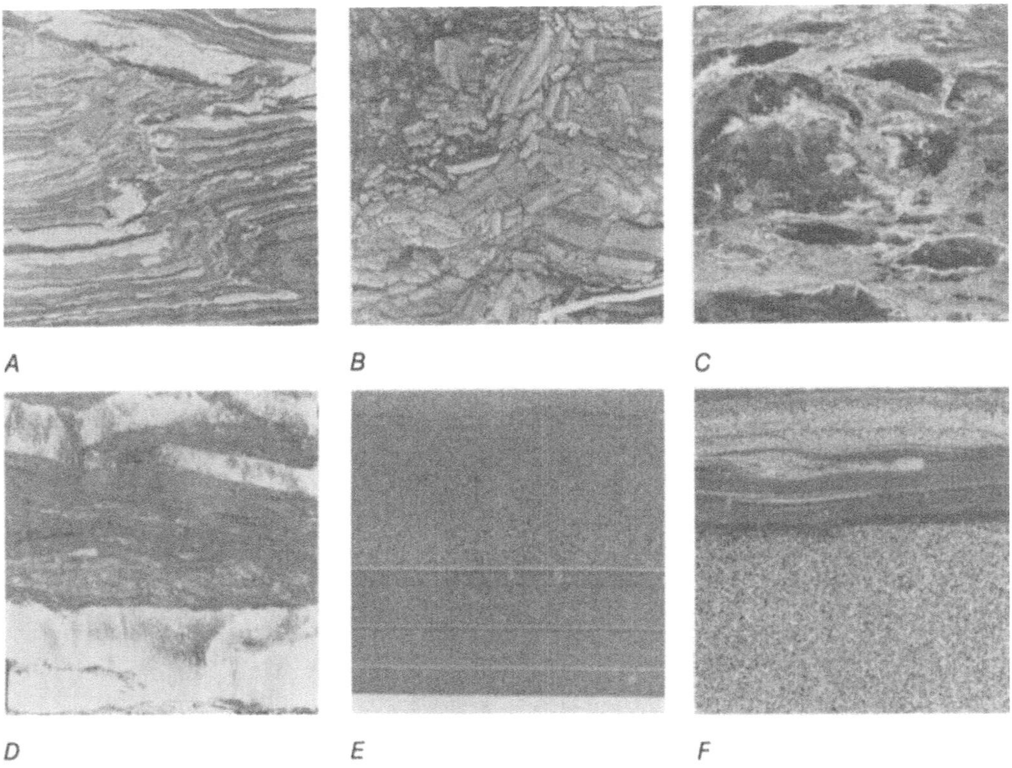

Figure 36-4. Numbered sedimentary units are as described in Figure 36-2; × 1.5.

A. Unit 3 (−188 meters). Micritic dolostone, showing more abundant darker organic laminae and paler microconcretionary layers, apparently causing crinkly disruption wedges, in part clearly compressional. Environment as for 1?

B. Unit 2 (−220 meters). Disrupted dolomicrite breccia; local collapse of paler platy and blocky lithoclasts in darker brown, silty, organic-stained dolomitic matrix. Supratidal lagoon?

C. Unit 2 (−224 meters). Brecciated, irregularly bedded to concretionary, dolomitic gypsum with minor satin-spar vein Restricted, supratidal lagoon?

D. Unit 2 (−228 meters). Finely laminated, dark, gypsiferous, silty dolostone; satin-spar veins. Fine, gray, gypsum laminae and lenses, in part concretionary, alternate with silty dolomicrite. Restricted, supratidal lagoon?

E. Unit 2 (−233 meters). Fine-grain, greenish-gray, micaceous subarkose. Pale laminae are reddish dolomicrite. Middle 1.5 cm bed is graded, flecked with biotite plates. Satin spar at base. Playa lake?

F. Unit 1 (−250 meters). Fine-grain, subarkose, overlain by laminated dolomitic mudstone with minor, thin sandstone layers and lenses. Microcurrent flaser bedding and microcompaction structures. Initial crater-lake stage.

stricted to relatively few beds. During deposition of units 7, 8, and 9, the repeated alternation of red shaly dolostone layers (and similar gray and greenish layers, some of which may also have been reddish at the time of deposition) with gray fossiliferous marine limestone and dolomitic limestone is probably influenced by wind tides, as well as by glacial and tectonic causes. In some layers, an abrupt change to coarse sediments and fossil debris, upwards, is suggestive of storm-generated tides. Almost throughout, the crater beds are characterized by repeated bed discontinuity with superimposed minor laminations (Figs. 36–3 and 36–4), which, over the total thickness of 260 meters, average about 1 cm per bedding unit and 1 mm per lamina in the lower part.

Interpretation

Neither uniform layering by an extensive shelf sea nor sedimentation in a deep depression fits the stratigraphy. A model involving a moderate (meteorite-induced) excavation in the upper zone of a broad tidal flat, subsequently covered by shallow transgressive seas, is in agreement with all the data presently available. Intermittent flooding by exceptional spring tides or hurricane tides, through and/or over the porous wall of the 3,000-meter-wide crater, deposited the successive laminae of dolomite and gypsum at the bottom of the sequence (Fig. 36–4).

Many aspects of these beds are remarkably similar to the Recent wind-tidal flat deposits of Laguna Madre (Miller, Paper 8). The lower 110 meters suggests a supratidal to upper intertidal zone, with the initial crater floor slightly below the normal sea level. Compaction of the underlying, highly porous impact-breccia made room for the subsequent tidalites. The middle 30-meter sand unit (unit 5) is interpreted as the stratigraphic record of the breaching of the crater wall. The uppermost 120 meters of predominantly fossiliferous limestones and dolomitic limestones ranges from subtidal biomicrites, biopelmicrosparites, and oosparites to intertidal and supratidal biomicrites, micrites, dolomicrosparites, and red beds.

Sedimentation close to mean sea level is implicit in the model proposed for the entire sedimentary sequence of the Brent Crater. This interpretation implies a delicate equilibrium among sedimentation, reflux, and compaction subsidence, so that repeated lithological responses to even minor marine fluctuations were preserved, and saline brines could from time to time withdraw to the ocean.

All but vestiges of Ordovician sedimentation have been eroded from the surrounding area. The sedimentary section of the Brent Crater, the upper units in particular, owe their preservation to the fact that compaction of the underlying highly porous impactite, due to deposition of overlying beds since removed, carried the crater beds below the general level of the surrounding Precambrian terrain. The section is an exceptional case of the "preservation index" discussed by Mountjoy (Paper 44).

Major tidal fluctuations and superimposed climatic, tectonic, and storm events are thought to account for the laminated beds. In addition, the long period maxima of lunar tides would affect deposition, particularly if the moon was closer to the earth in lower Paleozoic time. With a closer moon, tides would have been greater and, therefore, at least marginally, geologically more effective on the low relief topography of lower Middle Ordovician times, especially when combined with low pressure fronts or wind action.

Certainly the episodes of dolomite and gypsum precipitation must have required more time than the daily tidal maxima would permit. Assuming that the laminae reflect no more than annual events and possibly up to 20-year events (18.6 year cyclicity of longer term lunar tide maxima, and approximate local frequency of subtropical hurricanes), a very rough estimate of the crater filling up to unit 5 would be from 125,000 to about 2.5 million years. The overlying beds are early Middle Ordovician in age, so that the "splash down" was also presumably an early Middle Ordovician event.

Global Significance

The broader significance of the unmetamorphosed sedimentary fill of the Brent Crater is that it is an instantaneous cortical event as far as the earth's crust is concerned. It appears to have occurred on a stable continental platform (then situated in a probable tropical climatic zone) that had an elevation close to the mean sea level prevailing in Middle Ordovician times. Assuming that the southern Ontario and Hudson's Bay Ordovician, Silurian, and Devonian outcrops were once continuous, and therefore covering the Brent area, the crater was possibly depressed to about 500 meters below sea level by Upper Devonian times. This whole sequence, since removed, was in fact almost certainly deposited close to sea level on a gently subsiding continent. Subsequent to the Acadian orogeny (Upper Devonian, Paleo-Atlantic closing), erosion has prevailed, but the area was probably never much higher than its present elevation of about 300 meters, or the chances of its preservation would have been small.

Determination of precise ancient continental levels relative to their mean sea level is pertinent to the continued evaluation of theories concerning plate tectonics and the perplexing vertical components of continental movements. The common occurrence of tidalites throughout the post-Archean stratigraphic record of all continents constitutes a severe restraint on, for example, expanding or contracting earth theories that involve more than nominal changes in the mean radii of the earth.

Finally, from a paleoclimatologic point of view, the almost ubiquitous occurrence of apparently warm climatic conditions on all continental platforms, and the relative rarity of ancient cold climatic indicators, must prompt continuing review of the criteria we use for identifying climates. We presently favor the interpretations of tropical conditions commonly made on a basis of evidence such as that gathered at Brent, and explain the abundance of "warm" sediments by global geometry. Thus, for example, the surface area of the tropical belt is about 40 percent of the total area of the earth, the area between 33°N and 33°S latitude comprises about 55 percent, and between 45°N and 45°S is about 70 percent of the whole surface. There is much more room for continents in random relative motion to find space in equatorial positions rather than in polar positions.

Reference

DENCE, M. R. 1968. Shock zoning at Canadian craters: Petrography and structural implications. *In* French, B. M., and Short, N. M., eds., *Shock Metamorphism of Natural Materials*. Baltimore, Mono Book Corp., pp. 339–362.

37

Mississippian Tidal Deposits, North-Central New Mexico

Augustus K. Armstrong

Occurrence

The Mississippian Arroyo Penasco Group (Armstrong and Mamet, 1974; Armstrong, 1955, 1958, 1967) of north-central New Mexico contains the Espiritu Santo and Tererro Formations of Baltz and Read (1960). The Arroyo Penasco Group is 3 to 42 meters thick, rests on a peneplain of Precambrian rock, and is unconformably overlain by strata of Pennsylvanian age. Mississippian rocks crop out in the San Pedro, Nacimiento, Jemez, Sandia, Manzanita, Manzano, and Sangre de Cristo Mountains of north-central New Mexico (Figs. 37–1 and 37–2).

The Del Padre Sandstone Member, 1.5 to 20 meters thick, at the base of Espiritu Santo Formation, rests on Precambrian rock and is composed of quartz conglomerate, sandstone, and thin shale. Within the Espiritu Santo Formation there is a nearly complete carbonate depositional cycle consisting of dolo-

Publication authorized by the Director, U.S. Geological Survey.

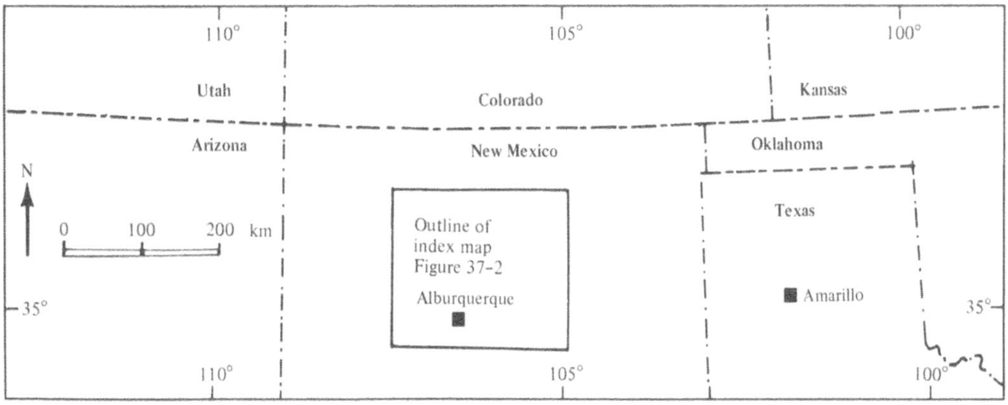

Figure 37-1. Index map showing the location of north-central New Mexico. For detailed location of outcrops see Figure 37-2.

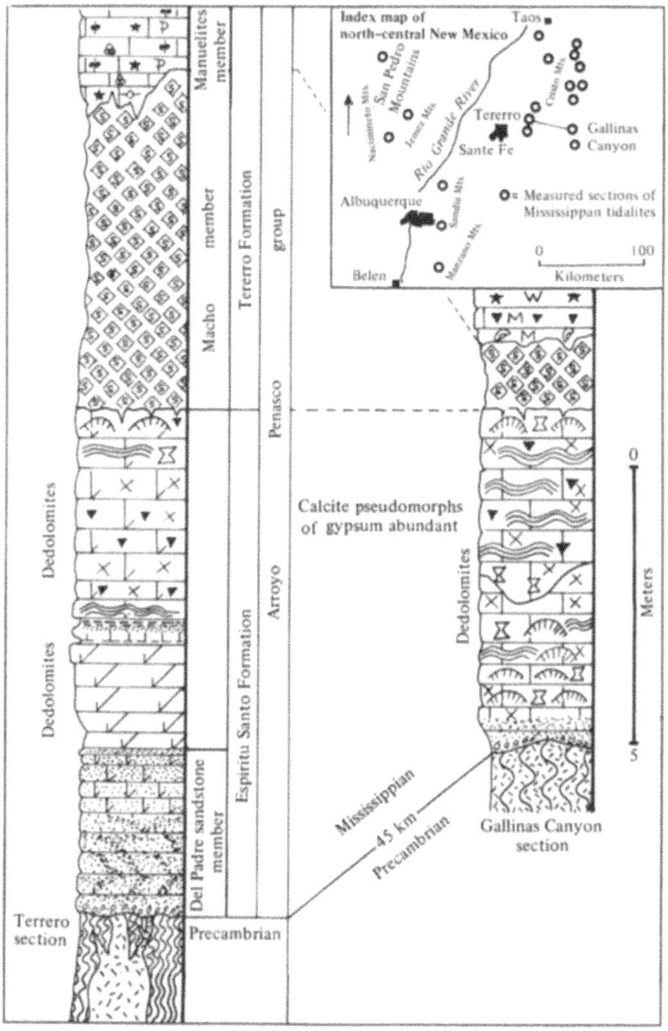

Figure 37-2. Lower part of the Mississippian Arroyo Penasco Group, the Espiritu Santo Formation sections at Tererro and Gallinas Canyon, Sangre de Cristo Mountains, New Mexico, and an index map of north-central New Mexico.

Figure 37-3. Agal mats (indicated by arrow), lithoclasts, and lamination preserved a few feet above the Precambrian contact in dedolomites of the late Osagean Espiritu Santo Formation; Gallinas Canyon outcrop, Sangre de Cristo Mountains.

mite, dedolomite, and coarse-grain, poikilotopic calcite with corroded dolomite rhombs. These rocks contain gray nodular chert that preserves the original microfacies and microfauna of late Osagean age. The sediments of the cycle are pellet, ostracode, lime mudstones overlain by 1- to 3-meter-thick beds containing birdseye structure, stromatolitic structure, 1 to 3-meter-wide, 1-meter-deep channels, lithoclasts, and mud cracks (Fig. 37–3)—all of which indicate deposition in an intertidal environment. In the Sangre de Cristo Mountains, the stromatolite is overlain by a 2- to 10-meter-thick breccia.

A disconformity representing all of Salem age, Meramecian time, separates these tidalites from the overlying echinoderm wackestones to lime mudstone and dolomite that have microfauna of St. Louis equivalent, Meramecian age.

The St. Louis equivalent, Meramecian age limestone beds

Figure 37-4. Mississippian outcrop at Jacks Creek, showing dolomites and dedolomites that have a flat surface in contact with the overlying collapse breccia. The breccia is composed of limestone fragments ranging in size from silt to angular blocks. Some of the blocks resemble the overlying limestones in having similar ooid facies and fauna. The contact between the breccia and overlying limestone is irregular.

Figure 37-5. Mississippian outcrop at Tererro, showing the late Osagean dedolomites; collapse breccia composed of late Osagean and Meramecian rocks, and overlying Meramecian carbonate strata. Mississippian-Pennsylvanian unconformity is shown.

are separated by a disconformity and hiatus from the lower Chesterian, which consists of wackestone to arenaceous oolitic to ooid-echinoderm packstone grading upward into lime mudstone and dolomite.

Late Mississippian and early Pennsylvanian uplift and erosion resulted in extensive erosion and removal of the Mississippian Arroyo Penasco Group carbonate rocks. A solution limestone collapse breccia, 2 to 10 meters thick, rests on a smooth surface of stromatolite dedolomite in the Sangre de Cristo Mountains (Figs. 37–4 and 37–5). The breccia resulted from movement of meteoric groundwaters in late Mississippian or early Pennsylvanian time. These dissolved a 2- to 10-meter-thick gypsum bed and caused subsequent collapse of interbedded late Osagean limestones and dolomites and adjacent overlying Meramecian carbonate rocks. Solution activity was extensive, and sinkholes developed. After becoming saturated with calcium sulfate, these waters flowed through the underlying anhydrite or gypsiferous dolomite and calcitized or dissolved the gypsum or anhydrite and partially calcitized the dolomite.

Evidence for Tidal Origin

The sedimentary and microfabric traits of the intertidal-supratidal environment of the late Osagean Espiritu Santo Formation have been obscured by the dolomitization and dedolomitization process. Only the gross sedimentary structures persist in the dedolomites: stromatolites (Fig. 37–6), abundant birdseye structure, calcite pseudomorphs after gypsum, lithoclasts, and 1- to 3-meter-wide, 1-meter-deep channels. The cherts preserve molds of the algal filaments, worm burrows, and desiccation cracks within the stromatolites.

Discussion

The abundant evidence of the former existence of gypsum associated with the carbonate rock of the Espiritu Santo Formation, together with the 2-meter beds of stromatolites, all suggest deposition in a hot arid climate similar to that described by Illing et al. (1965) for the Persian Gulf.

Earlier interpretation of the geologic history by Baltz and Read (1960) and Sutherland (1963) involved a sequence of major unconformities to explain the breccias and supposed pre-Mississippian age, Devonian or older, for the Espiritu Santo Formation. Study of the microfauna and sedimentary structures, outlined above, lead to a somewhat different interpreta-

Figure 37-6. Stromatolite and lamination structures preserved in dedolomite in the basal late Osagean Mississippian Espiritu Santo Formation of the Sandia Mountains. A 15-cm-long scale is shown at top of photograph.

tion. The microfauna of the bedded carbonates and the clasts and blocks in the breccia are of late Osagean, and in the upper part Meramecian, age. Because the sedimentary structures indicate deposition near sea level in an arid climate, the channels and collapse breccia are a result of the depositional environment rather than a long hiatus and extensive subaerial erosion.

References

ARMSTRONG, A. K. 1955. Preliminary observations on the Mississippian System of northern New Mexico. *New Mexico Bur. Mines Mineral Resources Circular 39.*

——— 1958. Meramecian (Mississippian) endothyrid fauna from the Arroyo Penasco Formation, northern and central New Mexico. *J. Paleontol. 32*, 970–976, pl. 127.

——— 1967. Biostratigraphy and carbonate facies of the Mississippian Arroyo Penasco Formation, north-central New Mexico. *New Mexico Bur. Mines Mineral Resources Memoir 20*, 80 p.

ARMSTRONG, A. K., and MAMET, B. L. 1974. Biostratigraphy of the Arroyo Penasco Group, Lower Carboniferous (Mississippian), north-central New Mexico. *New Mexico Geol. Soc., 25th Ann. Field Conf. Guidebook*, pp. 145–157.

BALTZ, E. H., and READ, C. B. 1960. Rocks of Mississippian and probable Devonian age in Sangre de Cristo Mountains, New Mexico. *Am. Assoc. Petrol. Geol. Bull. 44*, 1749–1774.

ILLING, L. V., WELLS, A. J., and TAYLOR, J. C. M. 1965. Penecontemporary dolomite in the Persian Gulf. *In* Pray, L. C., and Murray, R. C., eds., *Dolomitization and Limestone Diagenesis: A Symposium. Soc. Econ. Paleontol. Mineral. Spec. Publ. 13*, 89–111.

SUTHERLAND, P. K. 1963. Paleozoic rocks. *In* Miller, J. P., Montgomery, A., and Sutherland, P. K. Geology of part of the southern Sangre de Cristo Mountains, New Mexico. *New Mexico Bur. Mines Mineral Resources Memoir 11*, 22–46.

Section VI

Recognition of Ancient Carbonate Examples

Sedimentary Features and Facies Patterns

Each of the eight papers in Section VI illustrates how sedimentary features and/or facies patterns are used to recognize ancient tidal deposits. The variety of sedimentary features includes: reversals in cross-bed dip direction, mud cracks, fenestral fabrics (birdseyes), stromatolites, nodular evaporites, fossils and ichnofossils, and distinctive types of stratification. In several examples, the interpretations based on sedimentary features are strengthened substantially by the indications of environment deduced from the vertical or lateral facies pattern, an application of Walther's Law. Some examples are underlain by shallow marine deposits and overlain by terrestrial sediments; in others, the indications of prolonged subaerial exposures are more subtle: vadose cementation of shallow marine sands, an abrupt decrease in marine fossils, and upward-decreasing grain size with parallel changes in mineralogy (calcite to dolomite and/or evaporites). A combination of this kind of larger-scale facies sequence with the sort of internal sequence described in Section IV makes the strongest argument for deposition within the zone of fluctuating sea level.

38

Tidal Sediments and Their Evolution in the Bathonian Carbonates of Burgundy, France

B. H. Purser

Occurrence

The Middle Jurassic sediments of Burgundy have been deposited in three major, regressive, carbonate sequences totaling ca. 200 meters (Fig. 38–1). The middle sequence, essentially of Bathonian age, grades from *Ostrea* marlstones at the base, through calcisiltstones and bioclastic and oolitic calcarenites to terminate in the 60-meter-thick Comblanchien Formation. This formation consists of interbedded carbonate mudstones with oncoids, pellet calcarenites and calcirudites, and dolomites. Birdseyes, stromatolites, and desiccation features are common in certain mudstone layers, while small-scale cross-bedding, birdseyes, and syn-sedimentary dripstone cement are common in pelletoidal sands and gravels.

Facies

The Comblanchien Formation in the vicinity of Dijon is a light beige, completely cemented limestone renowned both for its building and wine-producing qualities. Its sedimentary tex-

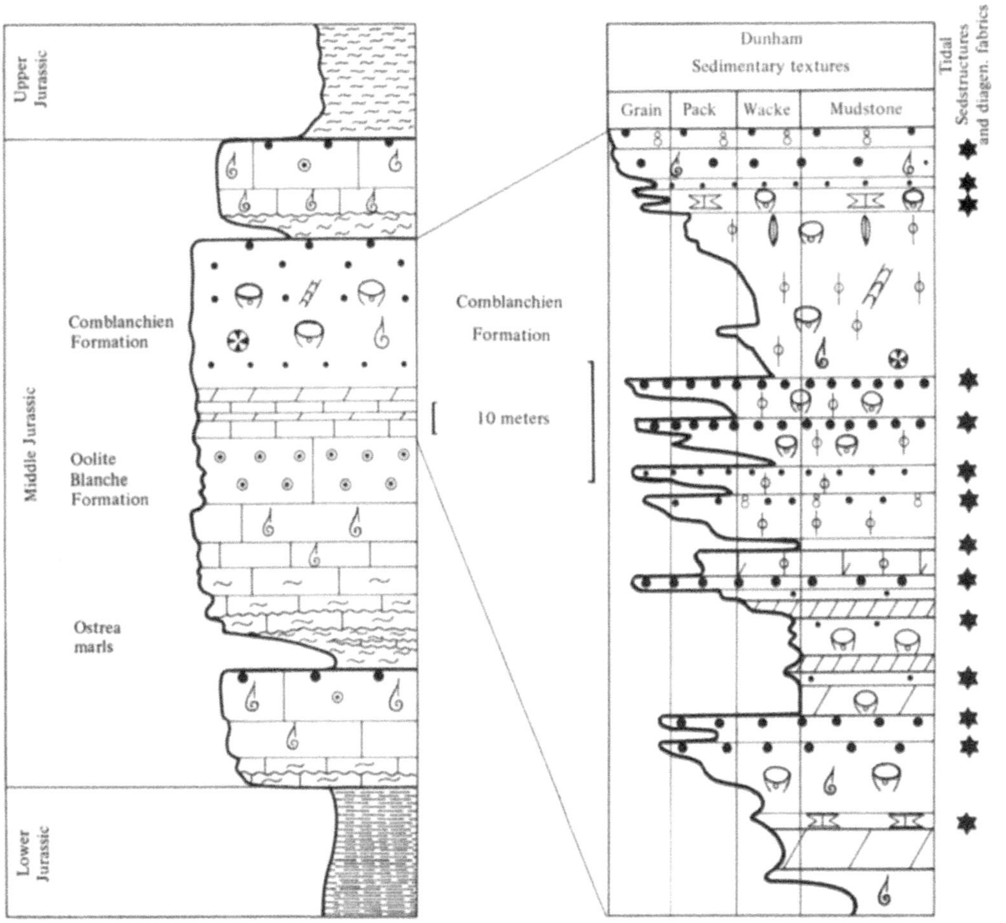

Figure 38-1. Middle Jurassic sequence in Burgundy, France, showing the distribution of tidal sediments.

tures and grain types are extremely variable, although micritic textures, pelletoidal grains, and various forms of algae are particularly characteristic. The vertical sequence of textures and sedimentary structures is, nevertheless, systematic, and comprise a series of minor sedimentary cycles within this major Bathonian cycle. Two main types of cycle occur.

MUDDY CYCLES. Muddy cycles (Fig. 38–2) average 3 to 4 meters in thickness and consist typically of 1- to 3-meters-thick dolomite or lime mudstone at the base. This dolomite is massively bedded and includes numerous, partially dolomitized oncoids. It grades upward into partially dolomitized, bioturbated lime wackestone, also with numerous algal oncoids, occasional whispy, "cryptalgal" laminae, and rare domed stro-

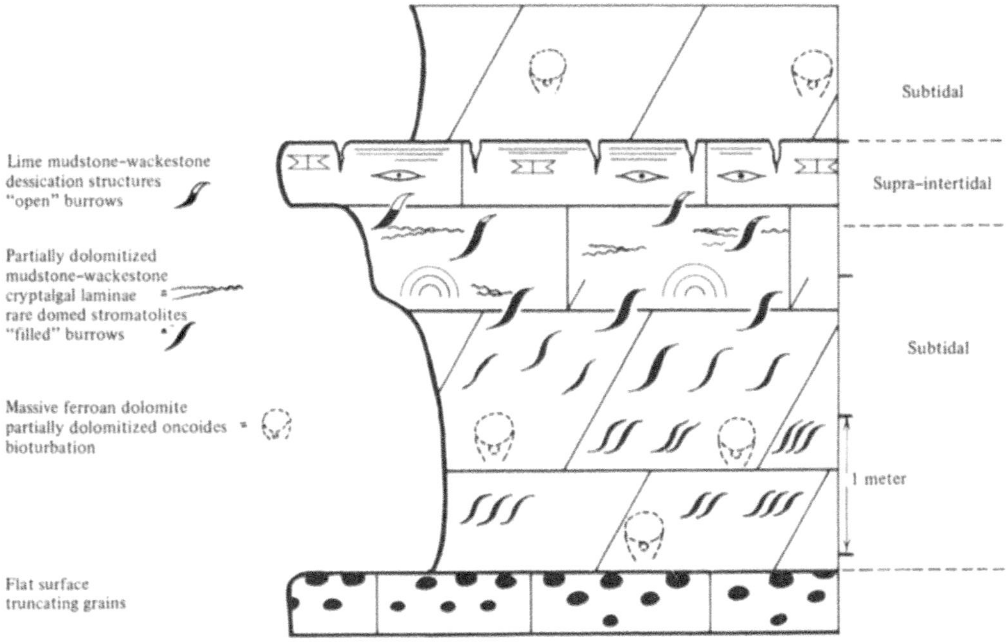

Figure 38-2. The essential aspects of a muddy (tidal-flat) cycle.

matolites 2 to 20 cm high. The top 50 cm of most muddy cycles also have mud-supported textures but are essentially nondolomitic. Vaguely defined burrows, filled completely with dolomitized sediment (Fig. 38–3*A*), grade rapidly upward into burrows that are partially filled with geopetal sediment or dolomite and sparry calcite (Fig. 38–3*B*); the transition from filled to open burrows almost invariably takes place 30 to 50

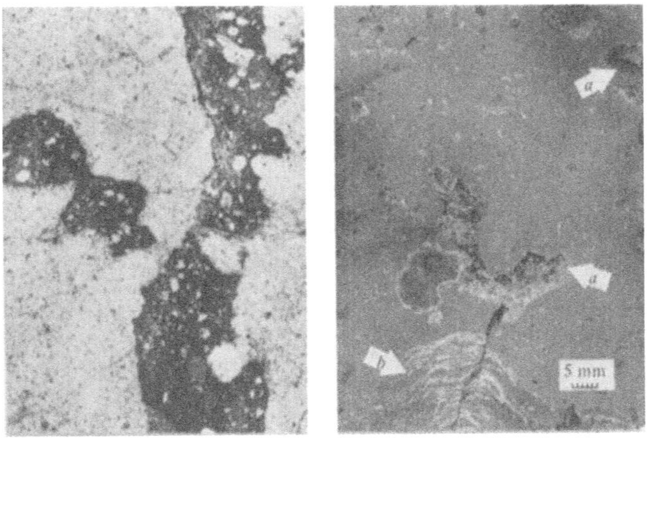

Figure 38.3. *A.* Filled burrows containing dolomitized sediment; polished surface; × 1. *B.* Open burrows partially filled with geopetal sediment (*a*); domed stromatolite (*b*); polished surface.

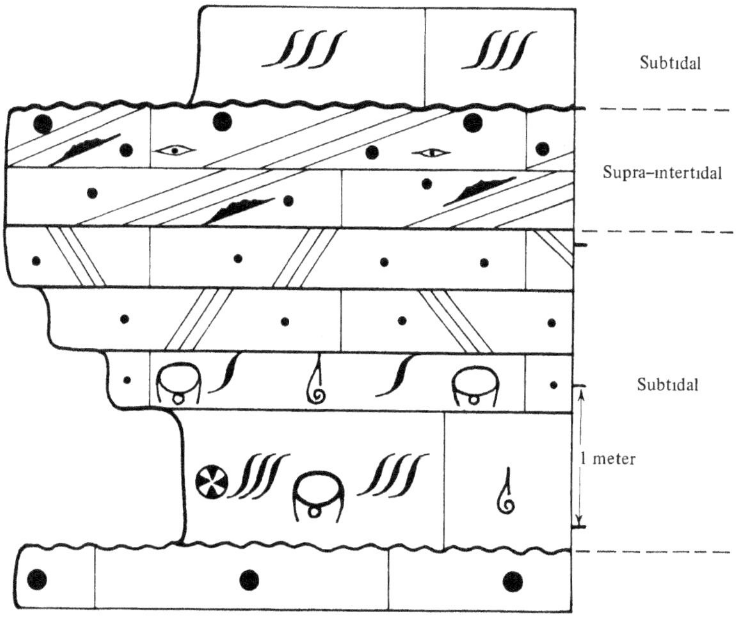

Figure 38-4. The essential aspects of a grainy (beach) cycle.

cm below the top of each cycle. The open burrows in the top 30 cm of the cycle are generally associated with birdseyes (both planar and ovoid), desiccation cracks and related sedimentary breccias, possible root casts, and traces of horizontal laminae. The top of each cycle is sharply defined.

GRAINY CYCLES. Grainey cycles (Fig. 38–4) average 3 to 4 meters in thickness, although one in the upper parts of the formation is 12 meters thick (Fig. 38–3). Each cycle consists of a muddy basal unit, generally with homogenized textures, often rich in large (2 to 5 cm) oncoids and other algal forms, scattered solitary corals, and mollusks. It is rarely dolomitized. These lime mudstones grade rapidly upward into grainier sediments, the uppermost 1 to 2 meters consisting of pelletoidal calcarenites and calcirudites which coarsen upward (see Fig. 38–4). In contrast to the muddy cycles, the sequence terminates in

Figure 38-5 (opposite). *A.* Macrostructure of a microcavern (*a*) situated immediately below a bedding plane (*b*) that truncates grains (*c*); polished surface (scale in mm). *B.* Microstructure of a microcavern showing stalactitic (*a*) and stalagmitic (*b*) layers and geopetal (*c*) and sparry calcite (*d*) infillings. Note that the stalactitic layer has been eroded locally (*e*), probably by crustaceans inhabiting the cavern prior to the sediment infilling; thin slide; × 5.

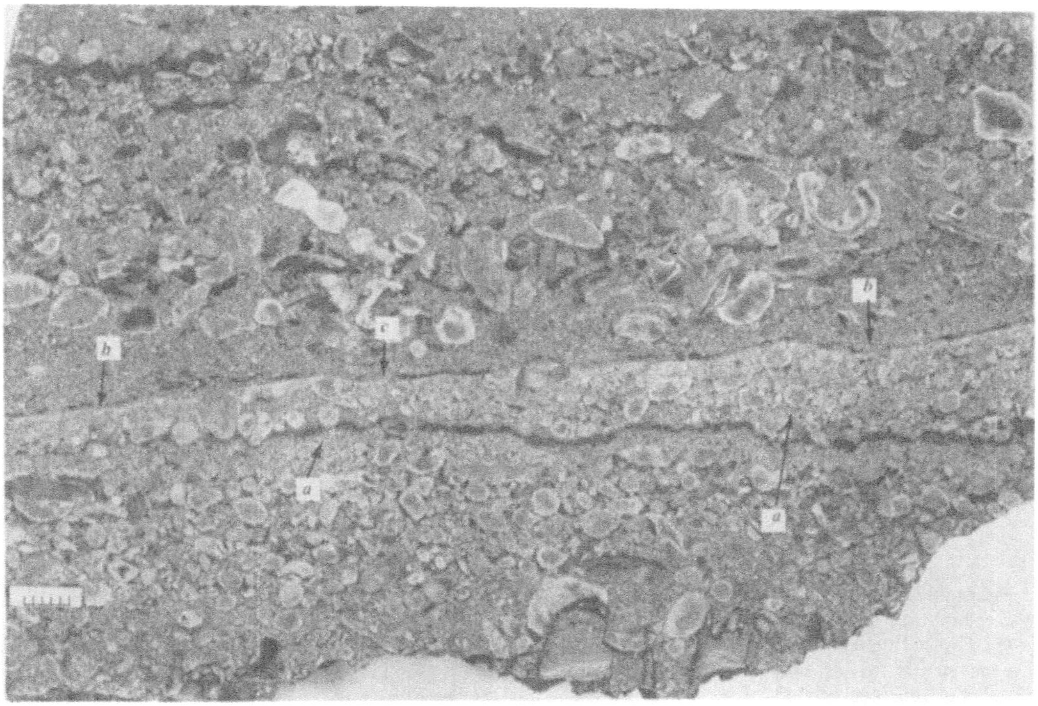

A

B

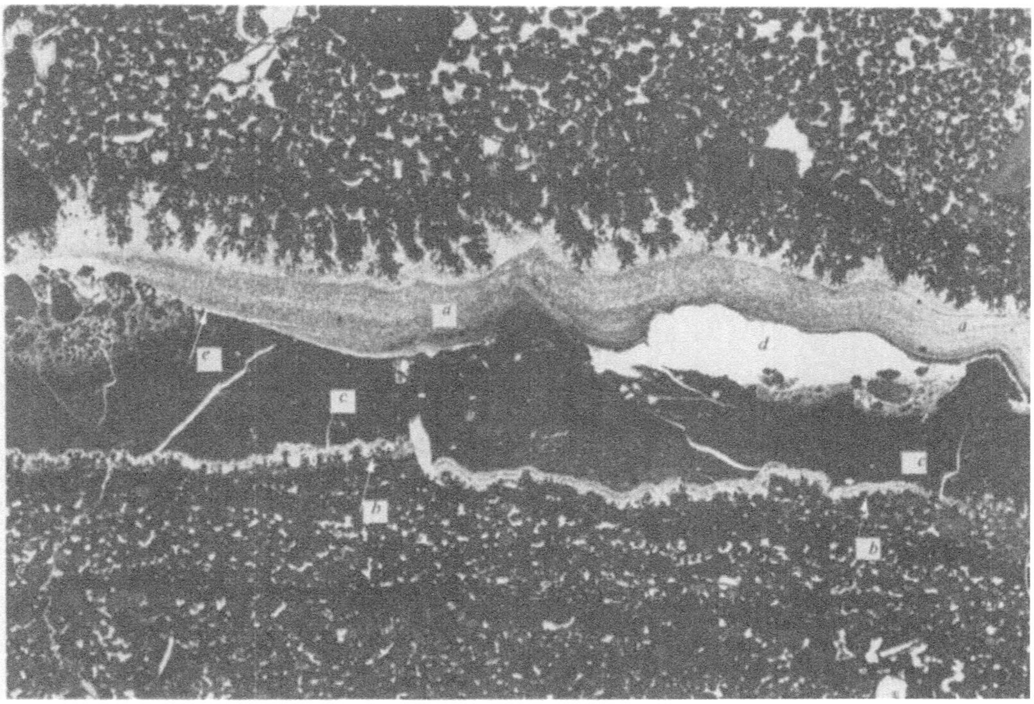

micrite-free, pelletoidal calcarenite or rounded oncoidal calcirudite.

The top meter of each granular cycle exhibits various types of inclined bedding, open-space structures, including numerous large birdseyes, and microcaverns. The latter are lenticular structures 2 to 20 mm high and 5 to 20 cm long, and are generally oriented parallel to the inclined stratification (Fig. 38–5*A*). They are lined with an irregular layer of early fibrous calcite, often with a dripstone morphology, and are partially filled with geopetal sediment that overlies the fibrous cement (Fig. 38–5*B*). Late sparry calcite fills the remaining space. These features have much in common with the structure termed "stromatactis." A similar fibrous calcite has lithified the top meter of many grainy cycles and, where cementing coarse sand

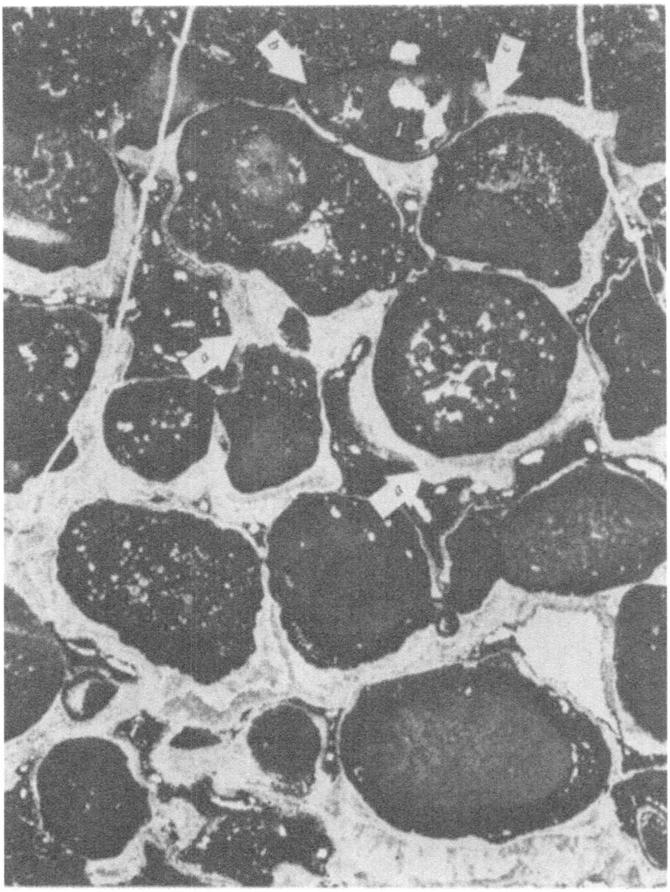

Figure 38-6. Beach-rock cement: carbonate gravel coated with microstalactitic cement thickest under grains (*a*). An erosion surface truncates grains (*b*) and the beach-rock cement (*c*); thin slide; × 7.5.

or gravel, is frequently thickest on the undersides of individual grains. That this "microstalactitic" calcite is syn-sedimentary is clearly demonstrated by its erosion along the top of the cycle (Fig. 38–6).

Grainy cycles occur throughout the Comblanchien Formation, but are more typical of the upper half; muddy cycles with dolomite occur only in the lower half of the formation (Fig. 38–3).

Evidence for Tidal Origin

In muddy cycles (Fig. 38–2) birdseyes, mud cracks, and associated breccias, by analogy with modern environments, strongly suggest desiccation and therefore emergence. The presence of a water table (sea level) may also be reflected in the vertical transition from open to filled burrows. Both types occur in muddy sediment and it is possible that drying above the water table facilitated the preservation of burrows in an open state—a feature characteristic of many modern tidal flats. Sediments permanently below the water table, because they are more fluid, would be *less* favorable for the preservation of open burrows, although there are many obvious exceptions in modern environments. However, the fact that the transition from open to filled burrows (the former associated with desiccation features) is repeated from cycle to cycle gives these structures added significance.

Because of their micrite textures and associated desiccation features, muddy cycles are interpreted as being the products of tidal-flat accretion over adjacent shallow marine muds, followed by episodic transgression. The associated dolomites are probably an integral part of this tidal-flat system.

That grainy cycles (Fig. 38–4) terminated above sea level is clearly indicated by the nature of their syn-sedimentary, fibrous calcite cements; the microdripstone fabric of this cement could not have formed below the water table (sea level). Similar microstalactitic aragonite cements are common in coarse-grained beach rocks along the southern shores of the Persian Gulf, and are particularly well developed in small burrows made by crabs beneath layers of beach rock (Purser and Loreau, 1973).

The nature of the syn-sedimentary cement, inclined stratification, and coarse textures at the tops of these Bathonian grainy cycles suggest that they are the product of prograding beaches or coastal spits.

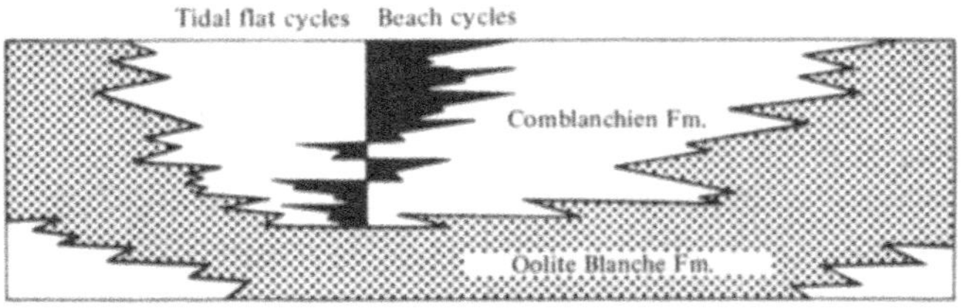

Figure 38-7. Profile showing schematic relationships between Bathonian tidal sediments and the oolitic barrier complex.

Discussion

Tidal-flat sequences (muddy cycles) occur only in the lower half of the Comblanchien Formation, while beach sequences (grainy cycles) dominate the upper parts (Fig. 38–1). This progressive change from a low energy, muddy shoreline to one of increased water agitation favoring beach accretion suggests changes in the adjacent offshore regime. These changes are readily understood when one examines the regional field relationships between the essentially pelleted mudstone Comblanchien facies and its lateral time equivalent, the "Oolite blanche" illustrated schematically in Figure 38–7. In Burgundy (particularly in the vicinity of Dijon) the lower parts of the Comblanchien, with their tidal-flat sequences, were situated relatively close to the oolitic barrier complex during their deposition. Conversely, because this carbonate bank had widened progressively in upper Comblanchien time, the shoreline facies were more remote from the protective oolite barrier. This led to the progressive disappearance of tidal flats and the development of beaches or coastal spits.

A modern analog demonstrating these processes occurs along the south shore of the Persian Gulf. Wide tidal flats (sabkhas) with associated dolomite are best developed in eastern Abu Dhabi where the shoreline is often muddy, mainly because of the protection offered by the adjacent coastal barrier complex (Evans, 1964). However, because the protecting barrier is oriented obliquely to the regional shoreline, the decreasing coastal protection results in a lateral change from low energy, evaporitic sabkhas to higher energy beaches.

The Bathonian carbonates of Burgundy exhibit sedimentary and early diagenetic features that enable one to establish the fundamental paleogeographic limits between land and sea. However, because the character of these tidal deposits was closely related to that of the adjacent offshore environments, they have a much wider potential application. If interpreted correctly they may be valuable guides to the understanding of Bathonian subtidal and other offshore environments, and, perhaps, to the directions of regional winds and currents. These interpretations will be most effective when the number and types of tidal sequence within a given stratigraphic unit are mapped on a regional scale.

References

DUNHAM, R. J. 1962. Classification of carbonate rocks according to depositional texture. *In* Ham, William E., ed., Classification of carbonate rocks. *Am. Assoc. Petrol. Geol. Memoir No. 1,* 108–121.

EVANS, G. et al. 1964. Origin of the coastal flats, the sabkha, of the Trucial Coast, Persian Gulf. *Nature, 202*, 759–761.

PURSER, B. H., and LOBREAU, J.-P. 1972. Structures sédimentaires et diagénétiques précoces dans les calcaires bathoniens de la Bourgogne. *Bull. B.R.G.M.* Sect. IV, No. 2.

PURSER, B. H., and LOREAU, J.-P. 1973. Aragonitic, supratidal encrustations on the Trucial Coast, Persian Gulf. *In* Purser, B. H., ed., *The Persian Gulf; Holocene Carbonate Sedimentation and Diagenesis in a Shallow Epicontinental Sea.* Springer-Verlag, New York.

SHINN, E. A. et al. 1969. Anatomy of a modern carbonate tidal flat, Andros Island, Bahamas. *J. Sed. Petrol. 39,* 1202–1228.

TINTANT, H. 1963. Observations stratigraphiques sur le Jurassique moyen de la Côte d'Or. *Bull. Sci. Bourgogne 21,* 93–117.

39

Evidences of Tidal Environment Deposition in the Calcare Massiccio Formation (Central Apennines—Lower Lias)

R. Colacicchi, L. Passeri, and G. Pialli

Occurrence

The limy Calcare Massiccio Formation outcrops in wide areas in southern Liguria, Toscana, Umbria, Marche, northern Lazio, and Abruzzi. This formation, up to 400 meters thick, overlies a thin-bedded unit (*Avicula contorta* beds) that consists of micritic limestones and marls with isolated, very thin levels of gypsum. This member overlies a thick sequence of evaporitic anhydrites and dolomites, which represent the first marine sedimentation of the Mesozoic Apennine sequence together with the upper levels of the underlying Verrucano Formation. The top of the Calcare Massiccio is overlain by thin-bedded pelagic limestone. The age of the formation is defined at the base by the underlying *A. contorta* beds (Rhaetian Auct.), while on the top it varies, being Upper Hettanian in the Tosco-Ligure zone and Upper Pliesbachian in the Marche region.

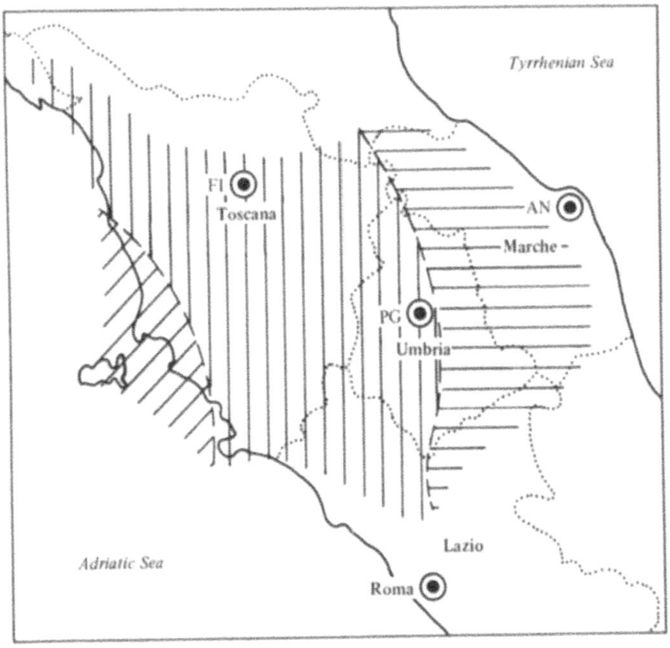

Figure 39-1. Facies distribution of the Calcare Massiccio. Vertical lines: lagoon; horizontal lines: tidal flat; inclined: offshore bar FI, Firenze; AN, Ancona; PG, Perugia.

Facies

Two fundamental lithologic characters typify the Calcare Massiccio: 1, a repeated, highly variable, tide-controlled facies occurs in the Umbro-Marchigiano region (Pialli, 1971) and crops out along the Tyrrhenian shoreline (Boccaletti and Manetti, 1972) (Fig. 39–1); 2, a more constant, homogeneous, muddy facies of a low energy environment (Passeri and Pialli, 1972) crops out in southern Liguria and in Toscana and western Umbria.

Calcare Massiccio of Eastern Umbria and Marche

The Calcare Massiccio of Umbria Marche is characterized by the following sedimentologic features:

Lagoon facies. Lagoon facies (Figs. 39–2 and 39–3) are white micrite, poor in sedimentary structures, and containing peloids, oncoids, whole fossils, and debris. Some peloids are pellet aggregates in micritic mud, an indication of binding algae and of weak currents along the sea bottom. The presence of ooids in a micrite matrix and discontinuous levels of oncolites suggests current transport from zones of higher energy; bioturbation is infrequent. These characteristics are considered

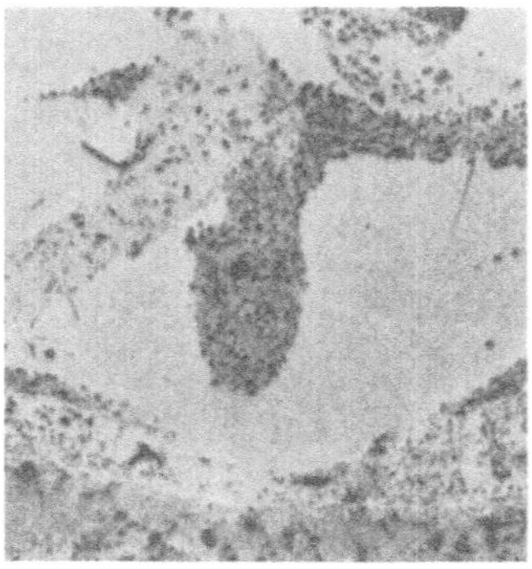

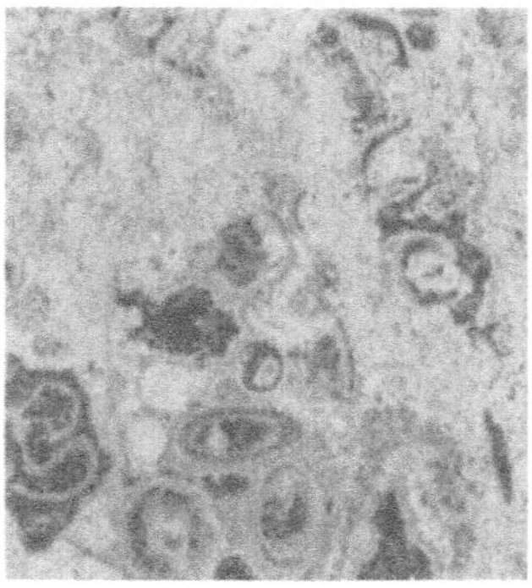

Figure 39-2 (left). Thin-section in plane-polarized light, × 2.5, of lagoon facies (Toscana) showing bioturbated fabric; negative print.

Figure 39-3 (right). Thin-section in plane-polarized light, × 5.5, of lagoon facies (Marche) showing bioclasts in a mud matrix and the effects of late diagenesis; negative print.

typical of a low energy environment with occasional tide-induced currents.

Subtidal channel facies. The main feature of the subtidal channel facies (Figs. 39–4 and 39–5) is the predominance of sand-sized grains over micrite matrix. The grains are oncolites, ooids, intraclasts, peloids, and very fine-grained fossil debris. The size distribution of grains is either bimodal or unimodal, but strongly asymmetrical. A sample from Frasassi consists of 6.5 percent oncolites 8 millimeters in diameter, and 90 percent ooids <2 millimeters in diameter, in a sparry calcite cement. The main sedimentary structures are low-angle, cross-lamination and ripple marks; the cross-lamination is asymmetrical with one set, probably the foreset, much more prominent than

Figure 39-4 (left). Thin-section in plane-polarized light, × 2.5, of subtidal channel facies showing biomodal size distribution; negative print.

Figure 39-5 (right). Thin-section in plane-polarized light, × 3, of subtidal channel facies showing bioclastic sand in a micrite matrix; negative print.

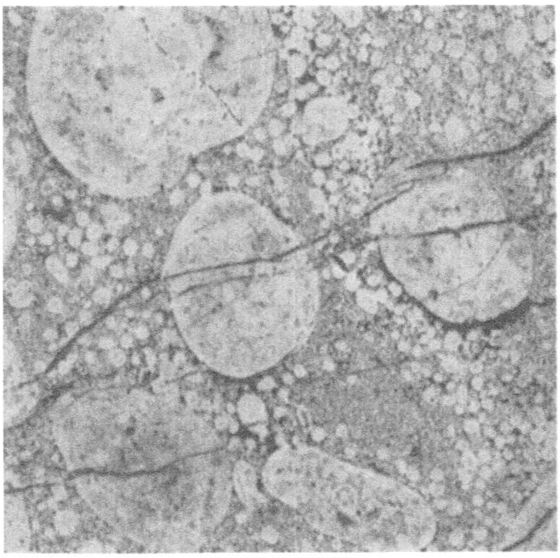

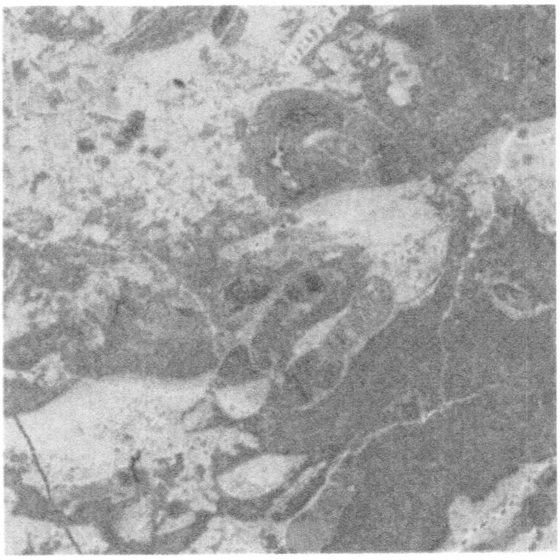

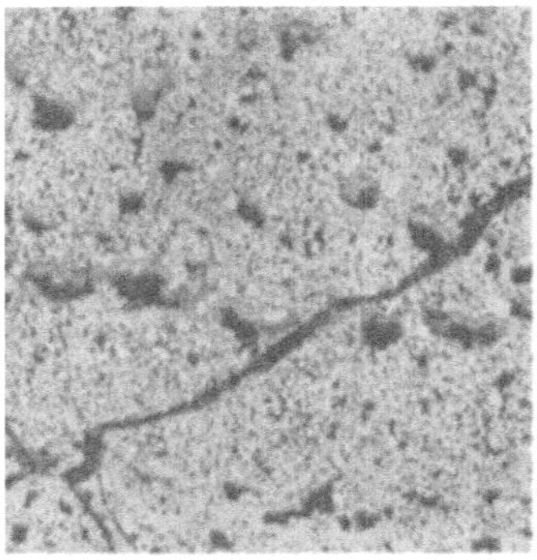

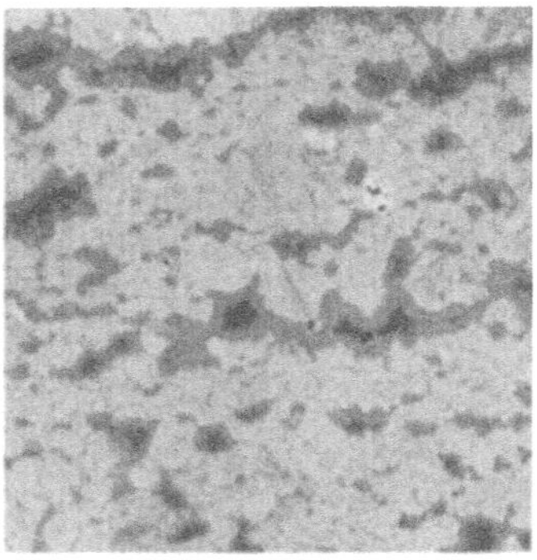

Figure 39-6 (left). Thin-section in plane-polarized light, × 3.5, of the beach facies showing keystone vugs with drusy calcite; negative print.

Figure 39-7 (right). Thin-section in plane-polarized light, × 3, of beach facies showing horizontally aligned vugs; negative print.

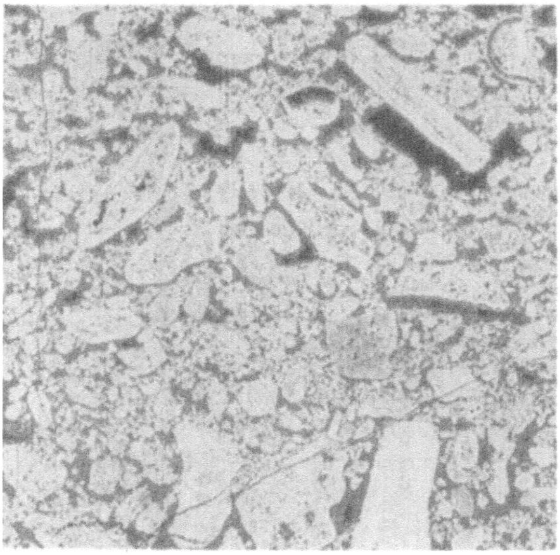

Figure 39-8. Thin-section in plane-polarized light, × 1.5, of channel slope deposits showing flat pebbles and mud aggregates with keystone vugs; negative print.

the other. Large grains are frequently imbricated, and the wave lengths of ripple marks range from a few millimeters to 40 cm.

The sedimentary environment is well characterized by the structures and the bimodal grain size distribution; it is probably the axial zone of tidal channels swept by fairly strong currents and always submerged because of the absence of desiccation structures, algal laminations, and so on.

Beach facies. Beach facies (Figs. 39–6 and 39–7) are fine-grained, oolitic and bioclastic, poorly sorted, lime sands with intraclasts, flat pebbles, peloids, and either sparry calcite cement or reddish micrite matrix. The main sedimentary structures are cavities similar in size to the largest grains or even larger, and lined by a ring of fibrous calcite that is succeeded by a filling of blocky calcite (Fig. 39–8), similar to "keystone vugs" of

Recent and Pleistocene beaches (Dunham, 1970). There are also laminations of mechanical origin and flat rock fragments, maximum 50 mm, similar in texture to the surrounding rock; these intraclasts of lime sand probably came from the erosion of beach rock.

INTERTIDAL-FLAT FACIES. Pellets, mud aggregates, and a few bioclasts (fragments of gastropods and forams) in a micrite matrix are the main constituents of this facies. Bioturbation is common as narrow vertical tubes infilled with pellets. In the higher parts, laminations of both mechanical and algal origin (Fig. 39–9) are interrupted by prism and sheet cracks. Some channel features with cut-and-fill and imbricated grains occur together with desiccation structures and flat pebbles of algal-laminated sediment derived from the supratidal zone.

LEVEE FACIES. Levee facies (Fig. 39–10) is characterized by typical structures: (1) very thin laminations of micrite alternating with laminae of graded pellets that pinch out laterally in a cross-laminated structure. These laminae were probably made by sheet flooding over an algal mat. (2) Desiccation structures—vugs aligned parallel with the laminations and filled with drusy and blocky calcite. Some vugs have reddish internal micrite as geopetal fills. Prism and sheet cracks with two generations of calcite fillings are frequent. (3) Laminated dolomitic crusts, centimeters thick, are interrupted by desiccation cracks.

Figure 39-9 (left). Thin-section in plane-polarized light, × 1.5, of the inter-supratidal facies showing algal mat structure; negative print.

Figure 39-10 (right). Thin-section in plane-polarized light, × 1.5, of levee facies showing laminae of pellets and lime mud deposited by sheet flooding; negative print.

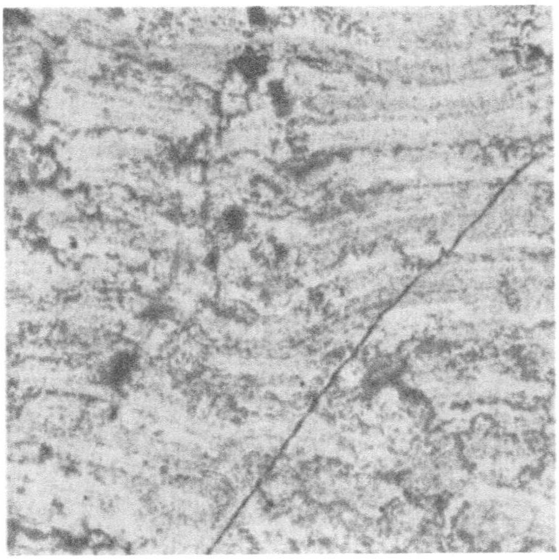

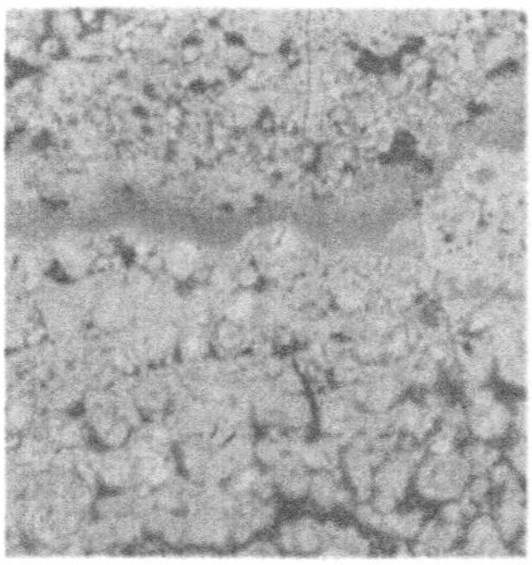

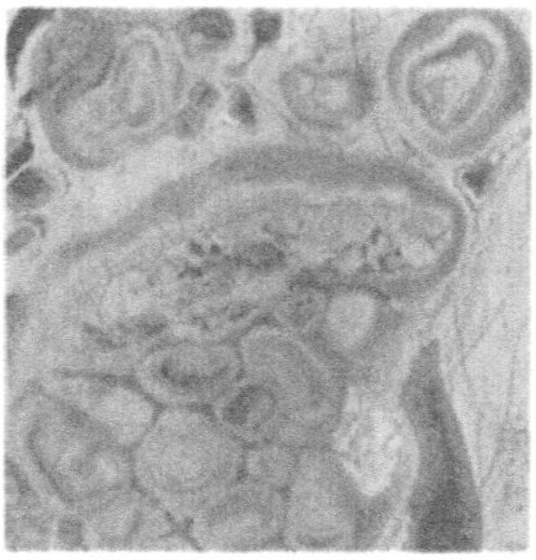

Figure 39-11 (left). Thin-section in plane-polarized light, × 1.5, of caliche showing vadose crusts and reverse grading; negative print.

Figure 39-12 (right). Thin-section in plane polarized light, × 2, of caliche showing polygonal vadose pisolites and asymmetrical growth; negative print.

SUPRATIDAL TO CONTINENTAL FACIES. Some thin levels of this facies (Figs. 39–11 and 39–12) are characterized by superimposed vadose textures and structures that at times completely obliterate the depositional textures. The main characteristics of these levels are: (1) vadose pisolites (Dunham, 1969) with nuclei of any grain; the covering laminae of microspar are thicker at the bottoms of the grains and the innermost laminae are sometimes dolomitized; laminar sheets frequently cover two or more grains, making a complete pisoid. (2) Laminar crusts made of fibrous calcite with a horizontal or undulating trend; the crusts attain thicknesses of some centimeters and they occur individually or repeatedly as a kind of zebra rock. (3) Karst cavities varying in dimension and shape with corroded walls and linings of polylaminated fibrous calcite that are succeeded by fillings of fine, brownish-red, well-cemented internal sediments similar to terra rossa of karst terrains.

Facies Sequence and Pattern

The lithofacies described above are interpreted as tide controlled because their structures, textures, and grain types are similar to those of Recent tidal deposits. In general, these lithofacies occur in cyclical sequences; the modal cycle (Fig. 39–13) has the following characteristics from bottom to top:

1. An erosional surface, more or less irregular
2. Transgressive intertidal bed: very thin, pink or buff, beach facies (high intertidal) and laminated structures
3. Subtidal bed: massive, white, thick, with numerous fossils and oncoids; the prevailing environment is lagoonal or a small tidal channel
4. Regressive intertidal bed: a little thicker than (2), but similar to it, with greater development of textural and structural features

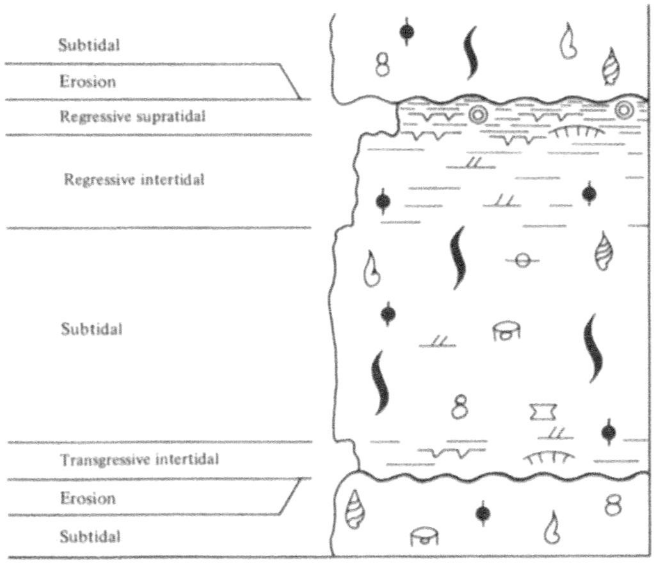

Figure 39-13. The modal cycle of the Calcare Massiccio in Umbria-Marche, standard symbols in this volume.

5. Supratidal bed: laminated, with pisoids, and crusts of the vadose environment. The last bed is almost always truncated by an erosion surface that is the base of the overlying cycle.

The cycles, as seen in outcrop, are different from the modal cycle because the transgressive intertidal bed (2) is normally missing through erosion that very often reaches down to the

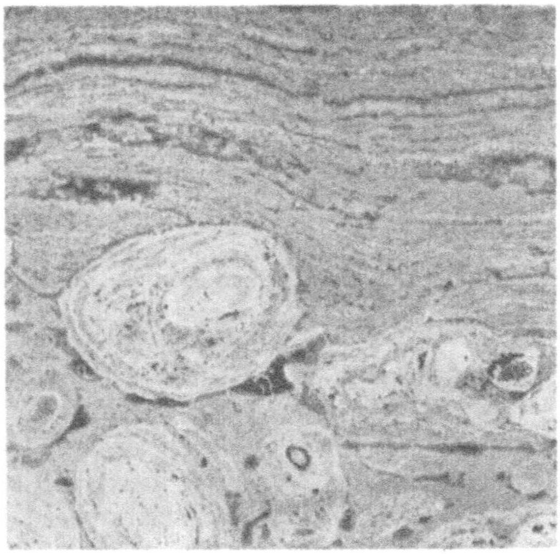

Figure 39-14. Thin-section in plane-polarized light, × 2, showing the transition from subtidal oncolites to supratidal laminated algal mats; negative print.

subtidal bed (3), (see Fig. 39–14). Each cycle, generally more than a meter thick and frequently 3 or 4 meters, cannot be the result of individual diurnal or monthly tide cycles, as such a rate of sedimentation is unrealistic. It seems more likely that the cycles are the result of regression produced by progradation.

Calcare Massiccio of Toscana and Western Umbria

The lithofacies of the western Calcare Massiccio is not directly controlled by tidal fluctuations, rather its environment is just below the tidal zone and only a few levels show tidal zone characteristics. The main lithofacies is gray micrite matrix with peloids, oncoids, micritized grains, rare bioclasts, and no ooids; stratification and desiccation features are absent and the only structure is bioturbation. The environment indicated by these characteristics is a quiet lagoon.

Farther east, closer to the Umbro-Marchigiano facies, another lithofacies is interbedded with the lithotype described above; it resembles the one from Umbria-Marche, characterized by keystone vugs, desiccation structures, stromatolites, vadose structures, and a lesser amount of lime mud. The occurrence of the second lithofacies points to a position close to the boundary between the lagoon and the lower part of the tidal zones where occasional progradations of the tidal flat could bring the lagoon up to tide levels.

The paleogeographic relationships between the two main types of Calcare Massiccio are not easy to detect because outcrops are rare and scattered, but they may be ascribed to a general pattern (Fig. 39–1) in which the Tosco-Umbro Calcare Massiccio represents a very large lagoon, deep enough not to be directly influenced by tidal currents, and bordered on the west by an offshore bar with ooids and other high energy sediments, and grading eastward to a widespread, highly dissected tidal flat. No reefs, platform edge, or continental shoreline have been detected, probably they were far off the modern Italian shorelines.

Comparison with Modern Environments

The lithofacies and depositional environments the Calcare Massiccio may be compared with some Recent carbonate sedimentary environments such as Florida Bay (Ginsburg, 1956) and Andros Island (Purdy, 1963; Shinn et al., 1969). With respect to the Andros tidal flat, the Umbria-Marche Calcare Massiccio contains more ooids, a more winnowed sediment, a lesser concentration of laminated structures, and dolomitic crusts, while beach facies, channel features, and traces of subareal action are more frequent. On the basis of these characters we may conclude that the Calcare Massiccio tidal flat was narrower than the Andros one, and was more subdivided by a diffuse channel system. Probably the Liassic tidal flat developed behind and between a series of small islands separating small channels, in a situation like the lower Florida Keys.

In the Ligurian and Tosco-Umbra facies the percentage of lime mud in the Calcare Massiccio is similar to that of recent lagoonal environments. The percentage of skeletal grains and peloids is nearer that of the Andros tidal flats than of the sediments of Florida Bay, although ooids are missing in the Calcare Massiccio of Tosco-Umbro, as they are in Florida Bay.

We conclude that the environment of the Calcare Massiccio has textures, structures, and grain types like those of Recent tidal deposits on Andros Island and in Florida Bay, but that in detail the pattern and sequence of features are different.

References

BOCCALETTI, M., and MANETTI, P. 1972. Caratteri sedimentologici del Calcare Massiccio della Toscana a Sud dell'Arno. *Boll. Soc. Geol. Ital. 91*, 559–582.

DUNHAM, R. J. 1969. Vadose pisolite in the Capitan reef (Permian), New Mexico and Texas. *In* Friedman, G. M., ed., Depositional environment in carbonate rocks. *Soc. Econ. Paleontol. Mineral. Spec. Publ. 14*, 182–191.

——— 1970. Keystone vugs in carbonate beach deposits. *Am. Assoc. Petrol. Geol. Bull. 54*, 845.

GINSBURG, R. N. 1956. Environment relationship of grain size and constituent particles in some South Florida carbonate sediments. *Am. Assoc. Petrol. Geol. Bull. 40*, 2384–2427.

PASSERI, L., and PIALLI, G. 1972. Facies lagunari nel Calcare Massiccio dell'Umbria occidentale. *Boll. Soc. Geol. Ital. 91*, 345–364.

PIALLI, G. 1971. Facies di piana cotidale nel Calcare Massiccio dell'Appennino Umbro-Marchigiano. *Boll. Soc. Geol. Ital. 90*, 481–507.

PURDY, E. G. 1963. Recent calcium carbonate facies of the Great Bahama Bank. 1. Petrography and reaction groups. *J. Geol. 71*, 334–355.

SHINN, E. A., LLOYD, R. M., and GINSBURG, R. N. 1969. Anatomy of a modern carbonate tidal-flat, Andros Island, Bahamas. *J. Sed. Petrol. 39*, 1202–1228.

40

Upper Jurassic Oolite Shoals, Dorset Coast, England

R. C. L. Wilson

Occurrence

Oolitic sediments described here form part of the Osmington Oolite Series of the Corallian Beds, which are Upper Oxfordian in age (units do not conform to standard stratigraphic usage, but are firmly embedded in existing literature). The examples described here are exposed in cliff sections northwest of Weymouth, Dorset. The Osmington Oolite Series consists of a series of lens-shaped oolite beds interleaved with phyllosilicate clays and quartz silts, resting on quartz sands of probable intertidal origin (Wilson, 1968). Above the Oolite horizons occurs a series of *Rhaxella* biomicrites; *Rhaxella* are bean-shaped sponge spicules usually replaced by calcite in the Corallian Beds (Wilson, 1966). The general sequence of the Corallian Beds is shown in Figure 40–1. A field guide is given by Wilson (1969).

Facies

Lateral variation of oolitic units of the Osmington Oolite Series is shown in Figure 40–2. Four main facies assemblages are distinguished:

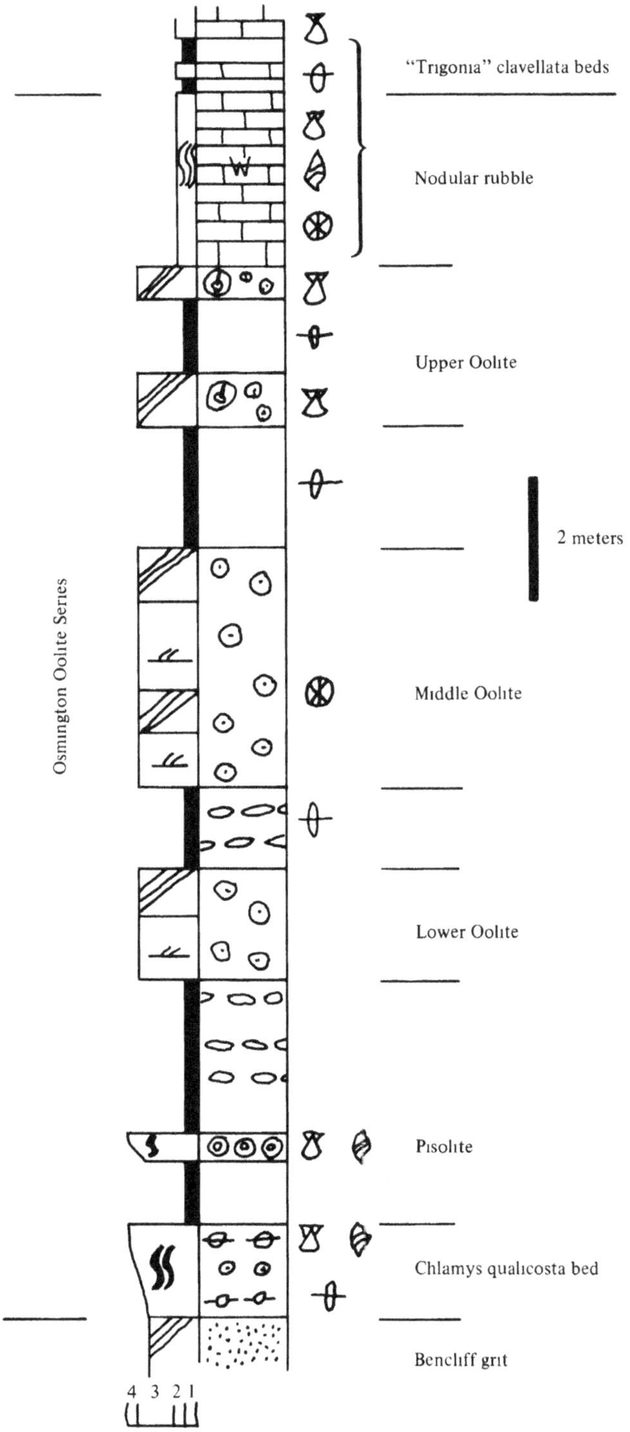

Figure 40-1. Generalized succession of the Osmington Oolite Series. "Brick" symbol is only included for micritic limestones. The left side of the column indicates variation in grain size. *1*, clay; *2*, silt; *3*, sand; *4*, gravel; black bar indicates phyllosilicate clay; *W*, wackestone; remaining carbonates are dominated by packstones.

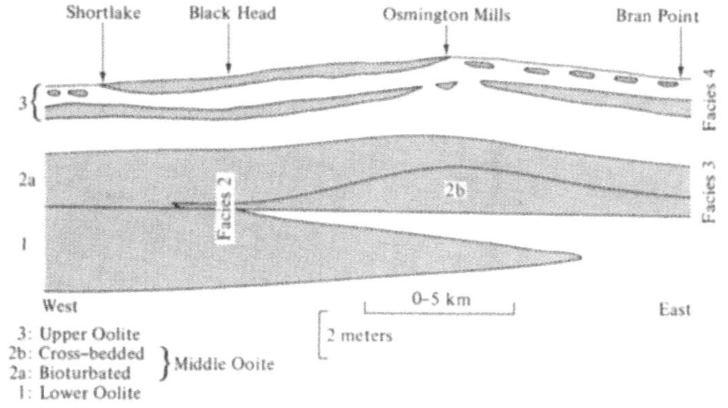

Figure 40-2. Lateral thickness variations in oolite horizons of the Osmington Oolite Series.

1. Coarsening-upward units shown by two beds of the Osmington Oolite Series: (1) *Chlamys qualicosta* bed: intramicrite-oomicrite-oosparite-poorly washed biosparite. (2) Pisolite: quartz sands and phyllosilicate clay-intramicrite-oomicrite-oosparite-oncolites (mode R of Logan et al., 1964).

2. Cross-bedded sets of oosparite showing 20° to 25° dips and sharp contacts either with phyllosilicate clays with nodular micrites (burrow fills described by Fursich, 1972), or bioturbated oolites (Fig. 40–3*A*). Some minor flaser bedding and clay drapes over current ripples also occur. Paleocurrent data for this assemblage are shown in Figure 40–4.

3. Association of *Rhaxella* biomicrites; intramicrites and oosparite are summarized in Figure 40–3 and Table 40–1.

4. Sheet deposits (5 to 10 cm) and large accretion sets (30 cm) of oomicrite and biomicrite with subsidiary oosparite and biosparite. Shell debris often shows imbricate structure, and the oomicrites are texturally inverted sediments, being a mixture of extremely well-sorted oolites in a micrite matrix. Some sets showing alternating current directions occur (see Figs. 40–3*D* and 40–4*C*).

Interpretation

Facies 1. The coarsening-upward units record increasing energy of environment through time, from intramicrite with intraclasts resembling Freeman's (1962) quiet water oolites, through oolitic sediments, and, finally, either algal-coated shells or mode R oncolites. These deposits were interpreted as the result of tidal-flat regression (Wilson, 1968), with high energy carbonates building out over finer grained, low energy ones. However, the tidal part of this description can only be included

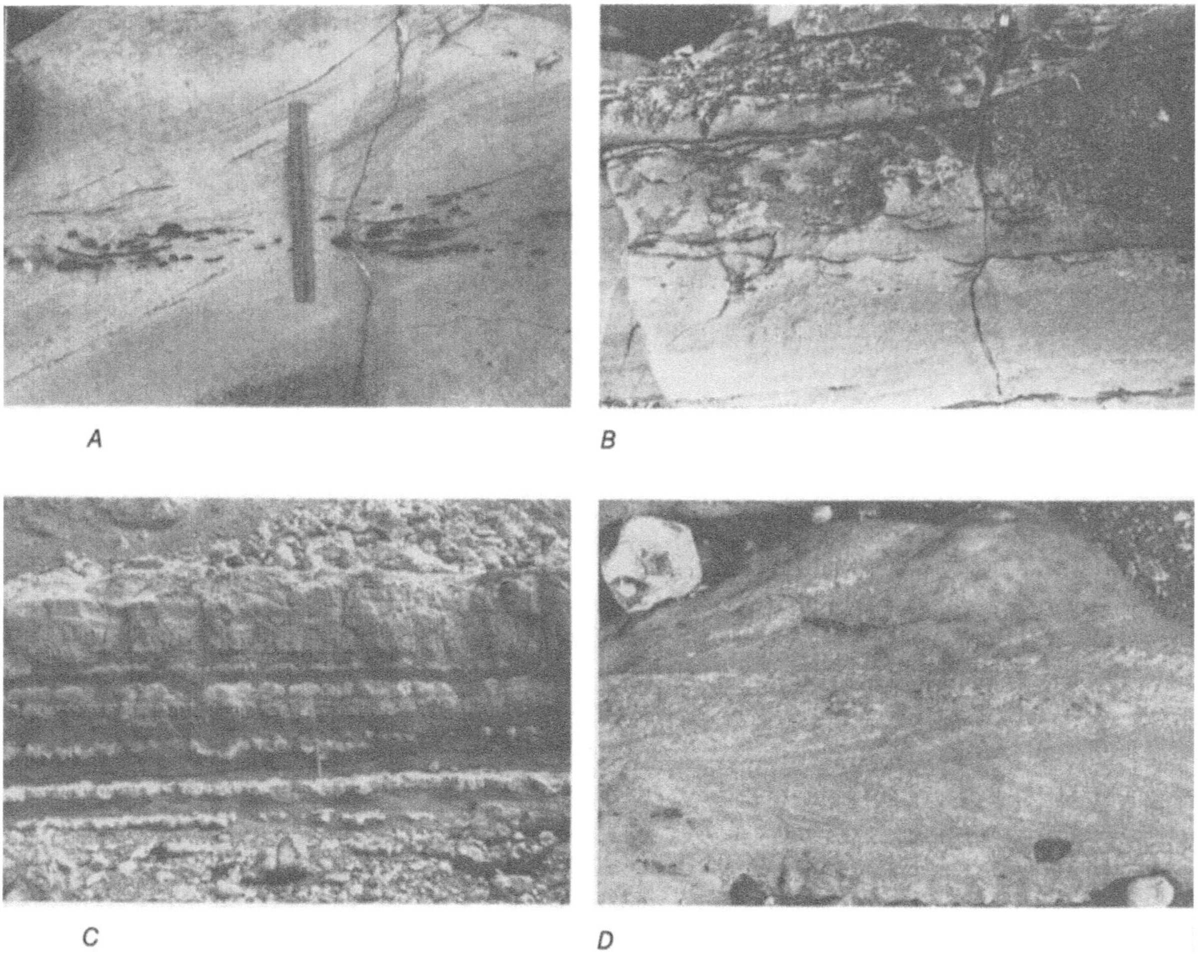

Figure 40-3. Sedimentary structures. (Scales shown in parts *A*, *B*, and *C* are all 30 cm long.) *A*. Large-scale, cross-bedded sets in oosparite of facies 2, with clay pebble lag in bottom sets. *B*. Low-angle cross-stratification and symmetric ripples with clay-filled troughs in oosparite of facies 2. Cross-bedding dips at right angles to that shown in *A* (compare Figs. 40-4*A* and 40-4*B*). *C*. Facies 3. Succession of intrasparites and nodular *Rhaxella* biomicrites formed as burrow fills (Furisch, 1972), overlain by oosparites containing small-scale, trough cross-bedding, *Arenicolites* and *Diplocraterion*, and traversed by large-scale cross-stratification produced by lateral accretion toward the right (east). *D*. Alternating dip direction of cross-bedding, developed in biomicrites and oolitic sediments of facies 4.

by association with the interpretation of facies 2 and 3. Considerable bioturbation and consequent loss of cross stratification in this facies suggest a low intertidal or subtidal environment.

Facies 2. The cross-bedded sets developed in the oolites represent avalanching down the steep advancing slopes of sand

Table 40–1 *Summary of features of facies 3 at Bran Point, Dorset*

Sediment type	Sedimentary structures	Trace fossils	Shelly fossils
Oosparite	Large-scale sets traversing total 1-meter thickness of unit (see Fig. 40–3*B*) produced by lateral accretion eastward crossing from top to bottom of 1-meter bed in distance of 15 meters, with intrasets of small-scale trough cross-bedding.	*Diplocraterion* at top *Arenicolites* at base	*Nucleolites* (small burrowing echinoid) Bivalve debris
Intramicrite	Bioturbated	*Rhizocorallium*	*Nucleolites*
Rhaxella biomicrite		Fills *Thalassinoides* burrows	

waves on top of an oolite shoal in a more rapid current regime than assemblage 3. Although paleocurrent data (Fig. 40–4) is unimodal for one locality, inaccessible portions of the cliff sections show cross-stratification dipping east, as well as west. In addition, at Black Head, thick (up to 1 meter) sets of cross-bedded oolite dipping east are succeeded by thinner sets (30 cm) orientated at 90° as reflected in two plots of directions in Figure 40–4*B*. (see also Figs. 40–3*A* and 40–3*B*).

FACIES 3. This facies is interpreted as having accumulated on the margins of an oolite shoal. A facies model is presented in Figure 40–5, based on accounts of Recent carbonate sedi-

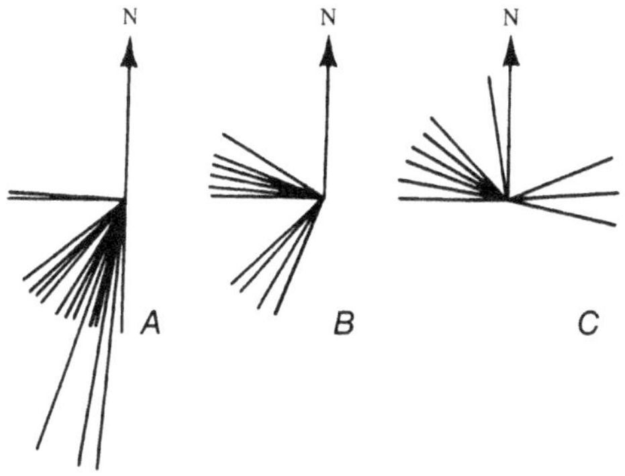

Figure 40-4. Paleocurrent data for the Osmington Oolites. *A*. Lower Oolite (Shortlake), 17 readings; facies 2. *B*. Lower Oolite (Black Head), 10 readings; facies2. *C*. Upper Oolite (Black Head), 10 readings; facies 4.

ments and fauna (such as Imbrie and Buchanan, 1965; Farrow, 1971). The bioturbated sediments represent intershoal "lagoonal" sediments, which, in the absence of marine grasses in the Jurassic, were probably stabilized by *Rhaxella* sponges. The general increase in grain size and proportion of oolitic coating to the grains are comparable to those described by Purdy (1961) and Ball (1967) for Bahaman oolite shoals and associated bank sediments.

FACIES 4. The sheet deposits of texturally inverted sediments are analogous to beach and storm deposits deposited on high tidal or supratidal flats described by Hayes (1964), Imbrie, and Buchanan (1965), but possible tidal influence is also indicated by the reversal of cross-bed dip directions shown in Figures 40–3*D* and 40–4*C*.

GENERAL ENVIRONMENT. There is a general shallowing of environments up the sequence, from low intertidal (assemblage 1) to high intertidal (assemblage 4). Assemblages 2 and 3, although not part of the same oolite bed, are interpreted as being, respectively, at the front of and marginal to an oolite shoal.

Unfortunately, the two-dimensional nature of the cliff sections in Dorset prevent an accurate orientation of the facies model presented in Figure 40–5. However, it does seem that the geometry of the oolite sand bodies is largely influenced by east-west flowing tidal currents. Lack of data on the geometry of the oolites also prevents the distinction between Ball's

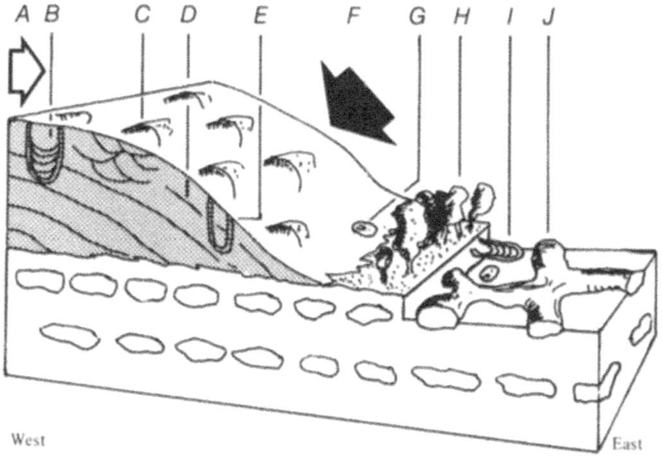

Figure 40-5. Environmental model for facies 3. Section (see Fig. 40-4*C*). Stipple: oosparite. Greytone: intramicrite. White: *Rhaxella* biomicrite. *A*, direction of lateral accretion; *B*, *Diplocraterion*; *C*, linguoid ripples producing trough cross bedding; *D*, large-scale cross-stratification marking lateral accretion; *E*, *Arenicolites*; *F*, principal current direction; *G*, *Nucleolites*; *H*, *Rhaxella* sponges (actual morphology is not known); *I*, *Rhizocorallium*; *J*, *Thalassinoides*.

Table 40–2 *Components of Klein's (1971) tidalite model identified in Corallian Beds*

	Facies of Osmington Oolites			
	1	2	3	4
A. 1. Cross stratification with sharp set boundaries		■		■
2. Herringbone cross-stratification			■	■
4. Parallel laminae	■			
5. Complex internal organization of dunes, etc.		■	■	■
B. 6. Reactivation surfaces		■		■
7. Bimodal distribution of set thicknesses		■		
9. Unimodal distribution of dip direction of cross strata		■	■	
10. Orientation of cross-strata parallels sand-body trend			?	
C. 13. Small current ripples to larger current ripples			■	
19. Symmetrical ripples			■	
D. Flaser bedding			■	
F. 35 & 36 channel log deposits			■	
H. Burrowing, etc.	■	■	■	

(1967) marine sand belts or tidal-bar belts. However, assuming an east-west trend to tidal currents, (see Fig. 40–4) a slope break (Ball, 1967, p. 557) trending roughly north-south is indicated, possibly due to shoaling of the Wessex Basin westward toward the Hercynian massif of Devon and Cornwall. The variation in current direction may be due to variations in tidal ebb and flood flow or the interaction of tides and wind-induced currents.

Tidal influence in a subtidal environment seems probable, as many of Klein's (1971) criteria for "tidalites" are present (see Table 40–2) in the Osmington Oolite Series. However, tidal range cannot be determined.

References

Ball, M. M. 1967. Carbonate sand bodies of Florida and the Bahamas. *J. Sed. Petrol. 37*, 556–591.

Farrow, G. E. 1971. Back reef and lagoonal environments of Aldabra atoll, distinguished by their crustacean burrows. In *Regional Variation in Indian Ocean Coral Reefs*. Symp. Zool. Soc. London, Vol. 28, pp. 455–500.

Freeman, T. 1962. Quiet water oolites from Laguna Madre, Texas. *J. Sed. Petrol. 32*, 475–483.

Fursich, F. T. 1972. *Thalassinoides* and the origin of nodular limestone in the Corallian Beds (Upper Jurassic) of southern England. *Neues Jahrb. Geol. Palaeontol. Abhandl. 140*, 33–48.

HAYES, M. O. 1964. Effect of hurricanes on nearshore sedimentary environments of coastal bend area of south Texas. *Bull. Am. Assoc. Petrol. Geol. 48*, 530 (abst.).

IMBRIE, J., and BUCHANAN, H. 1965. Sedimentary structures in modern carbonate sands of the Bahamas. *Soc. Econ. Paleontol. Mineral. Spec. Publ. 12*, 149–172.

LOGAN, B. W., REZAK, R., and GINSBURG, R. N. 1964. Classification and environmental significance of algal stromatolites. *J. Geol. 72*, 68–83.

PURDY, E. G. 1961. Bahamian oolite shoals. In *Geometry of Sandstone Bodies*. Am. Assoc. Petrol. Geol.

WILSON, R. C. L. 1966. Silica diagenesis in Upper Jurassic limestones of southern England. *J. Sed. Petrol. 36*, 1036–1049.

——— 1968. Carbonate facies variation within the Osmington Oolite Series in southern England. *Paleogeog. Paleoclimatol. Paleoecol. 4*, 89–123.

——— 1969. Field Meeting to south Dorset. Report by the Director. *Proc. Geol. Assoc. 80*, 341–352.

41

Some Examples of Shoaling Deposits from the Upper Jurassic of Portugal

R. C. L. Wilson

Occurrence

Over 5,000 meters of Jurassic sediments accumulated in the Lusitanian Basin. The Upper Jurassic contains a great variety of facies, ranging from reefs to fluviatile red beds. The basin contains a fully marine sequence (see Fig. 41–1 and Table 41–1), whereas the shelf area exhibits a great variety of shallow marine and continental deposits. The three examples of shoaling deposits described in this paper are from the top of the Montejunto Beds, the Vale Verde Beds and overlying *Pholodomya protei* Beds, and the top of the Alcobaça Beds.

Interpretation

Example 1 (Fig. 41–2) shows vertical variations in grain size and texture, which suggest variations of water depth with time, but contains no definitive evidence of tidal influence. This

This was work carried out with the aid of a Natural Environment Research council Grant.

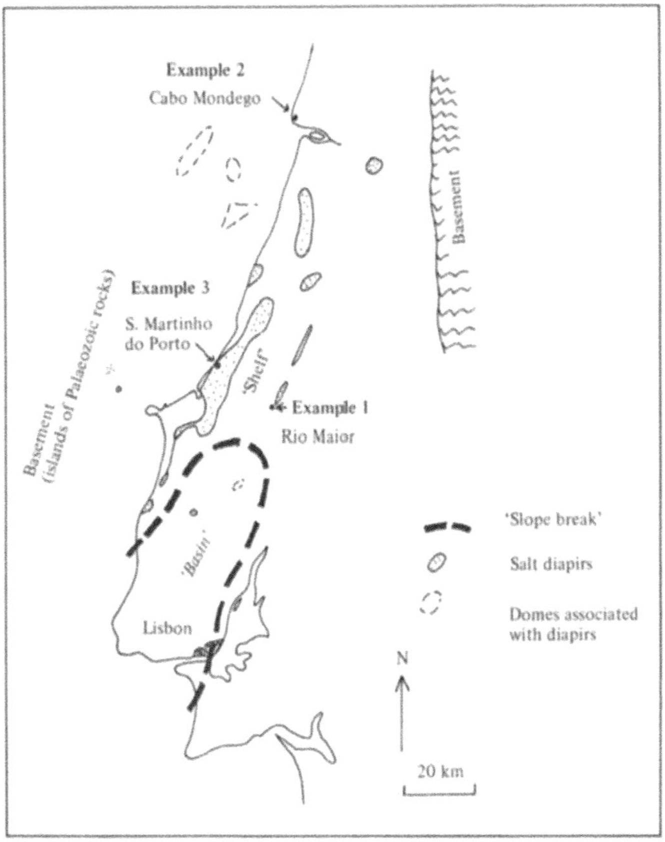

Figure 41·1. The Lusitanian Basin, showing the position of the main basin, separated from the shelf by a slope break. The diapirs shown in the shelf area were active during the Upper Jurassic.

Table 41–1 *Simplified Oxfordian and Kimmeridgian stratigraphy of the Lusitanian Basin*

Stage	Basin (meters) (see Fig. 41–1)		Shelf (meters)
Kimmeridgian (in French sense)	~100	*Lima pseudo alternicosta* Beds, Amaral Coral Lsts.	100–1,000 Alcobaça Beds
	800	Abadia Beds	
Oxfordian	350	Montejunto Beds	80–200 *Pholodomya protei* Beds
Upper	300–500	Cabaços Beds	100–200 Vale Verde Beds
Lower		M I S S I N G	
	C A L L O V I A N		

(After Ramalho, 1971; Ruget-Perrot, 1961.)

type of deposit is characteristic of the margins of the basin and was probably controlled by a slope break (Ball, 1967) interacting with wind-driven currents and, if present, tidal currents.

Example 2 (Figs. 42–3*A* and 41–4) shows a transition from open marine conditions in the underlying Callovian to fluviatile sediments overlying the *Pholodomya protei* Beds. As shown in Table 41–2, the laminated micrites and associated sediments of the Vale Verde Beds are analogous to Ginsburg and Hardie's (Chapter 23) Andros tidal flats. The remainder of the succession may be compared to sediments accumulating today in the Texas coastal bays (see Parker, 1959; Donaldson et al., 1970; Kanes, 1970). The coarse-grain carbonate up to 35 meters (in Fig. 41–3*A*) represents an offshore bar, and is overlain by a fauna (around 38 meters) reminiscent of Parker's (1959) inlet and high salinity oyster reef facies, followed by probable deltaic sands overlain by lignite (49 meters). The succeeding silts and calcareous silts (60 to 130 meters) probably accumulated in low salinity bay environments followed by the Andros-type laminated carbonates. The bivalve-rich biomicrites of the *Pholodomya protei* Beds represent bay sediments, and these are re-

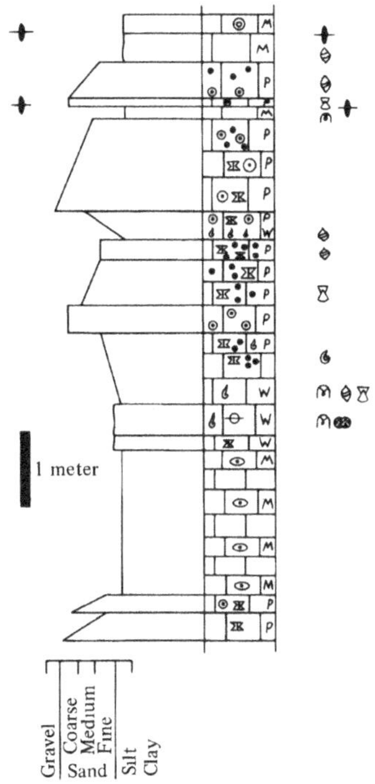

Figure 41-2. Example 1. Carbonate sequence at the top of the Montejunto Beds. Grain-size variation shown by variable width of left column. Textures indicated as follows: *M*, mudstone; *W*, wackestone; *P*, packstone.

placed laterally and vertically by fluviatile red beds, suggesting deltaic influence analogous to that in the Texas Gulf coast region.

Example 3 (Figs. 41–3*B* and 41–4) may be interpreted as a mixture of bay and inlet facies. In the southern part of the shelf area (see Fig. 41–1) bordering the slope break, the *Pholodomya protei* Beds are replaced by shallow water carbonates comparable to example 1, and the basal part of the Alcobaça Beds also contain oolitic and oncolitic sediments. These features

Figure 41-3. *A.* Example 2. Sequence in the Vale Verde and *Pholodomya protei* Beds. Black in left column indicates phyllosilicate clay; also see caption to Figure 41-2. Figures on right side indicate 10s of meters above junction with Callovian; wavy lines indicate laminated fenestral micrites. The junction of the Vale Verde and *Pholodomya protei* Beds is at 195 meters. *B.* Example 3. Sequence at the top of the Alcobaça Beds (figures at right indicate 10s of meters above top of Vale Verde Beds; black strip on right indicates red color, dashed lines indicates slightly red coloration.

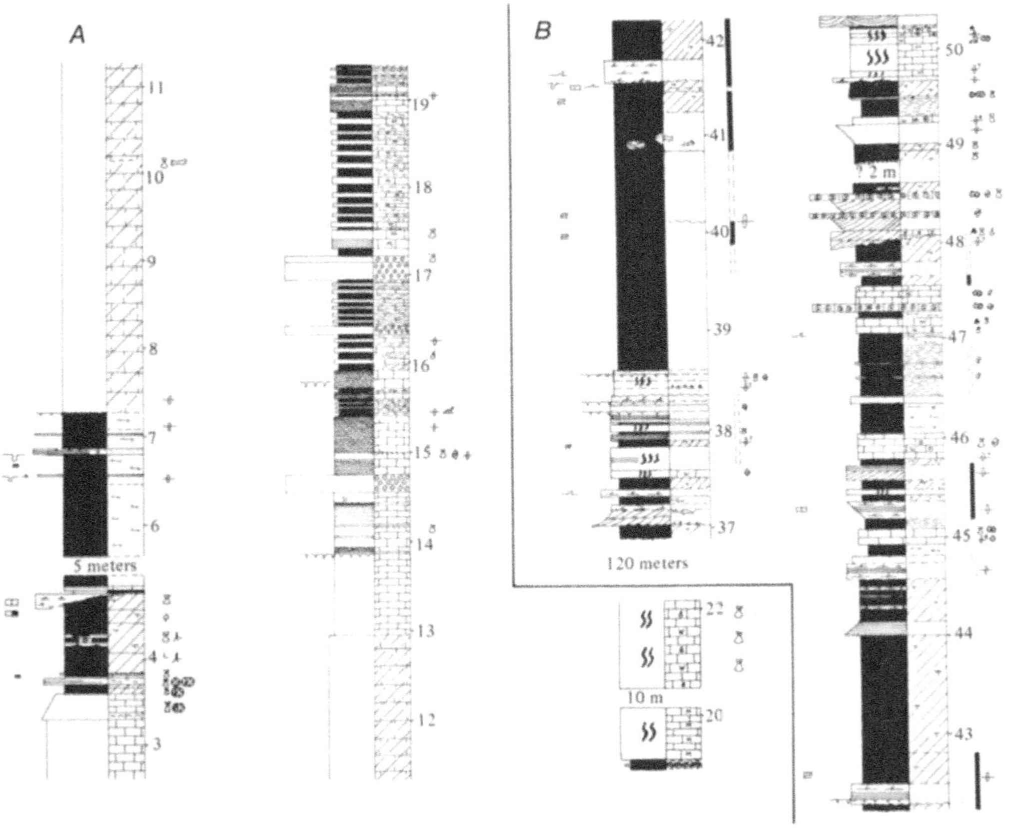

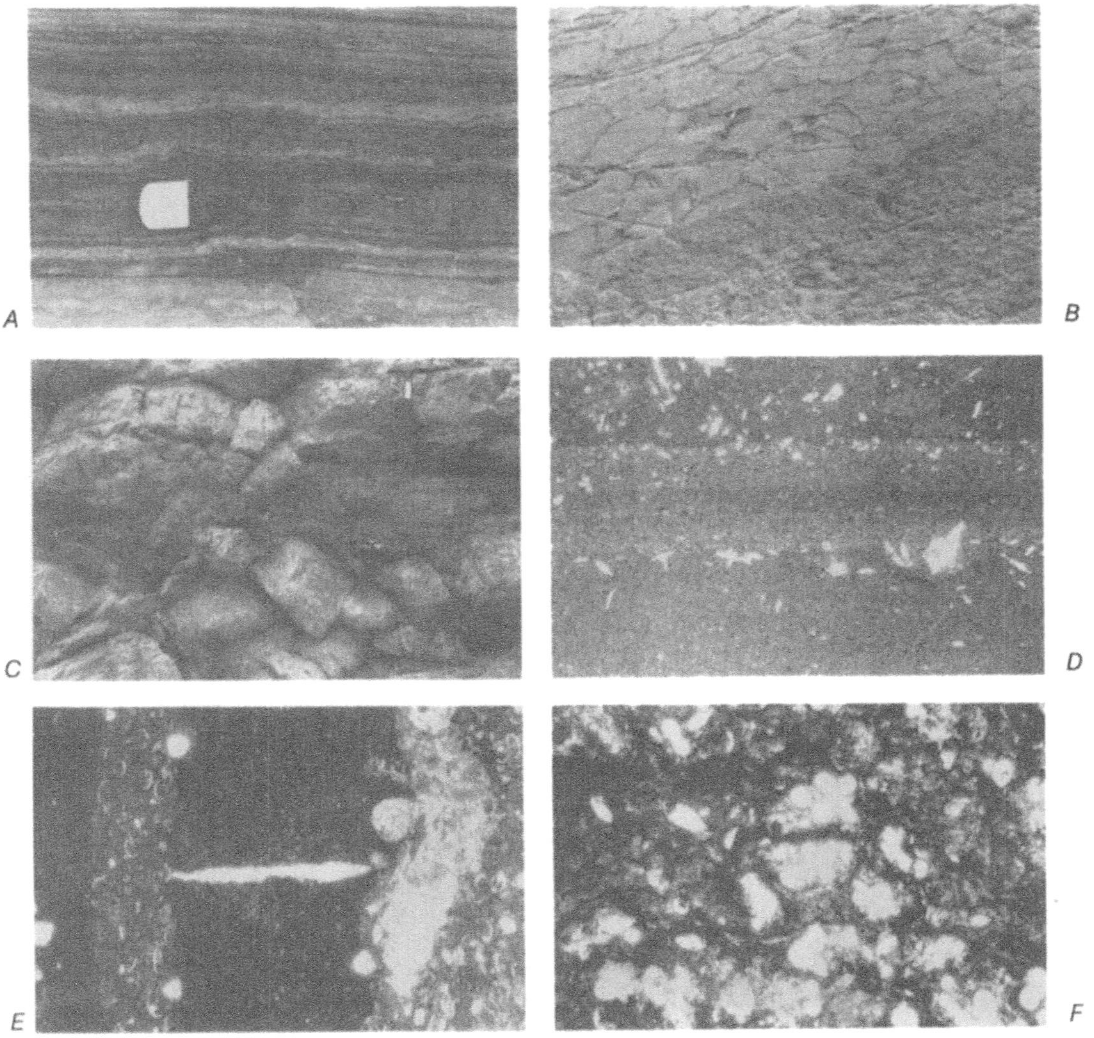

Figure 41-4. Features of example 2 (Vale Verde Beds). *A.* 152 meters (see Fig. 41-3*A*); laminated fenestral micrites, with crenulations similar to those in modern sediments described by Davies (1970). Lighter layers are extremely porous and crumbly in texture, whereas darker layers contain fine laminations (see *D* and *E*). Tape 4.5 cm. *B.* 139 meters; burrowed horizon overlying prominent desiccation-cracked surface. This micritic limestone contains small cerithid gastropods (cf. Ginsburg and Hardie, Chapter 23, Fig. 23-3*D*). *C.* 146 meters; breccia containing limestone blocks up to 1 meter across. *D.* 139 meters; laminated micrite containing calcite pseudomorphs after gypsum; × 50. *E.* 148 meters; detail of millimeter laminations, showing band of pellets and ostracod tests, intrastratal desiccation crack in micrite, and cavernous micrite; × 50 (cf. Ginsburg and Hardie, Chapter 23, Fig. 23-3*C*). Top is to right. *F.* 143 meters; sections through nodes of *Chara* stems; oogonians of *Chara* are found in the succession, but not usually associated with the remains of the main body of the algae; × 50.

A

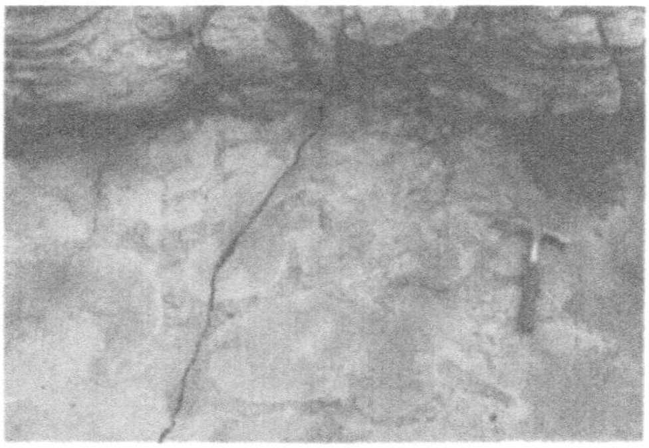

B

Figure 41-5. Features of Example 3 (top of Alcobaça Beds). *A.* 473 meters; oncolites: algal coatings on large nerineiid gastropods and coral fragments. Tape is 4.5 cm wide. *B.* 480 meters; *Thalassinoides* type burrow penetrating reddish phyllosilicate clay underlying channel fill of cross-bedded biosparite seen at top of photo. *C.* Fallen block of oncolites and underlying sandstone that contains clay flasers marking intrasets of smaller scale cross bedding. *D.* 485 meters; oncolites showing upward increase in grain size.

C

D

suggest the formation of marine sand belts under the influence of wind-generated, or possibly tidal, currents.

Discussion

There is clear evidence that all three examples accumulated around the land-sea interface and that sea level fluctuated. However, it is not possible to prove any tidal influence. If the

Table 41–2 *Criteria for environmental interpretation*

Example/ Stratigraphic unit	Features	Interpretation	General conclusions
Example 1: Top of Montejunto Beds (Fig. 41–2)	Oolites, oncolites, bored horizons. Fining-upward (oolite-micrite) and coarsening-upward units. Marine bivalve and coral fauna.	Shallow, turbulent water, probable emergence and lithification; change of energy of environment due to change in water depth caused by tidal-flat regression/transgression or migrating sandbars.	Clear evidence of shallow turbulent marine conditions, and sea-level changes, but no unequivocal evidence of tidal action.
Example 2: Vale Verde and *Pholodomya protei* Beds (Figs. 41–3*A*, 41–4)	Laminated fenestral micrites. Desiccation cracks. *Chara:* oogonians and stems. Abundant ostracods. *Isognomon, Cerithium,* burrows. Breccia horizons. Sequence from marine Callovian to overlying fluviatile sandstones (see Fig. 41–3*A*).	Assemblage very similar to Andros tidal flats and supratidal marshes described by Ginsburg and Hardie (Paper 23), with additional feature of breccias interpreted as basal channel deposits formed by collapse of channel wall. Coastal environment similar to Texas bays (see Interpretation section), from open marine to offshore bar, inlet/high salinity oyster reef, forest-swamp, low salinity bay, freshwater marsh, and fluviatile environments.	Clear evidence of coastal environment, but features comparable to Andros tidal flats could be produced by wind tides.
Example 3: Top of Alcobaça Beds (Figs. 41–3*B*, 41–5)	Oncolites. Oolites. Channel, with large burrow systems, and cross-bedded fill with flasers. Desiccation cracks. Caliche horizons, burrow-mottled clays and silts. Fining-upward sandstone units.	Alternation of fluviatile and shallow marine conditions, with fluctuating water table; possibly high salinity bay environment with some inlet influence	Fluctuating sea level proved, but no clear evidence of tidal influence.

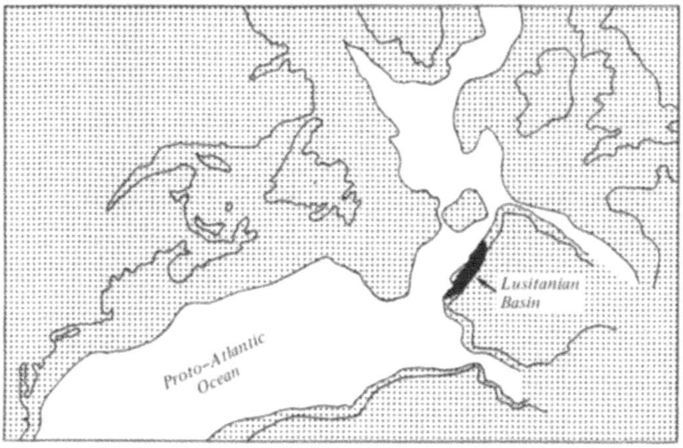

Figure 41-6. Location of the Lusitanian Basin in relation to the proto-Atlantic Ocean. (Modified after Pitman and Talwani, 1972.)

modern analogs quoted are a guide, then the tidal range in the shelf area would have been less than half a meter. The close similarities between part of example 2 and the Andros tidal flats suggest deposition under the influence of wind-generated tides.

Although the evidence for tidal influence on the deposition of the Upper Jurassic in Portugal is circumstantial rather than definitive, it is not improbable that there was some tidal influence, for the Lusitanian Basin opens south onto the Jurassic proto-Atlantic, which was presumably a mesotidal ocean (Fig. 41–6).

References

BALL, M. M. 1967. Carbonate sand bodies of Florida and the Bahamas. *J. Sed. Petrol. 37*, 556–591.

DAVIES, G. R. 1970. Algal-laminated sediments, Gladstone Embayment, Shark Bay, Western Australia. *In* Logan, B. W., Davies, G. R., Read, J. F., and Cebulski, D. E., eds., Carbonate sedimentation and environments, Shark Bay, Western Australia. *Am. Assoc. Petrol. Geol. Memoir 13*, 169–205.

DONALDSON, A. C., MARTIN, R. H., and KANES, W. H. 1970. Holocene Guadalupe Delta of the Texas Gulf Coast. *In* Morgan, J. P., Deltaic sedimentation. *Soc. Econ. Paleontol. Mineral. Spec. Publ. 15*, 107–137.

KANES, W. H. 1970. Facies and development of the Colorado River Delta in Texas. *In* Morgan, J. P., Deltaic sedimentation. *Soc. Econ. Paleontol. Mineral. Spec. Publ. 15*, 78–106.

PARKER, R. H. 1959. Macro-invertebrate assemblages of central Texas coastal bays and Laguna Madre. *Bull. Am. Assoc. Petrol. Geol. 43*, 2100–2166.

PITMAN, W. C., and TALWANI, M. 1972. Sea floor spreading in the north Atlantic. *Geol. Soc. Am. Bull. 88*, 619–646.

RAMALHO, M. M. 1971. Contribution à l'étude micropaléontologique et stratigraphique du jurassique supérieur et du crétace inférieur des environs de Lisbonne (Portugal). *Serv. Geol. Portugal Memoir 19.*

RUGET-PERROT, C. 1961. Études Stratigraphiques sur le Dogger et le Malm Inférieur du Portugal au nord du Tage. *Serv. Geol. Portugal Memoir 7.*

42

Carbonate-Sulfate Intertidalites of the Windsor Group (Middle Carboniferous) Maritime Provinces, Canada

Paul E. Schenk

Occurrence

During the Carboniferous, very thick postorogenic, mainly continental, red beds derived from intrabasinal horsts, accumulated in a complex rift-valley system called the Fundy Basin. During the Middle Carboniferous, generally hypersaline seas flooded perhaps a dozen times into the tortuously interconnected, intrabasinal grabens. Laterally, the resultant marine carbonates grade through nodular sulfate into increasingly coarser red beds, as well as lavas at basin margins. Red beds constitute approximately 80 percent of the Middle Carboniferous Windsor Group. Vertically, each carbonate blanket shows mainly offlap successions from shallow water through intertidal to supratidal environments. During the Late Carboniferous Maritime Disturbance, the assemblage was severely folded, faulted, locally metamorphosed, and intruded. Red beds flowed plastically or were metamorphosed locally to slate; carbonate units were folded and faulted; sulfates show spectacular metamorphic fabrics when not protected by adjacent carbonates.

Discussion

Many features of sedimentation, fauna, and diagenesis are strikingly similar between the Windsor Group of Atlantic Canada and the Pleistocene of Shark Bay, Western Australia (Logan et al., 1970; Schenk 1967a, 1967b, 1969). The Fundy Basin with its many semirestricted inlets appears identical to those of the bay. Specifically, the Antigonish Basin is surprisingly similar to Hamelin Pool both as to vertical and lateral relations of litho- and biofacies in depositional as well as diagenetic environments. Windsor-type evaporites, especially salt, is set-

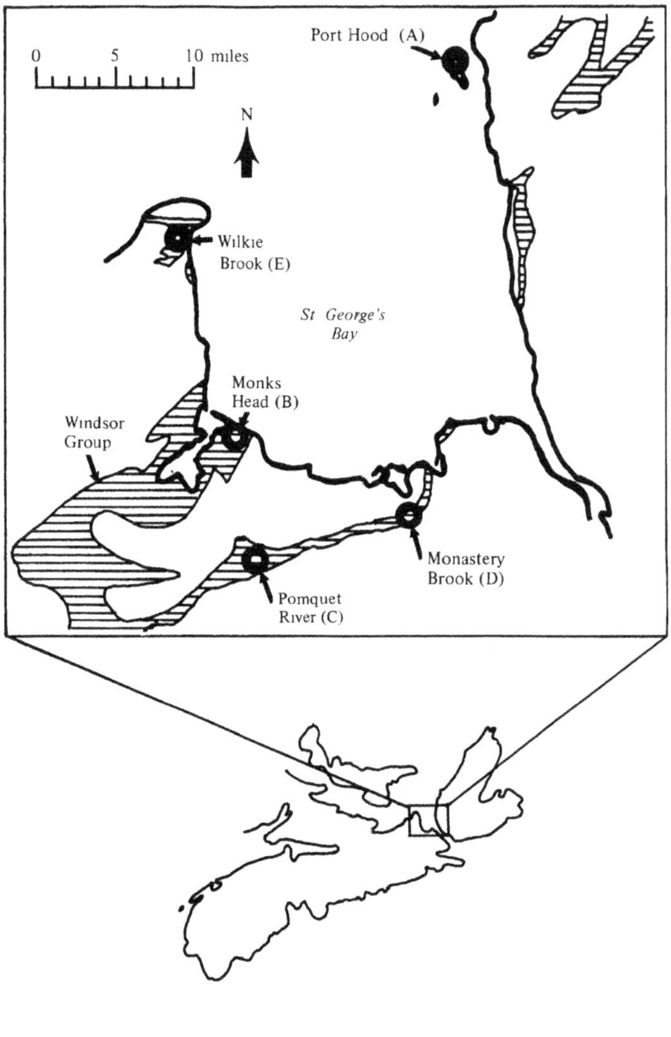

Figure 42-1. Antigonish basin, northeastern Nova Scotia, within the Carboniferous Fundy basin. Locations here are for Figure 42-2.

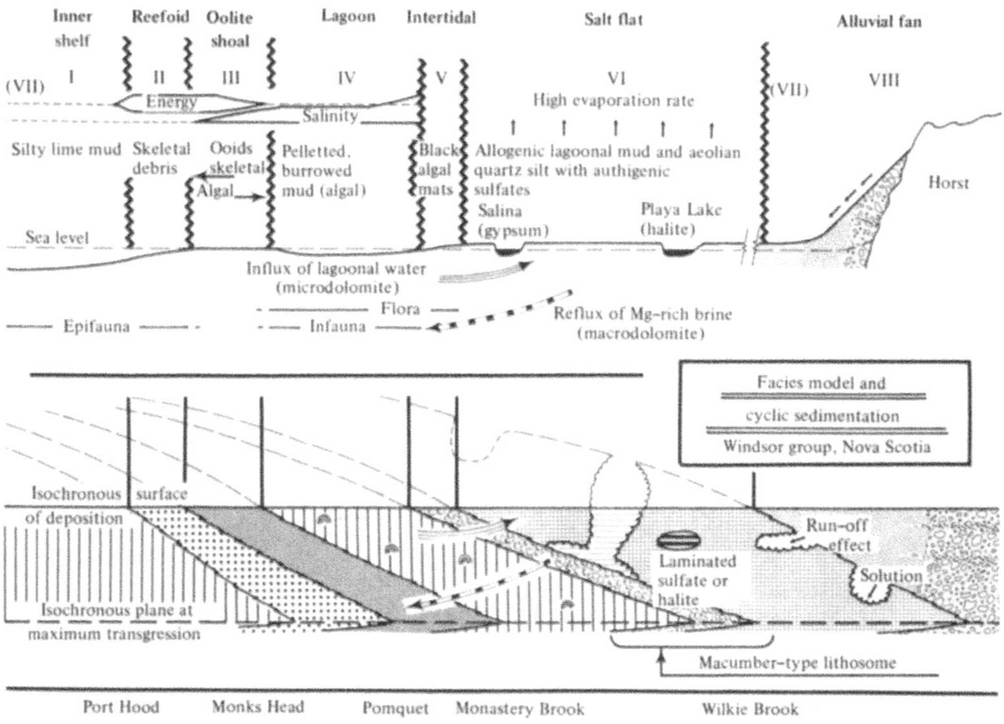

Figure 42-2. The upper diagram shows the inferred distribution of environments, processes, and the resulting facies types, numbered as in Table 42-1. The lower diagram shows the cyclical facies pattern developed by regression.

tling today in western Shark Bay as well as in the bay's probable future environment, directly north at Lake MacLeod. Although diagenetic gypsum precipitates now beneath arid salt flats of Shark Bay, the arid southwest coast of the Persian Gulf appears as a better analog for nodule growth (Shearman, 1966). Indeed, the carbonate-sulfate facies there mimic the Windsor. According to paleomagnetism, the Atlantic Provinces were at 10° South latitude during the Middle Carboniferous (Roy and Robertson, 1968); according to plate-tectonic reconstructions, the Windsor seas were the last dribbles of the Paleozoic Atlantic at this latitude as North America closed with Africa/Europe (Schenk, 1971). Therefore as to depositional, diagenetic, as well as tectonic settings, a third recent analog could be the southern Red Sea, specifically the Afar triangle (Hutchinson and Engels, 1972).

Table 42–1 *Facies of the Windsor Group*

Facies	Texture	Structures	Diagenesis	Fauna	Fig.
I Inner shelf	Biomicrite (wackestone)	Burrows or massive	Limestone (rarely macrodolomite), bituminous	Infauna, burrowers, diverse "normal"	42–3
II Reefoid	Biosparite-biolithite (grain or boundstone)	(Cross-stratification) geopetal.	Limestone (rarely cemented by macrodolomite)	Most diverse, brachiopods, corals, crinoids	42–4
III Ooid shoal	Oosparite (grainstone)	Cross-stratification, ripples, geopetal.	Dolomitic limestone to calcitic dolostone (macrodolomite cement)	Transported brachiopods, pelecypods	42–5
IV Lagoon	Biopelmicrite (wackestone)	Burrows or massive	Dolostone to calcitic dolostone (microdolomite)	Rare gastropods, ostracods, algae	42–6
V Intertidal	Biolithite (boundstone to pellet grainstone)	Laminated algal, stromatolites, channels, breccia sun cracks	Limestone to dolostone (microdolomite), bituminous	Blue-green algae, algal stromatolites, forams, gastropods	42–7—42–11
VI Salt flat	Sulfate nodules, microdolomite, biolithite, pelmicrite	Penemosaic sulfate nodules, laminated or fenestral carbonate	Diagenetic sulfate, blowing up algal, stromatolites, bituminous	Recrystallized skeletal fragments, blue-green algae	42–12
VII–VIII Alluvial fan	Red arkosic to lithic sandstone and conglomerate	Festoon cross-stratification, channels, laminated carbonate	Oxidizing, rotting of ferromagnesium; local reducing in lakes	Rare plants, ostracods	—

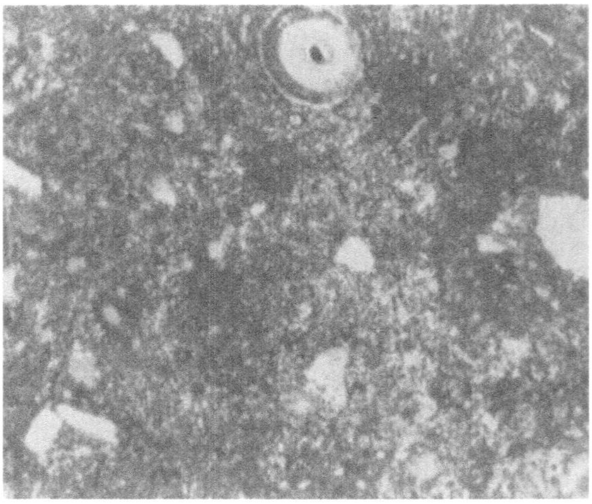

Figure 42-3. Facies 1. Poorly sorted brachiopod-crinoid biomicrite limestone, 16 percent quartz silt; × 6; Monk's Head.

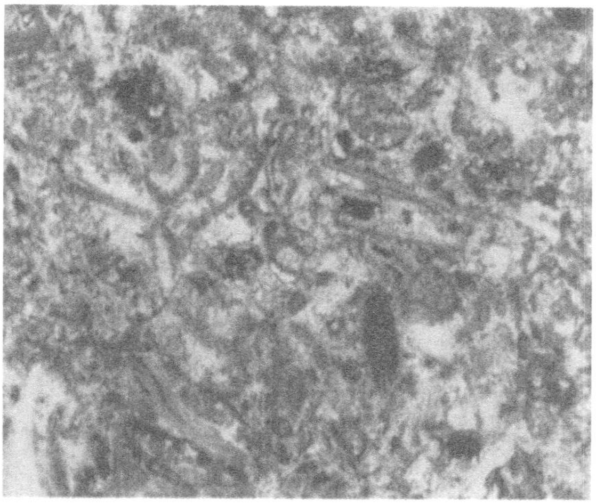

Figure 42-4. Facies 2. Well-sorted brachiopod-echinoid biosparite limestone; × 6; Port Hood.

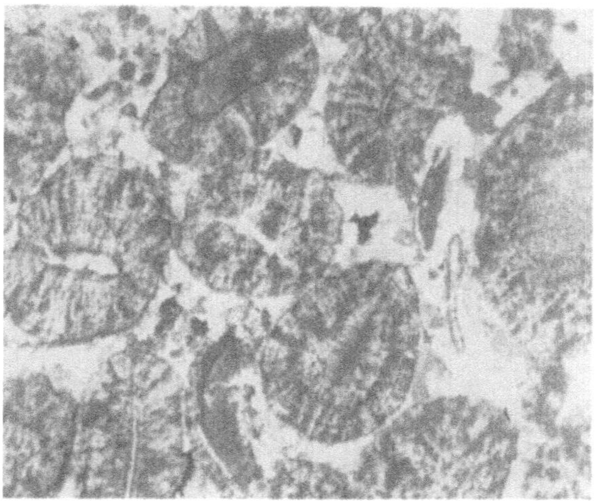

Figure 42-5. Facies 3. Well-sorted oosparite calcitic dolostone. Calcite ooids, macrodolomite cement; × 13; Port Hood.

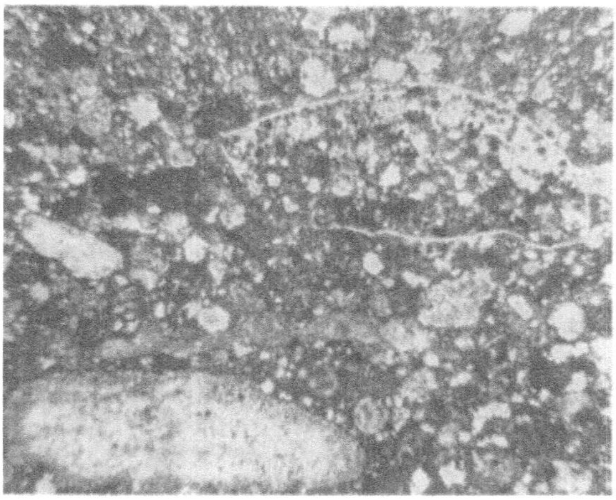

Figure 42-6. Facies 4. Massive biopelmicrite dolostone. Brachiopods, forams, intraclasts in microdolomite; × 8; Cumberland Basin.

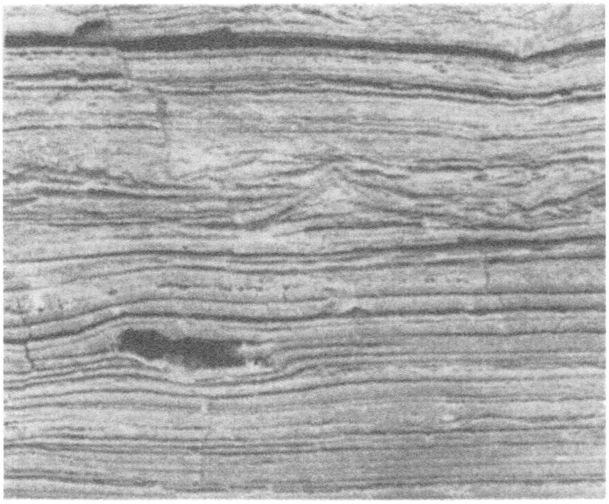

Figure 42-7. Facies 5. Biolithite limestone. Flat algal stromatolite with breccia, stromatactis, microteepee structure; × 1; Wilkie Brook.

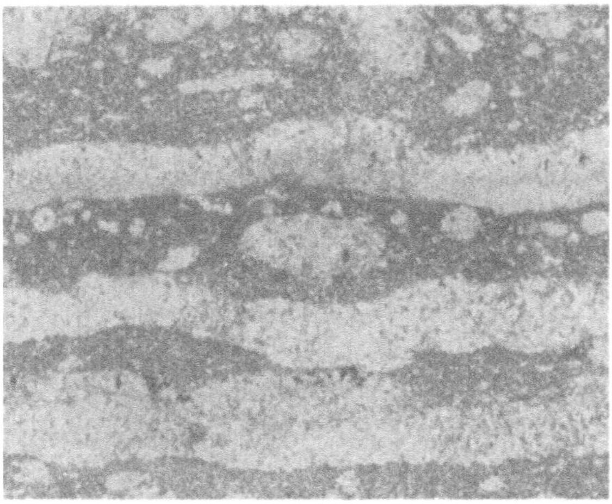

Figure 42-8. Facies 5. Photomicrograph of Figure 42-7. Light layers calcite with inverse grading; dark layers bituminous dolomicrite; × 8; Wilkie Brook.

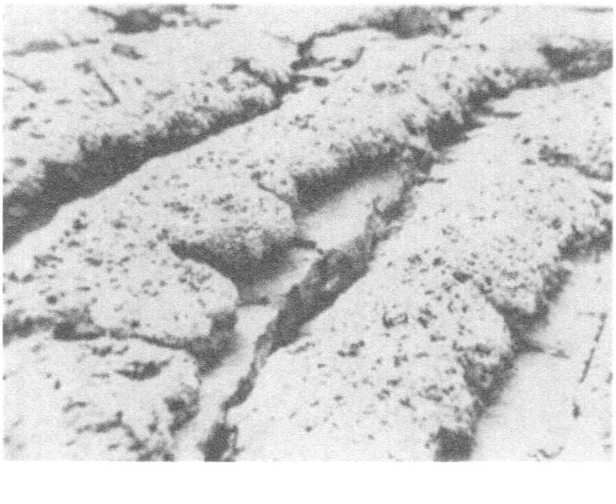

Figure 42-9. Facies 5. Linear algal stromatolites over facies 3 and under nodular gypsum. Rows 0 5 × 7 meters; Monk's Head.

Figure 42-10. Facies 5. Club-shaped algal stromatolites under sun-cracked polygons; Port Hood.

Figure 42-11. Facies 5. Sun-cracked polygons in flat algal stromatolite. Hammer is 28 cm long. Near Monastery Brook.

Figure 42-12. Facies 5. Core with anhydrite nodules exploding laminated algal stromatolite. Top to right; pencil point for scale. Cumberland Basin.

References

HUTCHINSON, R. W., and ENGELS, G. G. 1972. Tectonic evolution in the southern Red Sea and its possible significance to older rifted continental margins. *Bull. Geol. Soc. Am. 83*, 2989–3002.

LOGAN, B. W., DAVIES, G. R., READ, J. F., and CEBULSKI, D. E., eds. 1970. Carbonate sedimentation and environments, Shark Bay, Western Australia. *Am. Assoc. Petrol. Geol. Memoir 13.*

ROY, J. L., and ROBERTSON, W. A. 1968. Evidence for diagenetic remnant magnetization in the Maringouin Formation. *Can. J. Earth Sci. 5*, 275–285.

SCHENK, P. E. 1967a. The Macumber Formation of the Maritime Provinces—A Mississippian analogue to Recent strand-line carbonates of the Persian Gulf. *J. Sed. Petrol. 37*, 365–376.

——— 1967b. The significance of algal stromatolites to paleoenvironmental and chronostratigraphic interpretations of the Windsorian Stage (Mississippian), Maritime Provinces. *Geol. Assoc. Can. Spec. Paper 4*, 229–243.

——— 1969. Carbonate-sulfate-redbed facies and cyclic sedimentation of the Windsorian Stage (Middle Carboniferous) Maritime Provinces. *Can. J. Earth Sci. 6*, 1037–1066.

——— 1971. Southeastern Atlantic Canada, northwestern Africa, and continental drift. *Can. J. Earth Sci. 8*, 1218–1251.

SHEARMAN, D. J. 1966. Origin of marine evaporites by diagenesis. *Trans. Inst. Mining Met. 75*, 208–215.

43

Carboniferous Tidal-Flat Deposits of the North Flank, Northeastern Brooks Range, Arctic Alaska

Augustus K. Armstrong

Occurrence

In the Sadlerochit Mountains and adjacent areas (Fig. 43–1) the Carboniferous Lisburne Group, some 460 to 640 meters thick, was formed by a two-phase carbonate depositional cycle that began in late Mississippian time with the deposition of the basal Alapah Limestone and continued into Pennsylvanian time, when the Wahoo Limestone was deposited (Fig. 43–2). These carbonate sediments are part of a regional transgressive sequence. In the Sadlerochit Mountains above the Kayak(?) Shale, the basal late Mississippian Alapah limestones typically consist of well-sorted, pelletoidal-bioclastic grainstones and packstones and lesser amounts of ooid grainstones. Overlying these are beds of poorly sorted bryozoan-echinoderm packstones and wackestones. The environment of deposition for these carbonates is interpreted as open platform normal marine.

The lithology and sedimentary structures of the higher beds

Publication authorized by the Director, U.S. Geological Survey.

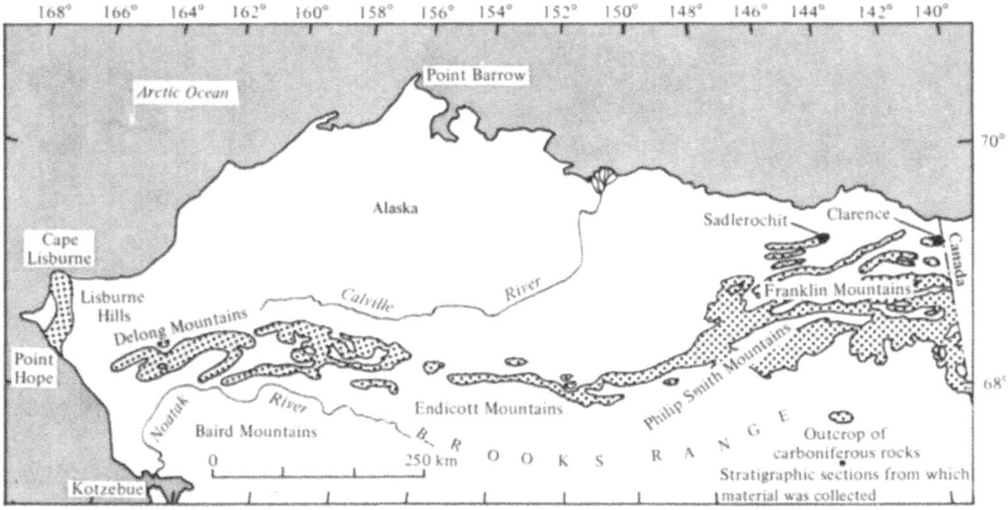

Figure 43-1. Index map showing outcrops in Brooks Range, arctic Alaska.

of the Alapah Limestone show a progressive decrease in biotic diversity, an increase in the amounts of pelletoidal packstones and lime mudstones, and an increase in the percentage of dolomite. This regressive sequence is very well developed in the east Sadlerochit Mountains section (Fig. 43–2). At 290 meters below the top of the section, the Alapah Limestone is a fine-grain, light brown-gray, cherty dolomite with well-developed stromatolites, calcite pseudomorphs after gypsum, and birdseye structure (Armstrong, et al., 1970; Mamet and Armstrong, 1972).

In the region of the Sadlerochit Mountains the second carbonate transgressive cycle of the Lisburne Group began in latest Chesterian time and continued across the Mississippian-Pennsylvanian boundary without interruption (Fig. 43–2). The earliest Morrowan-age carbonate rocks are lithologically similar to those of very latest Chesterian age; these are bryozoan-echinoderm wackestones and packstones that were probably deposited on an open marine platform. The outcrops of the Wahoo Limestone in the Sadlerochit Mountains indicate that, from the base of the Pennsylvanian to its top, the general trend in sedimentation is toward higher energy water, that is, shoaling water,

Figure 43-2 (opposite). Composite and idealized section of the Lisburne Group for the Sadlerochit Mountains showing interpreted shifts of carbonate depositional environments and stratigraphic location of major intertidal facies. The thick sequence of carbonate rocks in each cycle is believed to be the result of slow regional subsidence and slightly greater rates of carbonate deposition.

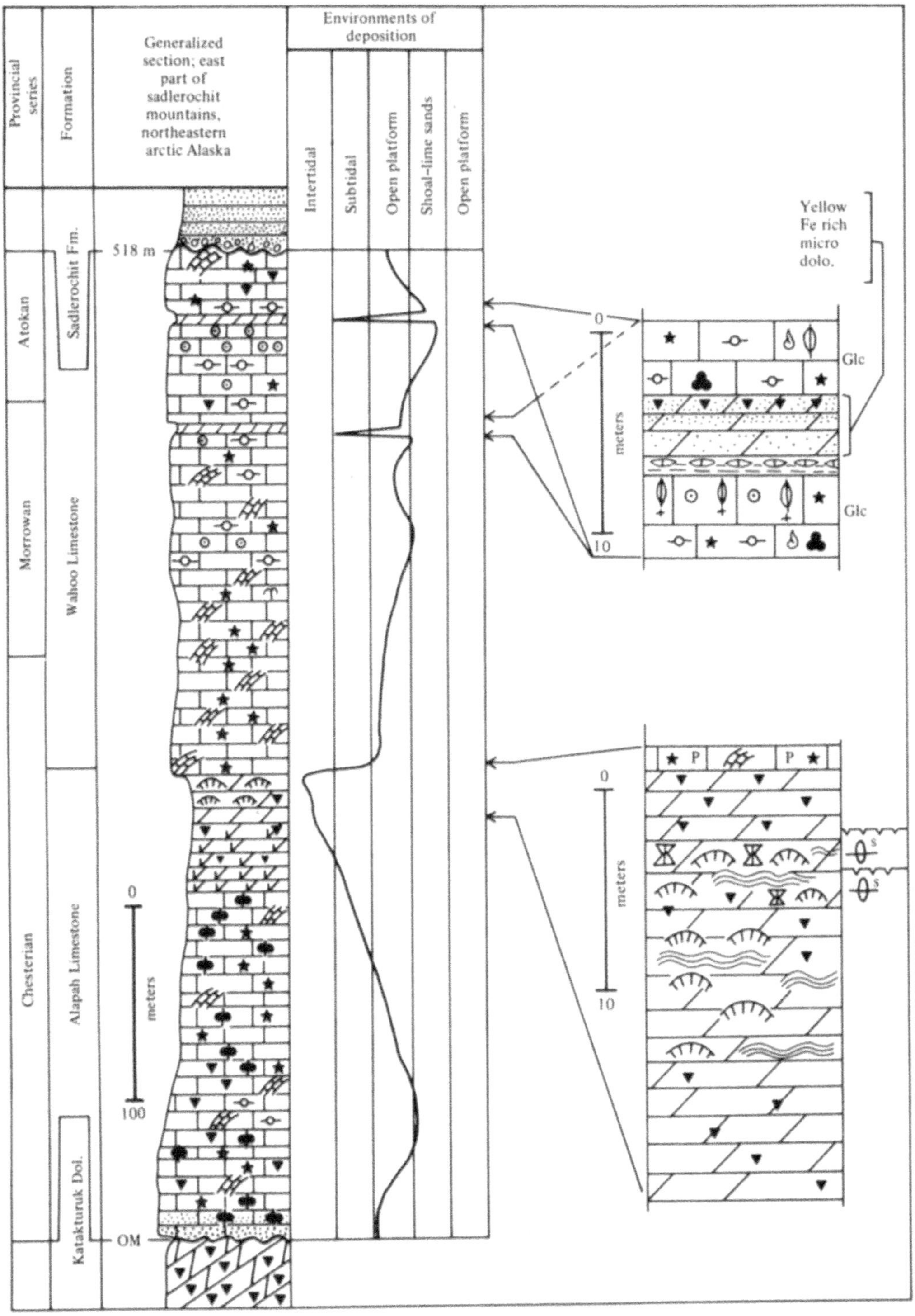
Provincial series
Formation
Generalized section; east part of sadlerochit mountains, northeastern arctic Alaska
Environments of deposition
Intertidal
Subtidal
Open platform
Shoal-lime sands
Open platform
Atokan
Morrowan
Chesterian
Sadlerochit Fm.
Wahoo Limestone
Alapah Limestone
Katakturuk Dol.
518 m
OM
0
meters
100
Yellow Fe rich micro dolo.
Glc
Glc
0
meters
10
0
meters
10
P
P
s
s

oolitic environment of deposition (Armstrong, 1972; Armstrong and Mamet, 1974). This trend is interpreted from the stratigraphically higher beds, which contain a lower content of micrite, a more diversified biota, glauconite, better sorting of fossil fragments, and well-developed oolites.

The oolitic grainstones, which are found in association with the glauconitic grainstones, are well stratified, generally 2 to 3 meters thick, and poorly cross bedded. The beds are often capped by 0.3- to 2-meter-thick argillaceous, arenaceous, limonotic, pale-yellowish-orange-weathering dolomites. This rock is formed by 30-μ dolomite rhombs with upward of 30 percent silt-size detrital quartz. Ball's (1967) description of modern carbonate sand bodies indicates that the Wahoo oolite grainstones were probably formed in a tidal-bar belt environment, transgressive over the underlying, open platform, normal marine, probably slightly reducing, glauconite-forming environment.

Interstratifications of the pale-yellowish-orange-weathering arenaceous dolomites and thick-bedded oolitic grainstones are interpreted as the record of very shallow, lime, mud tidal flats developed directly over oolitic tidal bars. This close physical relationship of oolitic grainstones and thin-bedded lime mudstones and dolomites is not unique to the Wahoo Limestone. Wilson et al. (1967, p. 81) report from Pennsylvanian carbonate rocks of southwestern New Mexico a similar sequence of oolitic grainstones and unfossiliferous mudstones. J. L. Wilson (written communication, 1970) states that similar "dolomite marker beds are also common in the Mississippian section of Montana where they are almost certainly tidal-flat and sebkha deposits with the former sulfate minerals leached out. In Montana they are also associated with oolitic grainstones but are separated from these by a transitional zone of birdseye pelletoidal mudstone and grainstone."

A similar sequence of lithologies can be seen in older beds of the Alapah Limestone at the Clarence River, near the Canadian border (Armstrong and Mamet, 1975).

The Lisburne carbonates are some 600 meters thick at the Clarence River section (Fig. 43–1) near the Canadian boundary. Although the Alapah Limestone is poorly exposed, its upper beds, which are late Meramecian or early Chesterian age, contain some 60 meters of dolomite with well-developed stromatolites, birdseye structure, intraclasts, and mud cracks, representing intertidal to supratidal environments of deposition.

This regressive suite of carbonate rocks, culminating in an intertidal-restricted marine facies, is overlain by a marine transgressive carbonate deposit of echinoderm-bryozoan-wackestones-packstones containing microfossils of earliest Chesterian age (Armstrong and Mamet, 1975).

Facies

The carbonate rocks within the Alapah Limestone, which are considered to represent deposition in an intertidal-supratidal environment, are generally microdolomites and have birdseye structure, stromatolites, mud cracks, intraclasts, and pseudomorphs of calcite after gypsum.

Discussion

The stromatolites in the Lisburne Group are characteristically low relief; those in the Alapah Limestone, which are best developed in the Clarence River outcrops, are less than 2 to 3 cm high. These stromatolites closely resemble those described from the Shark Bay by Davies (1970, Figs. 10–12) associated with shrinkage, polygons, graded beds of indurated crust, and storm deposits.

In the uppermost Alapah Limestone the low-relief stromatolites, laminations, and the lack of evidence of abundant primary evaporites (gypsum, pseudomorphs of gypsum) suggest deposition in the intertidal-supratidal zone in a relatively humid climate similar to that of present-day Andros Island, Bahamas, described by Shinn et al. (1965).

The evidence for a subtidal to an almost intertidal origin of the 0.3- to 2-meter-thick argillaceous, arenaceous, limonitic, orange-weathering dolomites in the Wahoo Limestone is subjective. These microdolomites directly overlie thick oolitic grainstones, lack fossils or sedimentary structures, and contain silt-size quartz.

Well-developed stromatolites, chips, laminae, channel fills, and microdolomites are also known in the Lisburne Group of late Mississippian age in the subsurface of the north slope (Armstrong and Mamet, 1974) and west-central Brooks Range (Armstrong, 1970).

The recognition of an intertidal-supratidal facies near the top of the Alapah Limestone and the subtidal dolomites in the Wahoo Limestone permit an analysis (Fig. 43–2) of the cyclic sedimentation and environments of deposition within the Lisburne Group of the Sadlerochit Mountains. This cyclic sedimentation can be used with the microfossils for detailed regional studies of carbonate stratigraphy and diagenesis of the Lisburne Group of northwestern arctic Alaska.

References

ARMSTRONG, A. K. 1970. Mississippian dolomites from Lisburne Group, Killik River, Mount Bupto Region, Brooks Range, Alaska. *Am. Assoc. Petrol. Geol. Bull. 54*, 251–264.

——— 1972. Pennsylvanian carbonates, paleoecology, and colonial corals north flank, eastern Brooks Range, Arctic Alaska. *U.S. Geol. Surv. Prof. Paper 747.*

ARMSTRONG, A. K., and MAMET, B. L. 1974. Carboniferous biostratigraphy, Prudhoe Bay State 1, to northeastern Brooks Range, Arctic Alaska. *Am. Assoc. Petrol. Geol. Bull. 58*, 646–660.

ARMSTRONG, A. K., and MAMET, B. L. 1975. Carboniferous biostratigraphy, northeastern Brooks Range, Arctic Alaska. *U.S. Geol. Surv. Prof. Paper 884.*

ARMSTRONG, A. K., MAMET, B. L., and DUTRO, J. T., JR. 1970. Foraminiferal zonation and carbonate facies of the Mississippian and Pennsylvanian Lisburne Group, central and eastern Brooks Range, Alaska. *Am. Assoc. Petrol. Geol. Bull. 54*, 687–698.

BALL, M. M. 1967. Carbonate sand bodies of Florida and the Bahamas. *J. Sed. Petrol. 37*, 556–591.

DAVIES, G. R. 1970. Algal-laminated sediments, Gladstone Embayment, Shark Bay, Western Australia. *In* Carbonate sedimentation and environments, Shark Bay, Western Australia. *Am. Assoc. Petrol. Geol. Memoir 13*, 169–205.

MAMET, B. L., and ARMSTRONG, A. K. 1972. Lisburne Group, Franklin and Romanzof Mountains, northeastern Alaska. *U.S. Geol. Surv. Prof. Paper 700-C*, C127–C144.

SHINN, E. A., GINSBURG, R. N., and LLOYD, R. M. 1965. Recent and supratidal dolomite from Andros Island, Bahamas. *In* Pray, L. C., and Murray, R. C., eds., Dolomitization and limestone diagenesis, a symposium. *Soc. Econ. Paleontol. Mineral. Spec. Publ. 13*, 112–123.

WILSON, J. L., MADRID-SOLIS, A., and MALPICA-CRUZ, R. 1967. Microfacies of Pennsylvanian and Wolfcampian strata in southwestern U.S.A. and Chihuahua, Mexico. *New Mexico Geol. Soc. Twentieth Field Conf. Guidebook*, 80–90.

44

Intertidal and Supratidal Deposits Within Isolated Upper Devonian Buildups, Alberta

Eric W. Mountjoy

Occurrence

Limestones with excellent intertidal features, including fenestral fabrics (Tebbutt et al., 1965; Choquette and Pray, 1970), occur in poorly developed cycles in the interior of many of the isolated Upper Devonian carbonate buildups (geologic reefs) of western Canada. Two representative buildups are Miette (in Rocky Mountains) long. 117°40′ W and lat. 53°03′ N (Fig. 44–1) and Golden Spike (oil-bearing, subsurface) long. 113°50′ W and lat. 53°28′ N (Tp51, R27, W4, Fig. 44–2).

General Setting

The buildups occur above a widespread stromatoporoidal biostromal platform (Flume and Cooking Lake Formations). The lower half to two-thirds of these restricted buildups normally consists of stromatoporoid carbonates ("lower" Leduc, upper Cairn). Limestones with intertidal features fenestral tex-

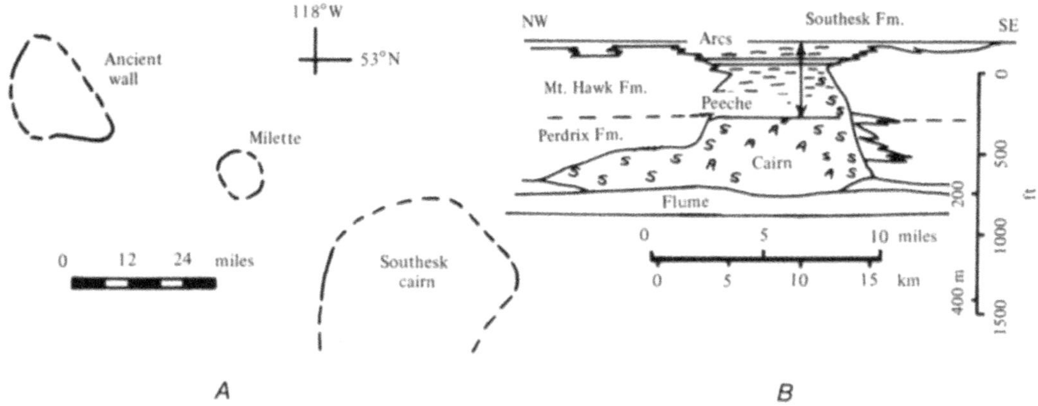

Figure 44-1. *A*. Map showing location of Miette and adjacent reef complexes in Alberta Rocky Mountains. *B*. Cross section of Miette complex.

tures are mostly restricted to the upper and interior portions of these buildups ("upper" Leduc, Peechee Member of Southesk). In turn, these are overlain by basin carbonate muds (Ireton and Mount Hawk Formations) or a coral unit (Grotto Member) in the mountain buildups (Figs. 44–1 and 44–2). In the Rocky Mountain buildups an additional sequence with intertidal limestones (Arcs Member) occurs above the Grotto (Fig. 44–1*B* and 44–3; Mountjoy, 1967).

Facies

The facies occur in a well-bedded sequence in which the beds range from 15 to 150 cm thick. The thickness of individual facies is considerably less, varying from 0.5 cm to 30 cm. The main facies present are:

1. Well-laminated fenestral packstones and wackestones (Fig. 44–4*A*, *B*, *F*)
2. Disturbed (bored?) fenestral wackestones (Fig. 44–4*C*, E)
3. Dense wackestones (Fig. 44–4*E*)
4. Intraformational rudites (Fig. 44–4*D*)

A fifth facies of rare green argillaceous carbonate mud occurs in discontinuous laminations, thin partings, between beds, along stylolites, and filling some small irregular cavities and vugs. These green carbonate muds (fill fractures) are more evident in the subsurface, and probably represent basin muds not winnowed from the buildups.

The packstones and wackestones consist of poorly to unsorted peloids (mostly pellets but also altered skeletal frag-

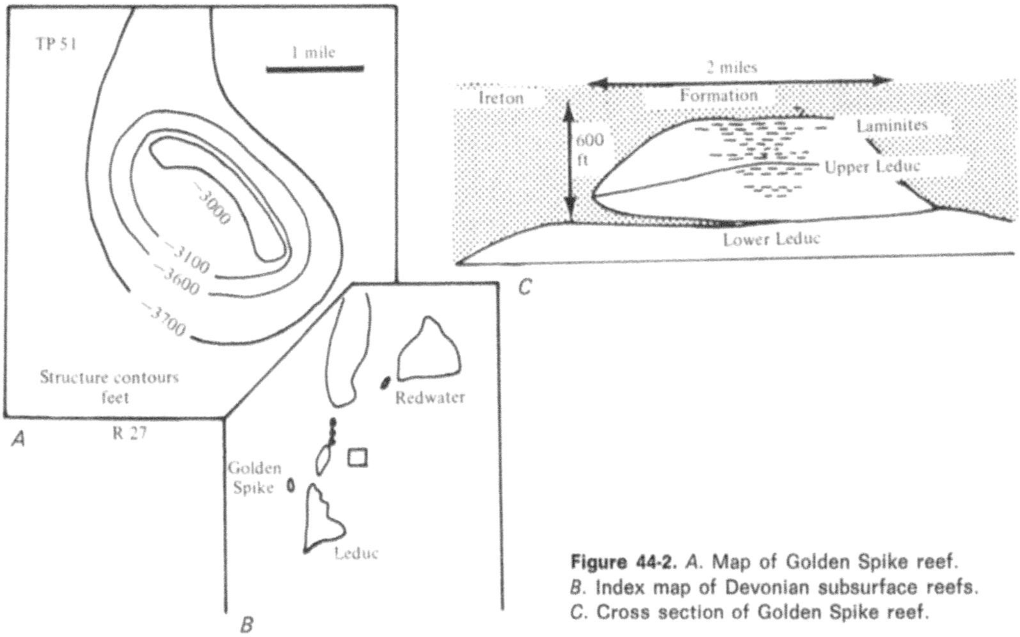

Figure 44-2. *A.* Map of Golden Spike reef. *B.* Index map of Devonian subsurface reefs. *C.* Cross section of Golden Spike reef.

ments), intraclasts, calcispheres, and rare Foraminifera. Some of the laminated fenestral rocks contain closely spaced, dense laminations that are likely of algal-mat origin (Fig. 44–4*A*, *B*). Other than very rare small or crinkled laminations a few centimeters across which possibly resulted from desiccation (only two observed in surface outcrops), no domal stromatolites are evident. Only rarely are beds with abundant *Amphipora* or stromatoporoids associated with the fenestral, dense, and megalodont packstones and wackestones. They are mostly near the base of the succession or within and near the Grotto Member (Fig. 44–3). The following subaerial or shallow water structures have not been observed: caliche crusts, desiccation cracks, pisolites, and cross bedding. Megalodont pelecypods (Fig. 44–3) occur sporadically within the dense wackestones and appear to occupy a similar position to megalodonts in the Lofer cycles (Fischer, 1964).

Interpretation

The main facies all have counterparts in Modern carbonate sediments reported from Florida, Bahama Banks, the Persian Gulf, and Shark Bay. The presence of fenestral fabrics, disrupted laminations, dark micritic laminations suggestive of algal mats, rip-up clasts, and the associated facies supports a shallow

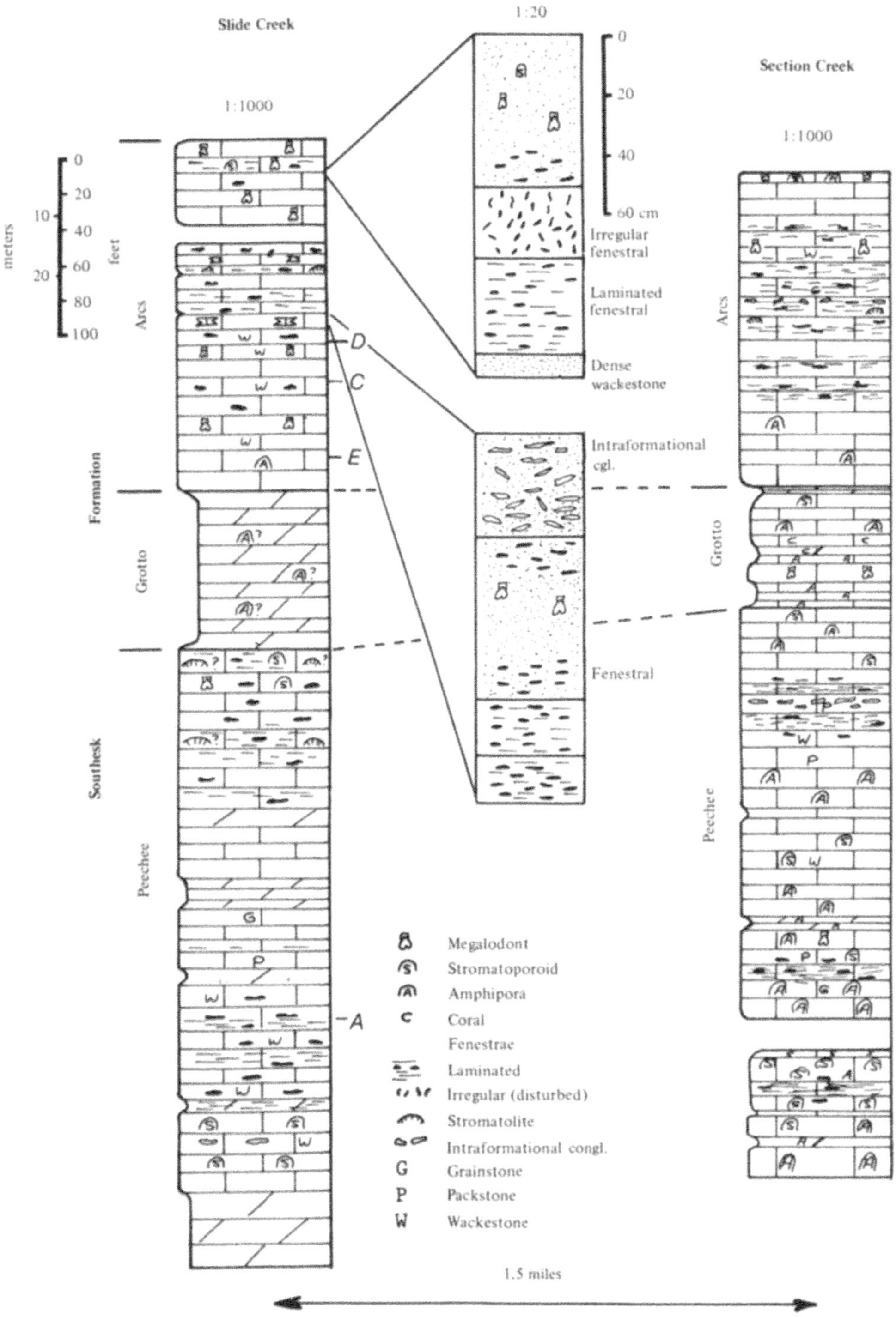

Slide Creek
1:1000
1:20
Section Creek
1:1000
meters
feet
Arcs
Formation
Grotto
Southesk
Peechee
0
20
40
60 cm
Irregular fenestral
Laminated fenestral
Dense wackestone
Intraformational cgl.
Fenestral
Megalodont
Stromatoporoid
Amphipora
Coral
Fenestrae
Laminated
Irregular (disturbed)
Stromatolite
Intraformational congl.
Grainstone
Packstone
Wackestone
1.5 miles

Figure 44-3 (opposite). Stratigraphic columns from Slide and Section Creek sections in central part of Miette reef complex (see Mountjoy, 1965). Center column gives representative details of well-exposed cycles. Letters refer to photographs in Figure 44-4.

intertidal or supratidal origin. However, only a small part has been preserved of the complete high intertidal and supratidal sequence that is observed in Modern environments (Ginsburg, et al., 1970; Roehl, 1967; Shinn et al., 1969).

Read (Chapter 29) has noted that the three types of fenestrae in Shark Bay tidal-flat sediments are characteristic of the type of overlying algal mats present, the type of algal mat being controlled by position within the intertidal and supratidal zones. All three types occur in the Devonian examples with the fine laminoid fenestrae (lower intertidal, Fig. 44–4*B*), the commonest; irregular fenestrae (middle and upper intertidal, Fig. 44–4*C*, 4*D*, 4*F*), occurring less frequently and tubular fenestrae very rarely (upper intertidal to supratidal). These data from Shark Bay, together with other features, suggest that the fenestrae in the buildups examined probably formed in the intertidal zone. Although fenestral fabrics alone are not diagnostic of an intertidal or supratidal environment, their association with definite subaerial features and their occurrence above shallow subtidal facies clearly indicate that they represent sediments deposited near a strand line.

These fenestral facies are widespread in lagoons of isolated buildups, land-fringing buildups, and broad carbonate shelves. These Devonian examples are similar to the well-known Lofer cycles (Fischer, 1964; Zankl, 1971) in terms of general cycle, fenestral textures, and megalodont facies.

Discussion

No distinctive cycles of the type noted by Read (Chapter 29) have been recognized, although these facies are locally interbedded with *Amphipora* and stromatoporoid limestones. These facies also lack the excellent desiccation and other features indicative of subaerial exposure. The lack of these cycles and subaerial features in these Devonian sediments suggests little change of sea level (or considerable reworking) compared to the Lofer and Pillara cycles. Hence, these buildups developed relatively constant facies patterns that are vertically continuous over several 10s of meters. If sea-level fluctuations occurred they were apparently of short duration and left little

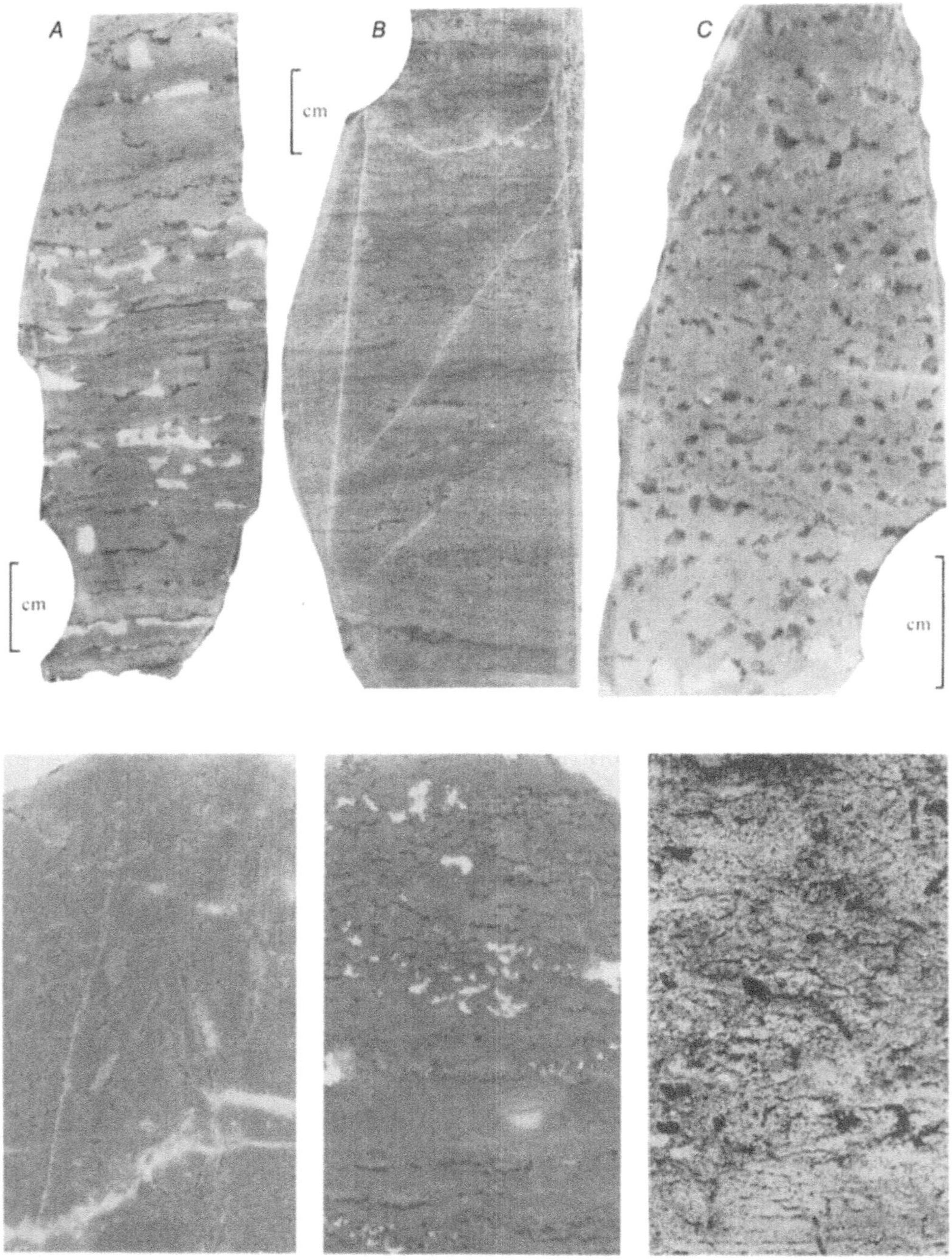
A
B
C
cm
cm
cm
D
cm
E
cm
F
cm

Figure 44-4 (opposite). Fenestral textures, Miette and Golden Spike reef complexes. *A, C, D,* and *E* from Slide Creek; *B* from Section Creek; and *F* from Golden Spike field. Bar scale is 1 cm. All oriented with tops up except *D.*

A. Laminated-fenestral wackestone, with minor disturbed fenestrae. Most of pores are filled (early?) with clear spar and remainder with dolomite (white). *B.* Finely laminated-fenestral packstone and wackestone locally disturbed, darker micritic layers may have resulted from algal mats. *C.* Irregular (disturbed) fenestral wackestone; some pores filled with white dolomite. *D.* Irregular fenestral, peloid wackestone to grainstone with ripped up fragments of micrite and peloid wackestone. Shelter pores occur beneath large clasts. Top to left. *E.* Laminated micrite to peloid wackestone with elongate fenestrae filled with an early clear spar and other fenestrae later filled with white dolomite. In middle, dense wackestone layer contains algal? encrusted circular void (presumed dissolved fossil). *F.* Open laminar and irregular to tubular fenestrae, south-central part of Golden Spike reef, near top of upper Leduc. Core slab from well 11-23-51-27W4, 5,414 ft depth.

or no evidence in the geologic record. Longer term sea-level fluctuations are clearly evident in the various formations, members, and so on (Mountjoy, 1965, 1967; McGillivray, 1970).

A weakness of many studies (including this one) of fenestral textures and associated facies is that generally the vertical sequences are insufficiently documented in terms of small-scale textures and variations. Even less is known about the lateral continuity of these facies. The lateral extent of the lower two portions of the Lofer cycles rarely exceeds 1,000 meters (Zankl, 1971, p. 162). The extent of the subfacies in the Miette buildups is unknown, but is at least a few hundred meters. Lateral extent is greater in the Golden Spike buildup, between 500 and 1,000 meters.

Most features of the exposure index of Ginsburg et al. (1970) are not preserved and hence cannot be used. In terms of the geologic record, one must consider what sediments and textures are most likely to be preserved; that is, can withstand exposure, reworking by physical and biological processes, and so on. A reconstruction of the stratigraphy in the Miette buildup (Mountjoy, 1965, 1967) indicates that subsidence relative to sea level was very slow, permitting considerable reworking by organisms and physical processes. Those sediments that are bound or trapped by organisms, buried rapidly, or cemented early would tend to be preferentially preserved. This preservation supports the concept that fenestral-laminated sequences are more easily preserved because of a protective algal-mat cover, lack of burrowing and local cementation—amply confirmed by most examples of Modern tidal flats. Adjacent sediments in Modern carbonate environments are generally more mobile and subject to reworking because they lack stabilizers, except for the baffling and trapping of sediment by *Thalassia* patches and banks.

What is the length of time represented by each fenestral lamination and by the thickness of facies? Do these deposits represent relatively continuous sedimentation or periodic deposits during storms of very high tides? One approach is to determine the sedimentation rates for the Frasnian portion of the Upper Devonian, as follows.

Assume that the length of the Frasnian is half the Upper Devonian, or 8.5 million years. Thicknesses vary from 400 to 1,000 meters for most of the Frasnian, hence sedimentation rates are approximately between 0.4 to 1.5 mm/year. This rate is comparable to Holocene sedimentation rates of 0.5 mm/year for the last 5,000 years in the Three Creek area west of Andros Island reported by Ginsburg et al. (1970) and to 0.6 mm/year at Crane Key for the last 3,300 years. These deposits therefore represent very slow and probably periodic accumulations, and hence could not possibly represent a tidal or even monthly cycle.

Although sedimentation rates can be greater in a few areas and at certain intervals, a sufficiently high rate of sedimentation for deposition of a complete subtidal to supertidal sequence is not attained in most areas of carbonate deposition in the past, except under conditions of relatively rapid sea-level fluctuations or sudden subsidence followed by a shoaling-upward sequence due to sediment accumulation. These facies represent periodic sedimentation presumably related to times of abnormally high seasonal or wind tides when sediment was stirred up and deposited as thin layers over the extensive intertidal flats within the lagoons of these Devonian buildups, as occurs today in most areas where smooth algal mats occur.

References

BALL, M. M. 1967. Carbonate sand bodies of Florida and the Bahamas. *J. Sed. Petrol. 37*, 556–591.

CHOQUETTE, P. W., and PRAY, L. C. 1970. Geologic nomenclature and classification of porosity in sedimentary carbonates. *Am. Assoc. Petrol. Geol. Bull. 54*, 207–250.

FISCHER, A. G. 1964. The Lofer cyclothems of the alpine Triassic. *Kansas Geol. Surv. Bull. 169*, 107–149.

GINSBURG, R. N., BRICKER, P. O., WANLESS, H. R., and GARRET, P. 1970. Exposure index and sedimentary structures of a Bahama tidal flat. *Geol. Soc. Am. 3*, 744–745 (abst.).

LEAVITT, E. M. 1968. Petrology, palaeontology, Carson Creek North reef complex, Alberta. *Bull. Can. Petrol. Geol. 16*, 298–413.

MCGILLIVRAY, J. G. 1970. Lithofacies control of porosity trends, Leduc Formation, Golden Spike Reef Complex, Alberta. Unpublished M.Sc. thesis, McGill Univ.

MOUNTJOY, E. W. 1965. Stratigraphy of the Devonian Miette Reef complex and associated strata eastern Jasper National Park. *Geol. Surv. Can. Bull. 110.*

——— 1967. Factors governing the development of the Frasnian Miette and Ancient Wall reef complexes (banks and biostromes) Alberta. *Alberta Soc. Petrol. Geol. Intern. Symp. Dev. System 2*, 387–408.

READ, J. F. 1973. Carbonate cycles, Pillara Formation (Devonian), Canning Basin, Western Australia. *Bull. Can. Petrol. Geol. 21*, 38–51.

ROEHL, P. O. 1967. Stony Mountain (Ordovician) and Interlake (Silurian) facies analogs of Recent low-energy marine and subaerial carbonates, Bahamas. *Am. Assoc. Petrol. Geol. Bull. 51*, 1979–2032.

SHINN, E. A. 1968. Practical significance of birdseye structures in carbonate rocks. *J. Sed. Petrol. 38*, 215–223.

SHINN, E. A., LLOYD, R. M., and GINSBURG, R. N. 1969. Anatomy of a modern carbonate tidal-flat, Andros Island, Bahamas. *J. Sed. Petrol. 39*, 1202–1228.

TEBBUTT, G. E., CONLEY, C. D., and BOYD, D. W. 1965. Lithogenesis of a distinctive carbonate rock fabric. *Wyoming Geol. Surv. Contrib. Geol. 4*, 1–13.

ZANKEL, H. 1971. Upper Triassic Carbonate Facies in the northern Limestone Alps. *In* Sedimentology of parts of central Europe. *Guidebook 8th Intern. Sediment Congr.* 147–185.

45

Carbonate Coastal Environments in Ordovician Shoaling-Upward Sequence, Southern Appalachians

Allan M. Thompson

Occurrence

Carbonate-rich rocks of shallow marine-coastal origin occur in the Upper Ordovician Inman, Leipers, and Sequatchie Formations in southeast Tennessee and northeast Alabama. The rocks average 60 meters total thickness, and occur in an area of complexly changing facies patterns developed through Late Ordovician time in response to shifting coastal positions on the distal margins of the Taconic clastic wedge (Thompson, 1970). These patterns generally involve thinning and increasing marine character to the west. Strata beneath the coastal rocks are open-marine shelf limestones of the Catheys Formation; strata above the coastal rocks are either Late Ordovician fluvial shales or Silurian marine rocks above an unconformity.

Facies

Four distinct lithofacies are recognizable (Thompson, 1971), shown schematically in Figure 45–1. Facies subdivisions were based on lithologic, petrographic, sedimentologic, and faunal

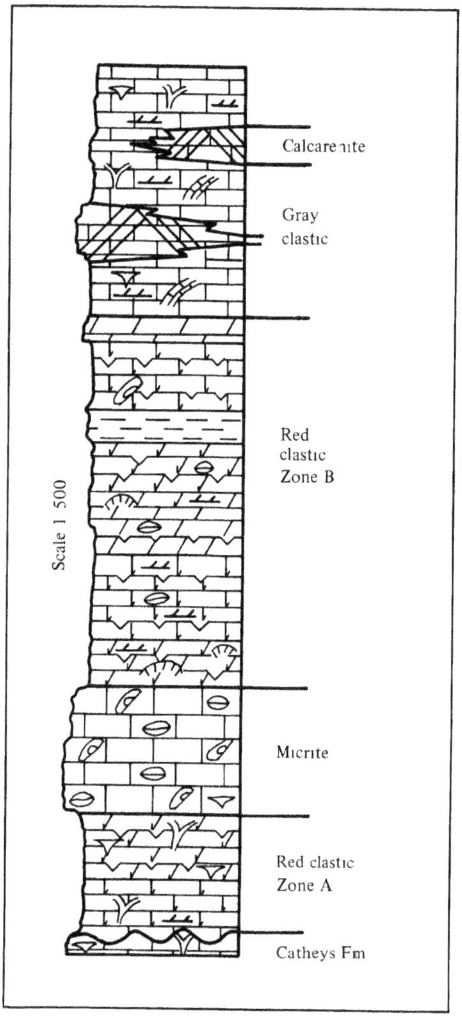

Figure 45-1. Schematic columnar section of coastal lithofacies and their composite vertical succession.

differences, and environments were interpreted by critical comparison with Modern carbonate environments. All facies contain considerable terrigenous sediment.

Red clastic facies. Rocks of the red clastic facies consist of up to 50 meters of red and gray, silty and clayey, pelletal biomicrite and calcareous shale, usually highly dolomitic. Bedding types include common parallel to irregular laminations, less common cryptalgal laminations with fenestral fabrics, and occasional microcross-laminations. Most laminations are alternately calcitic and dolomitic; small scours (relief 1 to 15 cm) frequently cut into dolomitic beds, and are filled with calcitic

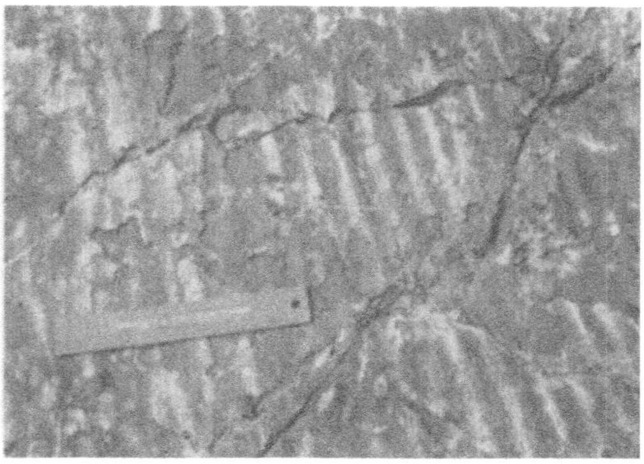

Figure 45-2 (left). Red (*dark*) mud-cracked surface in red clastic facies. Several thin red zones just beneath surface. Scale is 8 cm long.

Figure 45-3 (right). Rippled and mud-cracked surface in red clastic facies. Mud cracks lie close to fractures. Scale is 15 cm long.

sediment. Dolomitic intraclasts are abundant. Beds underlying laminations are commonly mud-cracked (Fig. 45–2); many are also rippled (Fig. 45–3). The sediment immediately beneath mud-cracked planes is often red and highly dolomitic, and grades downward into gray, less dolomitic rock (Fig. 45–4).

The red clastic facies is less fossiliferous than other facies, and contains two distinct faunal assemblages. The lower assemblage consists of brachiopods and bryozoans, with rare crinoids and ostracods, and characterizes zone A in Figure 45–1. The upper assemblage contains rare gastropods, ostracods, and small bivalves and rare bryozoan hash (zone B in Fig. 45–1). As a whole, the facies contains abundant nonskeletal organic material, including pellets, isolated vertical burrows (often filled with red sediment, Fig. 45–4), bedding-plane ichnofossils, and thin, dark organic films parallel to bedding in cryptalgal beds.

GRAY CLASTIC FACIES. Included in this facies are up to 30 meters of greenish-gray, silty and clayey micrite containing up

Figure 45-4. Red (*dark*) laminae, erosional above, grading down into buff calcitic laminae. Red zones truncating bedding are burrows. Note dark dolomitic intraclasts.

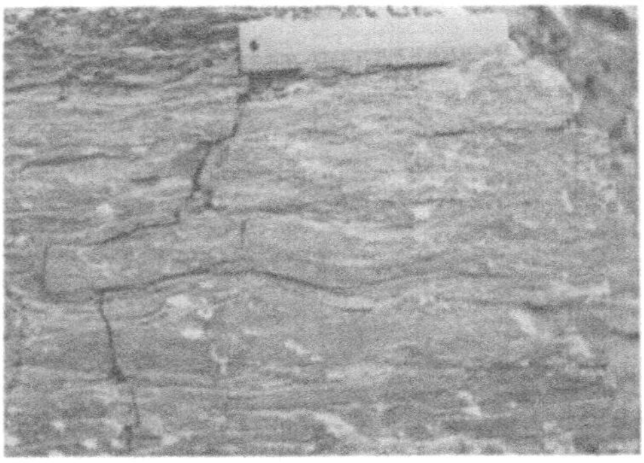

Figure 45-5 (left). Ripple bedding and preserved ripple forms in gray clastic facies. White spots are recrystallized bryozoan remains. Scale is 15 cm long.

Figure 45-6 (right). Gray clastic facies, showing current bedding and a small scour-and-fill structure in upper center. Exposed thickness is 2 meters.

to 70 percent clastic sediment. The major carbonate is calcite mud and silt; dolomite and pellets are absent. Bedding is dominated by wavy bedding, ripple bedding, and microcross-laminations (Fig. 45–5); current ripple trains are occasionally preserved, and small, shallow scours truncate ripples and other structures (Fig. 45–6). Large-scale cross-bedding is rare to absent. The sediment filling channels and scours is of the same grain size as that removed during channeling. Directional features suggest highly variable transport directions. The lowest meter is commonly glauconitic; red beds and mud cracks are absent. This facies contains an abundant open-marine fauna, dominated by bryozoans in more silty rocks and by brachiopods in more calcareous rocks. An infaunal population was apparently absent, as bioturbation and vertical burrowing are minimal.

Micrite facies. The micrite facies contains clayey biomicrite and minor calcareous shale, with rare terrigenous silt

Figure 45-7 (left). Thick bedding in micrite facies. Some thin bedding faintly visible locally. Thin-bedded zone at base is red clastic facies, zone A. Exposed thickness is 13 meters.

Figure 45-8 (right). Irregular, thick bedding of clayey micrite and skeletal layers in micrite facies. Scale is 15 cm long.

and dolomite. This facies is usually thick bedded (0.5 to 2 meters) to massive (Fig. 45–7), but occasionally shows gradations from thick bedded to thin bedded (2 to 4 cm) (Fig. 45–8). Primary bedding structures are exceedingly rare because of intense burrowing and bioturbation. Where preserved, stratification is as thin beds, rare microcross-lamination, and thin to thick alternations of skeletal, micritic, and argillaceous strata (Fig. 45–8). The skeletal bands contain abraded and comminuted brachiopods and bryozoans.

An abundant indigenous fauna includes gastropods, bivalves, and ostracods and rare brachiopods and bryozoans. Indigenous skeletal remains are whole; brachiopods and bryozoans are comminuted and broken and have been transported. Locally, the bioturbation has comminuted indigenous skeletal remains.

CALCARENITE FACIES. This facies consists of well-sorted, coarse-grained skeletal calcarenite (biosparite, grainstone) with little other sediment. Both the predominant calcite grains and rare quartz grains are exceedingly well rounded; micrite and most terrigenous clay are absent. The calcarenites are strongly cross-bedded in both planar-base, often solitary sets (Fig. 45–9) and trough-base, grouped sets of smaller scale but greater total thickeness (Fig. 45–10). Many sets show smoothly to irregularly rounded upper surfaces that are not apparently related to foreset evolution (Fig. 45–10), and herringbone patterns are common. Set orientations are bipolar, in roughly east-west directions. Small, shallow scours are cut in the calcarenites; filling sediments are slightly coarser, and grade upward into normal sediment.

Faunal elements of this facies include both micrite facies (mollusks, ostracods) and gray clastic facies (brachiopods,

Figure 45-9 (left). Large-scale, planar cross-strata in calcarenite facies. Paleocurrents to left (east-southeast). Hammer handle at left is 15 cm long.

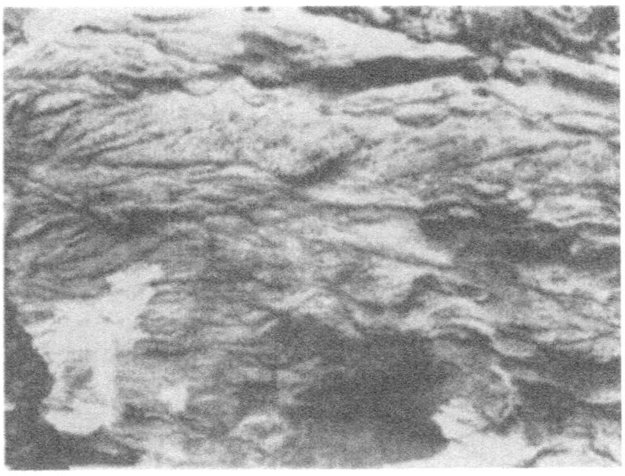

Figure 45-10 (right). Trough and planar cross-bedding in calcarenite facies. Ripples, megaripples, sand waves, and reactivation surfaces visible.

Figure 45-11. Interbedded calcarenite (*light*) and gray clastic (*dark*) facies. Two calcarenite zones are visible: one at left (5 meters thick), the other at upper right (0.4 meter thick).

bryozoans) taxa; brachiopods and bryozoa perdominate. All fossils are transported, and occur as sand-size grains; rarely, broken single shells are found. Burrowing is absent.

Rocks of this facies occur in intervals up to 6 meters thick, interlayered at random levels in the gray clastic facies (Fig. 45–11). These levels are not predictable or correlatable regionally.

Environmental Synthesis

The four facies described above occur in the vertical sequence shown in Figure 45–1. The sequence becomes progressively less fossiliferous, and contains more evidence of exposure, upward to the top of the red clastic facies; east of the study area, the red clastic facies is overlain by fluvial-floodplain shales and quartz sandstones. This pattern suggests a general shoaling and westward regression of shoreline-related environments. The fossiliferous gray clastic facies and calcarenite facies in the upper parts of the section represent a submergence and marine transgression, with the development of offshore marine environments above the former coastal zone.

Specifically, the red clastic facies is interpreted to represent coastal mud flats dominated by intermittent flooding, periodic deposition, low velocity traction currents of often local extent, considerable deposition from suspension, and considerable subaerial exposure. Rapid sediment addition as discrete layers is suggested by the prevalence of exposed bed tops and by vertical escape burrows generated by infaunal organisms attempting to adjust to a new sediment-water interface. Red colors are laminar, and consistently occur just beneath heavily mud-cracked bedding planes, suggesting a genetic relation between intense desiccation and ferric oxide production. Algal mats developed on newly deposited sediment layers, and promoted desiccation and possibly also the production of ferric oxide compounds. Dolomite generation within the environment is suggested by interlaminated dolomitic and calcitic laminae, and by dolomitic rip-up clasts in calcitic beds. The mud flats become progressively less red, less mud-cracked, and less dolomitic westward, suggesting decreasing exposure.

The gray clastic facies is interpreted as a nearshore-marine seafloor dominated by relatively low energy traction currents, whose velocities locally fluctuated considerably. Ripples and small sand waves were the dominant bedforms, and generated abundant ripple-bedding. Channels were scoured out during periods of high current competence, and were filled by later sediments deposited from currents of normal competence. Water circulation, aeration, and nutrient supply were sufficient to support a considerable benthonic fauna; the low tolerances of the involved taxa for salinity fluctuations suggest conditions of ample water mixing and minimal fresh-water dilution.

Data for the micrite facies indicate deposition of mixed carbonate and terrigenous mud in a quiet-water environment dominated by generally weak but sporadic traction currents. Occasional high current velocities were capable of winnowing mud from skeletal debris, and generated thin calcarenites. The indigenous fauna had wide salinity tolerance, suggesting considerable salinity fluctuations in the water mass. The consistently broken brachiopod and bryozoan skeletal remains may have been derived from adjacent environments (for example, gray clastic facies). The indigenous fauna included infaunal deposit-feeders, who thoroughly bioturbated and partially homogenized the sediment. The textures and fauna, combined with a stratigraphic position above zone A of the red clastic facies in a regressive sequence, suggest a lagoonal to estuarine environment.

The calcarenite facies is consistent with shallow-marine deposition from relatively high energy traction currents of reversing flow direction. The high rounding and sorting of all grains, especially of quartz grains, and the lack of mud indicate prolonged winnowing and reworking of sediment by consistently high velocity currents. The prevalence of normal-marine taxa in the skeletal remains, and the interbedding of calcarenites with gray clastic facies rocks carrying identical taxa, suggest a local, marine origin of the sand grains. Bipolar cross-bedding and herringbone cross-bedding were generated by reversing current flow, and the rounded upper set boundaries resemble reactivation surfaces (Klein, 1970), which also indicate reversing flow, probably in tidal environments. The prevalence of shore-directed cross-strata suggests a flood-dominated coast. The textures, structures, and fauna are consistent with current-swept shoals, marine sandbars, or tidal channels with large migrating sand waves. The environment was made up of many isolated and independent areas of sedimentation, and thus is more compatible with sandbar or tidal channel interpretations.

The array of environments existing during the regressive stage, then, included an intermittently exposed barrier mud flat (zone A, Fig. 45–1) fronting an open shallow marine shelf dominated by brachiopods and bryozoa. Behind the barrier lay a lagoonal-estuarine complex which graded landward into extensive coastal mud flats (zone B, Fig. 45–1). The array of transgressive environments included coastal mud flats (zone B) and a shallow, open shelf containing sandbars and/or tidal channels. The barrier flat had disappeared, and with it the distinction between lagoon and open shelf.

Evidence for Tidal Origin

The rocks contain no direct evidence for the existence of diurnal tides. The most convincing evidence for diurnal rise and fall of sea level—domal or columnar stromatolites—is conspicuously absent. Cryptalgal laminations in the red clastic facies do not connote daily water level fluctuations. However, the rocks contain abundant evidence for periodic rise and fall of sea level, even though such fluctuation may not have been diurnal.

1. Cryptalgal laminations and associated mud cracks indicate alternating flooding and exposure of the mud flats, and hence fluctuating water level.

2. Laminated dolomites, associated with mud-cracked surfaces and red laminae, suggest successive flooding, sediment addition, desiccation, and dolomitization, as seen in Modern high-intertidal zones (Shinn et al., 1965).

3. The local extent of the calcarenite facies and the presence of bipolar, herringbone cross-strata and reactivation surfaces document the existence of local, channel-like structures with reversing currents directed approximately onshore and offshore. These reversing currents strongly suggest periodic rise of water level, with resultant shore-directed current flow, and falling water, with sea-directed flow. This reversing flow could occur either in tidal channels or over the crests of intertidal sandbars.

Thus, this Ordovician coastal complex of open seafloor, barrier flat, lagoon, coastal mud flat, and later current channels was probably affected by periodic rise and fall of sea level, and the terminology associated with fluctuating water level (littoral or tidal) can probably be applied. However, the sea-level fluctuation was not necessarily diurnal, and may well have been of much longer and more irregular period.

References

KLEIN, G. DEV. 1970. Depositional and dispersal dynamics of intertidal sand bars. *J. Sed. Petrol. 40*, 1095–1127.

SHINN, E. A., GINSBURG, R. N., and LLOYD, R. M. 1965. Recent supratidal dolomite from Andros Island, Bahamas. *In* Pray, L. C., and Murray, R. C., eds., Dolomitization and limestone diagenesis. *Soc. Econ. Paleontol. Mineral. Spec. Publ. 12*, 112–123.

THOMPSON, A. M. 1970. Tidal-flat deposition and early dolomitization in Upper Ordovician rocks of southern Appalachian Valley and Ridge. *J. Sed. Petrol. 40*, 1271–1286.

——— 1971. Clastic-carbonate facies relationships in Upper Ordovician rocks of northeast Alabama. *In* Drahovzal, J. A., ed., *The Middle and Upper Ordovician of the Alabama Appalachians.* Alabama Geol. Soc., pp. 63–78.

Epilogue

Tidal Sedimentation: Some Remaining Problems

George deVries Klein

The preceding compilation of examples of tidalites has focused on their paleogeographic, stratigraphic, and facies relations. In a subsidiary way, this compilation has focused also on the potentially diagnostic features that may be utilized to identify other examples in the geologic record. These data are of value to students of historical geology.

Although the compilation has synthesized the first phase of tidal sedimentation studies through the establishment of some Holocene sedimentation models and the recognition of fossil counterparts, there are some remaining problems that require attention. It is the purpose of this note to review briefly some problems in tidal sedimentation that may prove to be fruitful areas of research in the immediate future.

First, the basic Holocene models used as reference standards for recognizing fossil analogs are surprisingly few. Well-studied clastic Holocene tidal flats are known from the North Sea coast of western Europe, the Minas Basin of the Bay of Fundy, the Coast of British Columbia, San Francisco Bay, the Gulf of

California (noted for its relatively sand-free tidal flats), the coast of Massachusetts, and the coast of Georgia. These seven areas share much in common in terms of textural distributions and structures, but also show important differences in facies patterns, preservation potential, hydrography, and climate. Well-studied carbonate models are known only from the Bahama Banks, the Persian Gulf, and Shark Bay (Western Australia). The major differences between these three carbonate tidal-flat models is the climatic regimen—tropical humidity, extreme aridity, and semiaridity being characteristic, respectively. One example of a mixed carbonate-silicaclastic tidal flat is known (Larsonneur, Paper 3). Clearly, seven clastic tidal-flat analogs, three carbonate tidal-flat analogs, and one mixed carbonate-silicaclastic analog is hardly sufficient to interpret examples from 3.2 x 10^9 years of the earth's sedimentary history. Future research is required concerning the nature of sediment variability of different Holocene intertidal areas under difference climatic regimens.

A particularly important topic for future intertidal silicaclastic research is the nature of depositional processes and sediments occurring on tidal flats flanking open coasts. Reineck (Paper 1) recognized this group of tidal-flat sediments, but most of the discussion at the associated tidal-flat conference focused on analogs of tidal flats flanking barrier islands, such as those known from the North Sea coast of the northern Netherlands and northwest Germany. It is noteworthy that only about 5,600 km of the world's coastlines are barrier island coastlines, and 3,200 km of this population occurs in the United States (Dickinson et al., 1972, p. 193). If so little of present-day coastal zones are characterized by barrier islands on a worldwide basis, it is perhaps more likely that open-coast tidal flats were more characteristic of tidal-flat sedimentation in the past. Hence, an understanding of such silicaclastic tidal flats is needed to further refine the facies models included in the preceding compilation.

This volume emphasized the recognition of intertidal sediments and their fossil counterparts. However, with the exception of Barnes and Klein (Chapter 19) and Klein (Chapter 17), almost no examples were cited of shallow, subtidal, tide-dominated sediments. These sediments, which include well-documented modern analogs from the North Sea, the China Sea, and the continental shelf of eastern North America (Sable Island, Georges Bank, Virginia Coast), owe their origin to the causal relation of shelf width and both tidal intensity and tidal range. It is conceivable that most epeiric sea sediments were dominated by tidal processes in both the intertidal *and* the shallow, subtidal, tide-dominated domain. Clearly, the shallow, subtidal, tide-dominated sediments require further study so as to discriminate them from intertidal sediments in studies of fossil analogs.

An area of particularly promising research in tidal sedimentation is the interaction of theoretitcal analysis of sediment transport, experimental studies, hydraulic studies, and fluid-flow sediment relations in a natural setting. The theoretical approach used by Komar (1969, 1972) in turbidities is potentially most useful for the study of tidalites. Field application of experimental data has been completed successfully in three cases (Boothroyd and Hubbard, 1971; Klein, 1970; Klein and Whaley, 1972), and their results permit preliminary estimates of the limiting quantitative variables operating during sediment transport by tidal currents.

Finally, although it is now recognized that tidal sediment transport processes may dominate shallow water areas and shorelines, tidal processes may also occur in deeper water zones. In two separate studies using a "deep-tow" instrument system, Lonsdale et al. (1972a, 1972b) have documented conclusively that tidal current systems operate in deep ocean basins at depths in excess of 2,000 meters. Reworking tidal currents at such depths redistribute and transport a considerable volume of sediment. Our understanding of abyssal tidal sedimentation is just emerging, but its implication for flysch sedimentation has been recognized already in one example (Laird, 1972) and may pose other implications as further results are reported.

In summary, this volume has synthesized where we stand in tidal facies studies, and has implicitly focused for us the future direction required for tidalite research.

References

BOOTHROYD, J. C., and HUBBARD, D. K. 1971. Genesis of estuarine bedforms and cross-bedding. *Geol. Soc. Am. 3*, 509–510.

DICKINSON, K. A., BERRYHILL, H. L., III, and HOLMES, C. W. 1972. Criteria for recognizing ancient barrier coastlines. *In* Rigby, J. K., and Hamblin, W. K., eds., Recognition of ancient sedimentary environments. *Soc. Econ. Paleontol. Mineral. Spec. Publ. 16*, 192–214.

KLEIN, G. DEV. 1970. Depositional and dispersal dynamics of intertidal sand bars. *J. Sed. Petrol. 40*, 1095–1127.

KLEIN, G. DEV., and WHALEY, MARGARET L. 1972. Hydraulic parameters controlling bedform migration on an intertidal sand body. *Geol. Soc. Am. Bull. 83*, 3465–3470.

KOMAR, P. D. 1969. The channelized flow of turbidity currents with application to Monterey deep-sea fan channel. *J. Geophys. Res. 74*, 4544–4558.

——— 1972. Relative significance of head and body spill from a channelized turbidity current. *Geol. Soc. Am. Bull. 83*, 1151–1156.

LAIRD, M. G. 1972. Sedimentology of the Greenland Group in the Paparoa Range, West Coast, South Island. *New Zealand J. Geol. Geophys. 15*, 372–393.

LONSDALE, P., NORMARK, W. R., and NEWMAN, W. A. 1972. Sedimentation and erosion on Horizon Guyot. *Geol. Soc. Am. Bull. 83*, 289–316.

LONSDALE, P., MALFAIT, B. T., and SPIESS, F. N. 1972. Abyssal sand waves on the Carnegie Ridge. *Geol. Soc. Am. 4*, 579–580.

Annotated Bibliography

Hans-Erich Reineck and R. N. Ginsburg

To explore the literature on any scientific subject, students and professionals need to find the key papers, those works that offer basic information and concepts. Finding the key papers on tidal deposits is no easy task; Straaten's bibliography[1], which concentrated on the North Sea region, lists 239 papers, and a more recent bibliography by Reineck[2] cites nearly 500 works. To help the neophyte with this voluminous literature we have selected 25 works on tidal deposits, 12 on siliciclastics and 13 on carbonates. We included both recent and ancient examples, and in listing the four categories we have indicated those papers that include introductory descriptions with an asterisk.

[1] Straaten, L. M. J. U. van. 1961. Sedimentation in tidal flat areas. *J. Alberta Soc. Petrol. Geol.* 9(7), 203–216.

[2] Reineck, H.-E. 1973. Bibliographie geologischer Arbeiten über rezente und fossile Kalk- und Silikatwatten (Bibliography of Recent and Ancient Tidalits). Cour. Forsch.-Inst. Senckenberg, 6, Frankfurt am Main, 57 pp.

Recent Siliciclastics

BAJARD, J. 1966. Figures et structures sédimentaires dans la partie orientale de la baie du Mont Saint-Michel. *Rev. Geog. Phys. et Geol. dyn. 8,* 39–112.

An atlas of sedimentary structures in carbonate-rich tidal deposits of a temporate sea with large tide range.

DÖRJES, J., GADOW, S., REINECK, H.-E., AND SINGH, I. B. 1969. Die Rinnen der Jade (Südliche Nordsee). Sedimente und Makrobenthos. *Senckenbergiana Maritima 50,* 5–62.

An integrated description of sediments, bedding types, and benthos from the subtidal zone of a classic region by a team of specialists.

HOUBOLT, J. J. H. C. 1968. Recent sediments in the Southern Bight of the North Sea. *Geol. Mijnbouw 47,* 245–273.

Strong tidal currents in the southern North Sea accumulate sand from the seafloor into ridges up to 30 meters high, which parallel the currents and have internal cross stratification.

KLEIN, G. DEV. 1970. Depositional and dispersal dynamics of intertidal sand bars. *J. Sed. Petrol. 40,* 1095–1127.

Shows how tidal currents and sandbar morphology control texture and internal stratification in an area of extremely high tide range. Valuable illustrations of bedforms and stratification.

*REINECK, H.-E. 1967. Layered sediments of tidal flats, beaches, and shelf bottoms of the North Sea. *Estuaries.* Am. Assoc. Advan. Sci. *83,* 191–206.

Concise, well-illustrated descriptions of bedding types, bioturbation structures, and their environments.

REINECK, H.-E. 1970, ed. *Das Watt, Ablagerungs- und Lebensraum.* Frankfurt am Main, Kramer, 142 pp.

A comprehensive, thoroughly illustrated summary of sedimentologic, geochemical, and organic processes and features, their areal distribution and vertical sequence, from a much-studied area.

*SCHÄFER, W. 1972. *Ecology and Palaeoecology of Marine Environments.* Edinburgh, Oliver & Boyd, 568 pp.

A remarkable synthesis of functional morphology, life habits, death, and preservation of marine animals in North

Sea sediments. Includes an illustrated analysis of trace fossils; maps and descriptions of biofacies of tidal flats.

*STRAATEN, L. M. J. U. VAN. 1954. Composition and structure of recent marine sediments in the Netherlands. *Leidse Geol. Mededel. 19,* 1–110.

A comprehensive description of morphology, hydrography, mineralogy, macro- and microstructures, transformation of iron compounds, and organic content, with a summary of sedimentary criteria for tidal zones.

THOMPSON, R. W. 1968. Tidal flat sedimentation on the Colorado River Delta, Northwestern Gulf of California. *Geol. Soc. Am. Memoir 107.*

The physiography, sediments, structures, and postglacial history in an arid climate with a large tidal range and a predominance of siliceous muds.

Ancient Siliciclastics

*RAAF, J. F. M. DE, AND BOERSMA, J. R. 1971. Tidal deposits and their sedimentary structures (seven examples from western Europe). *Geol. Mijnbouw. 50,* 479–504.

Critical analyses of the sedimentary structures of seven fossil examples with emphasis on the diagnostic bedding features and sequences.

STRAATEN, L. M. J. U. VAN. 1954. Sedimentology of Recent tidal flat deposits and the Psammites du Condroz (Devonian). *Geol. Mijnbouw 16,* 25–47.

A critical comparison of ancient and recent sedimentary structures and ichnofossils.

*SWETT, K., KLEIN, G. DEV., AND SMIT, D. 1971. A Cambrian tidal sand body. The Eriboll Sandstone of Northwest Scotland, an ancient-recent analog. *J. Geol. 79,* 400–415.

The assemblage of bedding types and bimodal dip directions of cross beds used to interpret a Cambrian sand as tidal.

Recent Carbonates

BALL, M. M. 1967. Carbonate sand bodies of Florida and the Bahamas. *J. Sed. Petrol. 37,* 556–591.

Includes the distinctive attributes of the morphology, sequence of structures, and grain kind in holocene tidal bars of ooid sands, with emphasis on their regional and local hydrography.

ILLING, L. V., WELLS, A. J., AND TAYLOR, J. C. M. 1965. Penecontemporary dolomite in the Persian Gulf. *In* Pray, L. C., and Murray, R. C., eds., Dolomitization and limestone diagenesis: A symposium. *Soc. Econ. Paleontol. Mineral. Spec. Publ. 13,* 89–111.

Dolomite in the upper half-meter of sabkha deposits (supratidal) forms by replacement of aragonitic sediment in brines with elevated Mg/Ca; illustrated descriptions of authigenic gypsum and algal-laminated sediments.

*KENDALL, C. G. St. C., AND SKIPWITH, P. A. D'E. 1969. Holocene shallow-water carbonate and evaporite sediments of Khor al Bazam, Abu Dhabi, Southwest Persian Gulf. *Am. Assoc. Petrol. Geol. Bull. 53,* 841–869.

Maps and illustrated descriptions of the carbonate and evaporite sediments and their environments with emphasis on grain type and occurrence of evaporite minerals.

KINSMAN, D. J. J. 1966. Gypsum and Anhydrite of Recent age, Trucial Coast, Persian Gulf. *Second Symposium on Salt.* Northern Ohio Geol. Soc., Cleveland, pp. 1302–1326.

The distribution of diagenetic minerals on a sabkha with emphasis on the occurrences and geochemistry of gypsum and anhydrite.

*LOGAN, B. W., DAVIES, G. R., READ, J. F., AND CEBULSKI, D. E. 1970. Carbonate sedimentation and environments, Shark Bay, Western Australia. *Am. Assoc. Petrol. Geol. Memoir 13,* 1–223.

A comprehensive, well-illustrated description of the environments, sediments, organisms, and Quaternary history of an embayment with progressively increasing salinity. Valuable descriptions of sea grass bank deposits, algal-laminated sediments, and postglacial facies.

LOGAN, B. W., REZAK, R., AND GINSBURG, R. N. 1964. Classification and environmental significance of algal stromatolites. *J. Geol. 72,* 68–83.

Relationships between growth forms of modern algal-laminated sediments and their environments provide basis for a simplified geometric classification of fossil stromatolites.

*SHINN, E. A., LLOYD, R. M., AND GINSBURG, R. N. 1969. Anatomy of a modern carbonate tidal flat, Andros Island, Bahamas. *J. Sed. Petrol. 39,* 1202–1228.

Descriptions and illustrations of the sedimentary structures, organisms, and their vertical sequence in subenvironments of a tidal flat where storm flooding is a major control.

Ancient Carbonates

*Bosellini, A., and Hardie, L. A. 1973. Depositional theme of a marginal marine evaporite. *Sedimentology 20,* 5–27.

Sedimentary structures and sequences in repeated cycles of dolomite and gypsum of Alpine Permian deposits are used to interpret the environment as a prograding lagoon-sabkha complex. Valuable comparisons of structures in recent and fossil evaporites.

*Fischer, A. G. 1964. The Lofer cyclothems of the Alpine Triassic. *In* Merriam, D. F., ed. Symposium on cyclic sedimentation. *State Geol. Surv. Kansas Bull. 169, 1,* 107–149.

Emphasizes diagnostic sedimentary structures in subtidal and "intertidal" elements; definitions and illustrations of loferites, prism, and shrinkage cracks; tectonic model for transgressive cycles.

*Laporte, L. 1967. Carbonate deposition near mean sea-level and resultant facies mosaic: Manlius Formation (Lower Devonian) of New York State. *Am. Assoc. Petrol. Geol. Bull. 51,* 73–101.

Well-illustrated example of the use of association of sedimentary structures and fossils to identify the deposits of tidal zones in an epeiric sea.

*Roehl, P. O. 1967. Stony Mountain (Ordovician) and Interlake (Silurian) facies analogs of Recent low-energy marine and subaerial carbonates, Bahamas. *Am. Assoc. Petrol. Geol. Bull. 51,* 1979–2032.

Sedimentary structures used to interpret subsurface oil reservoir rocks as tidal-flat deposits using Holocene Bahama example as reference. Reservoir properties related to subenvironments.

Walker, K. R. 1972. Community ecology of the Middle Ordovician Black River Group of New York State. *Geol. Soc. Bull. 83,* 2499–2524.

———, 1973. Stratigraphy and environmental sedimentology of Middle Ordovician Black River Group in the Type Area—New York State. *New York State Mus. Sci. Serv. Bull. No. 419.*

Two related papers that describe the lithologies, structures, and fossil communities used to identify tidal facies and map their distributions; valuable analysis of organism communities.

Contributors

Elizabeth A. Allen
Department of Geology
University of Delaware
Newark, Delaware

Augustus K. Armstrong
U. S. Geological Survey
345 Middlefield Road
Menlo Park, California

John J. Barnes
2307-B Wordsworth
Houston, Texas

F. W. Beales
Department of Geology
University of Toronto
Toronto, Canada

Moshe Braun
Department of Geology
Rensselaer Polytechnic Institute
Troy, New York

Raymond G. Brown
Department of Geology
University of Western Australia
Nedlands
Western Australia

Charles H. Carter
Ohio Geological Survey
(Lake Erie Section)
Sandusky, Ohio

R. Colacicchi
Instituto di Geologia
Università di Perugia
Perugia, Italy

Robert W. Dalrymple
Department of Geology
University of Delaware
Newark, Delaware

Richard A. Davis, Jr.
Department of Geology
University of South Florida
Tampa, Florida

Contributors

Graham Evans
Department of Geology
Imperial College
London, England

Alfred G. Fischer
Department of Geological and Geophysical Sciences
Princeton University
Princeton, New Jersey

Gerald M. Friedman
Department of Geology
Rensselaer Polytechnic Institute
Troy, New York

Robert N. Ginsburg
Rosenstiel School of Marine
and Atmospheric Sciences
University of Miami
Miami Beach, Florida

George M. Hagan
Department of Geology
University of Western Australia
Nedlands
Western Australia

Robert B. Halley
U.S. Geological Survey
Fisher Island
Miami Beach, Florida

Lawrence A. Hardie
Department of Earth and Planetary Sciences
Johns Hopkins University
Baltimore, Maryland

John C. Harms
Denver Research Center
Marathon Oil Company
Littleton, Colorado

Stanley Cooper Harrison
Exxon Company U.S.A.
3016 Casa del Norte Drive, NE
Albuquerque, New Mexico

Paul Hoffman
Division of Geological and Planetary Sciences
California Institute of Technology
Pasadena, California

Lubomir F. Jansa
Atlantic Geoscience Center
Geological Survey of Canada
Bedford Institute of Oceanography
Dartmouth, Nova Scotia
Canada

George deVries Klein
Department of Geology
University of Illinois
Urbana, Illinois

R. John Knight
Department of Geology
McMaster University
Hamilton, Ontario, Canada

John C. Kraft
Department of Geology
University of Delaware
Newark, Delaware

Naresh Kumar
Lamont-Doherty Geological Observatory of
Columbia University
Palisades, New York

Léo F. Laporte
Earth Sciences Board
University of California
Santa Cruz, California

Claude Larsonneur
Département de Géologie
Université de Caen
Caen, France

Brian W. Logan
Department of Geology
University of Western Australia
Nedlands
Western Australia

G. P. Lozej
Department of Geology
University of Toronto
Toronto, Canada

David B. MacKenzie
Marathon Oil Company
Post Office Box 269
Littleton, Colorado

James A. Miller
Union Oil Company of California
Research Center
Brea, California

Eric W. Mountjoy
Department of Geological Sciences
McGill University
Montreal, Quebec, Canada

A. Thomas Ovenshine
U. S. Geological Survey
Menlo Park, California

L. Passeri
Instituto di Geologia
Università di Perugia
Perugia, Italy

G. Pialli
Instituto di Geologia
Università di Perugia
Perugia, Italy

B. H. Purser
Laboratoire de Gélogie Historique
Centre d'Orsay
Université de Paris-Sud
Paris, France

J. F. Read
Department of Geological Sciences
Virginia Polytech Institute and
State University
Blacksburg, Va.

Hans-Erich Reineck
Institut für Meeresgeologie und Meeresbiologie
"Senckenberg"
Wilhelmshaven, West Germany

Antonio Rizzini
AGIP Direzione Mineraria
Milan, Italy

John E. Sanders
Department of Geology
Barnard College
Columbia University
New York, New York

Paul E. Schenk
Department of Geology
Dalhousie University
Halifax, Canada

Jean F. Schneider
Department of Geology
Swiss Federal Institute of Technology
Zurich, Switzerland

Bruce W. Sellwood
Sedimentology Research Laboratory
University of Reading
Reading, England

J. H. J. Terwindt
Rijkswaterstaat Deltadienst
's-Gravenhage,
The Netherlands

Allan M. Thompson
University of Delaware
Newark, Delaware

Robert W. Thompson
School of Natural Resources
California State University, Humboldt
Arcata, California

Glenn S. Visher
Department of Earth Sciences
University of Tulsa
Tulsa, Oklahoma

Roger G. Walker
Department of Geology
McMaster University
Hamilton, Ontario, Canada

Harold R. Wanless
Rosenstiel School of Marine and
Atmospheric Sciences
University of Miami
Miami, Florida

R. C. L. Wilson
Department of Earth Sciences
Open University, Walton
Bletchley, Bucks, England

Peter J. Woods
Department of Geology
University of Western Australia
Nedlands
Western Australia

Isabel Zamarreño
Depto. de Geomorfologia y Geotectonica
Universidad, Orviedo, Spain

Index

A

B

C

D

E

F

G

H

M

O

P

R

S

T

Rock Types and Grain Types

Rock Types

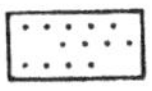
Quartz sand

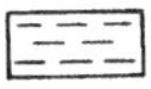
Shale

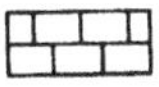
Limestone (undifferentiated)

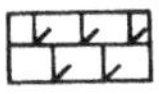
Dolomitic limestone

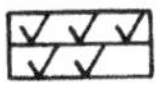
Calcareous limestone

Dolomite (undifferentiated)

DEPOSITIONAL

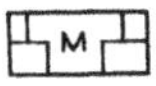

Mudstone

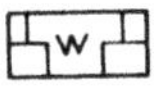

Wackestone

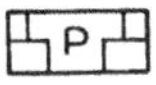

Packstone

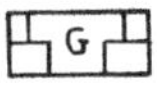

Grainstone

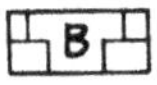

Boundstone

DIAGENETIC

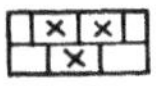
Crystalline

Shell Oil Company standard legend

Grain Types

SKELETAL GRAINS

General Symbols
- Fossils, undifferentiated
- Fossils, encrusting

Specific Symbols
- Algae, undifferentiated
- Algal plates
- Algal balls
- Algal stromatolites
- Brachiopods, undifferentiated
- Brachiopods, phosphatic
- Brachiopods, Productid
- Bryozoa, tube-like forms
- Bryozoa, fenestellid forms
- Corals, colonial
- Corals, solitary
- Crinoids
- Diatoms
- Foraminifera, undifferentiated
- Foraminifera, pelagic
- Foraminifera, small benthonic
- Foraminifera, large benthonic
- Foraminifera, Fusulinids
- Molluscs, undifferentiated
  - Belemnites
  - Gastropods
  - Pelecypods (clams)
    - Inoceramis
    - Oysters
    - Rudists undifferentiated
  - Tentaculites
- Ostracods
- Silicified wood
- Spicules
- Spines
- Spines, brachiopod
- Spines, echinoid
- Stromatoporoids, undifferentiated
- Stromatoporoids, Amphipora, undiff.

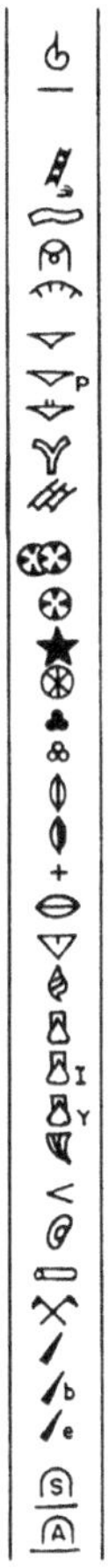

NONSKELETAL GRAINS

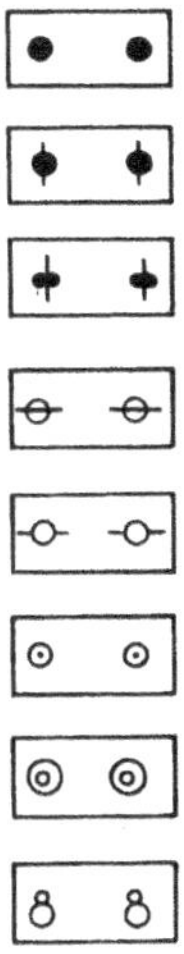

- Pelletoids
- Pellets (fecal particles)
- Micropelletoids (20–62μ)
- Coated grains
- Superficial oolites
- Oolites
- Pisolites
- Grapestone or aggregate grains

LITHOCLASTS

Fragments of previously lithified carbonate rock

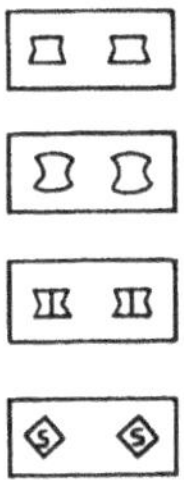

- Angular, undifferentiated
- Rounded, undifferentiated
- Intraclasts { locally derived / penecontemporaneous fragments
- Breccia, solution, collapse

GPSR Compliance

The European Union's (EU) General Product Safety Regulation (GPSR) is a set of rules that requires consumer products to be safe and our obligations to ensure this.

If you have any concerns about our products, you can contact us on ProductSafety@springernature.com

In case Publisher is established outside the EU, the EU authorized representative is:

Springer Nature Customer Service Center GmbH
Europaplatz 3
69115 Heidelberg, Germany

Batch number: 10838292

Printed by Printforce, the Netherlands